AF389307

The Effect of Diet and Nutrition on Postprandial Metabolism

The Effect of Diet and Nutrition on Postprandial Metabolism

Editors

François Mariotti
Dominique Dardevet

MDPI • Basel • Beijing • Wuhan • Barcelona • Belgrade • Manchester • Tokyo • Cluj • Tianjin

Editors
François Mariotti
AgroParisTech,
Université Paris-Saclay
France

Dominique Dardevet
Université Clermont Auvergne
France

Editorial Office
MDPI
St. Alban-Anlage 66
4052 Basel, Switzerland

This is a reprint of articles from the Special Issue published online in the open access journal *Nutrients* (ISSN 2072-6643) (available at: https://www.mdpi.com/journal/nutrients/special_issues/ Postprandial_Metabolism).

For citation purposes, cite each article independently as indicated on the article page online and as indicated below:

LastName, A.A.; LastName, B.B.; LastName, C.C. Article Title. *Journal Name* **Year**, *Article Number*, Page Range.

ISBN 978-3-03943-232-5 (Hbk)
ISBN 978-3-03943-233-2 (PDF)

Contents

About the Editors

François Mariotti (MEng, Ph.D., HDR, Prof.) is Professor of Nutrition in the Human Biology and Nutrition Department at AgroParisTech, the Paris Institute of Technology for Life, Food and Environmental Sciences (Paris, France), now part of the Université Paris-Saclay. He currently heads the 'PROSPECT' research group that studies the relations between dietary protein intake, nutritional security and cardiometabolic health, combining metabolism and physiology in clinical trials and the analysis of dietary intake of populations (at UMR Physiologie de la Nutrition et du Comportement Alimentaire, AgroParisTech, INRAe, Université Paris-Saclay, Paris, France). Prof. Mariotti has a strong background on postprandial metabolism and physiology. He has also developed and implemented multiple, multidisciplinary approaches to public health nutrition, and is the chairman of the standing committee on Nutrition at the French Agency (Anses).

Dominique Dardevet (Ph.D., HDR) is a Research Director at the INRAE, French National Research Institute for Agriculture, Food and Environment National Institute. His main research objectives aim at optimizing strategies to maintain the functional integrity of mobility/independency during muscle wasting (catabolic states or aging) by studying the nutritional control of muscle protein renewal and function, and the regulation of amino acid bioavailability. His approaches combine biochemistry, physiology, and cellular biology in preclinical studies and translational research leading to clinical trials. Since 2006, he has been the Editor of *Nutrition Research Reviews* and involved in several European consortia like The Joint Programming Initiatives (JPI) MaNuEl (Malnutrition in the elderly) and Intimic Knowledge Platform on Food, Diet, Intestinal Microbiomics and Human Health.

Preface to "The Effect of Diet and Nutrition on Postprandial Metabolism"

Humans feed themselves by discontinuous inputs that give an essentially dynamic characteristic to all nutritional processes. Metabolism is constructed according to a nutritional system: The individual's endogenous metabolism works on a fasting basis at a steady-state to ensure physiological functions with consistency, but this situation results in molecule/nutrient losses that must be compensated for by food intake. In the postprandial phase, conversely, the nutritional system must ensure the correct use of the influx of nutrients, by regulating flows while maintaining circulating concentrations within acceptable ranges. The postprandial phase is, therefore, the critical nutritional phase during which the body ensures its repletion while putting its homeostasis under pressure, and manages metabolic disturbance according to "hemodynamic" type processes. A decrease in good postprandial management capacity compromises the nutritional status of an individual, on the one hand, and, on the other hand, indicates alterations in metabolic health, i.e., a loss of metabolism flexibility and adaptability. The effectiveness of nutrient use in the postprandial phase appears to be a major determinant of long-term nutritional status for many nutrients. If we consider the resilience of the body to environmental stimuli as a central and modern definition of health, then we can view the smooth course of these postprandial processes as a definition of nutritional health. The efficiency of nutrient use and the time course of the activation of nutrient metabolism is modulated by the characteristics of the diet, patterns of intake, nature of the meals, individual genetics, phenotypes and health status, and all other lifestyle characteristics.

Changes in postprandial metabolism have been considered to be potential early markers in the pathophysiological course, leading to the risk of pathology development. They are sensitive to diets and the complex nature of meals, which can alter the allostatic load, and postprandial deregulations are predictive of the risk of chronic diseases.

With this Special Issue, we aimed to expand and add to the research on the importance of postprandial metabolism in nutrition. The book begins with two long reviews of the literature to understand the appearance of markers of altered cardiometabolic health in the postprandial phase, as well as what is known about its modulation by nutritional intakes, with a detailed review on the importance of modulation of postprandial lipemia. This is complemented by an original research paper on the nature of dietary lipids. Another article focused on the likely important mechanism of postprandial regulation, which is the appearance of low-grade endotoxemia, while another contribution has tended to rule out the possibility that arginine methylation processes, which lead to the appearance of cardiovascular risk factors, may be operating in the postprandial phase. This set provides a rather important overview of the presentation of postprandial processes and their determinants. The global and exploratory understanding of postprandial changes in metabolism is also addressed here, with metabolomics data allowing comparison between "normal" postprandial situations (i.e., after a standard meal) and situations resulting from an acute or prolonged nutritional challenge. Other data help to understand the re-organization of inter-organ flows in response to overnutrition, particularly during the postprandial phase. The last part of the book presents studies that account for the effects of meal composition on postprandial phenomena. It includes several studies that have focused on what has long been the poor relation of the nutritional regulation of postprandial metabolism: proteins. These studies illustrate that proteins themselves, or depending on their relationship with other nutrients (sucrose, lipids), may (1) have different levels of effectiveness in postprandial repletion in subjects with anabolic resistance, (2) modify postprandial metabolic responses in healthy or at-risk subjects, and (3) limit metabolic deregulation leading to ectopic lipid deposition—a highly pathogenic phenomenon. These studies illustrate the current vitality of this topic, which is decidedly modern because it offers a consistent conceptual framework for understanding the relationship between diet and health.

François Mariotti , Dominique Dardevet
Editors

Review

The Postprandial Appearance of Features of Cardiometabolic Risk: Acute Induction and Prevention by Nutrients and Other Dietary Substances

Laurianne Dimina and François Mariotti *

UMR PNCA, AgroParisTech, INRA, Université Paris-Saclay, 75005 Paris, France
* Correspondence: francois.mariotti@agroparistech.fr

Received: 7 July 2019; Accepted: 19 August 2019; Published: 21 August 2019

Abstract: The purpose of this review is to provide an overview of diets, food, and food components that affect postprandial inflammation, endothelial function, and oxidative stress, which are related to cardiometabolic risk. A high-energy meal, rich in saturated fat and sugars, induces the transient appearance of a series of metabolic, signaling and physiological dysregulations or dysfunctions, including oxidative stress, low-grade inflammation, and endothelial dysfunction, which are directly related to the amplitude of postprandial plasma triglycerides and glucose. Low-grade inflammation and endothelial dysfunction are also known to cluster together with insulin resistance, a third risk factor for cardiovascular diseases (CVD) and type-II diabetes, thus making a considerable contribution to cardiometabolic risk. Because of the marked relevance of the postprandial model to nutritional pathophysiology, many studies have investigated whether adding various nutrients and other substances to such a challenge meal might mitigate the onset of these adverse effects. Some foods (e.g., nuts, berries, and citrus), nutrients (e.g., l-arginine), and other substances (various polyphenols) have been widely studied. Reports of favorable effects in the postprandial state have concerned plasma markers for systemic or vascular pro-inflammatory conditions, the activation of inflammatory pathways in plasma monocytes, vascular endothelial function (mostly assessed using physiological criteria), and postprandial oxidative stress. Although the literature is fragmented, this topic warrants further study using multiple endpoints and markers to investigate whether the interesting candidates identified might prevent or limit the postprandial appearance of critical features of cardiometabolic risk.

Keywords: metabolic syndrome; postprandial; endothelial function; oxidative stress; nuts; berries

---

1. Introduction

This review focuses on the kinds of diets, food, and food components that affect postprandial inflammation, endothelial function, and oxidative stress, and which are related to cardiometabolic risk, including metabolic syndrome (MS), and ultimately, cardiovascular diseases (CVD) and type 2 diabetes. Although this review gathered a very large number of studies, it is not intended to be exhaustive; rather, it emphasizes the food and food components that have been studied the most, and the data that together help us to understand the impact of nutrition on cardiometabolic risk, as this can be studied during the postprandial period.

Metabolic syndrome refers to the clustering of a series of risk factors for CVD, whose prevalence is rising markedly at a global level [1–5]. Because MS is an important risk for CVD and type-II diabetes [6], considerable attention has been paid to analyzing its links with environmental factors and diet.

MS has been characterized from a clinical point of view using the following criteria: a high waist circumference; raised plasma triglycerides, plasma glucose, and systolic blood pressure; and lower

HDL-cholesterol concentration [5,7,8]. From a pathophysiological viewpoint, the heterogeneity of MS is considerable, but there is now consensus regarding the importance of a few related features that are major components of cardiometabolic risk. MS is mainly considered as being related to the development of resistance to the action of insulin in different tissues and on different metabolisms [9], linked closely to the onset of systemic low-grade inflammation, which in turn is associated with the development of abdominal fat [10]. The third element in the triad is the initiation of vascular endothelial dysfunction. Indeed, endothelial dysfunction is closely associated with insulin resistance and it is the manifestation of a pro-inflammatory and pro-atherogenic phenotype in the vascular milieu [8,11]. Nutrition, and in particular western diets, have been implicated in the onset of this cardiometabolic risk; for a review see [12–14]. Controlled studies in animals have provided further evidence that insulin resistance, systemic and adipose tissue low-grade inflammation, and vascular endothelial dysfunction, as promoted by western diets, are early features of this cardiometabolic risk cluster [15].

From a mechanistic standpoint, a growing body of evidence is tending to confirm the rationale for a close association between insulin resistance and endothelial function. Firstly, it has been suggested that endothelial dysfunction is the earliest manifestation of diet-induced cardiometabolic risk, even before the onset of insulin resistance and a systemic inflammatory state [15–18]. Secondly, endothelial dysfunction may be largely driven by an impairment of the action of insulin on the endothelium, so that this dysfunction could be considered as a vascular feature of insulin resistance, itself promoting a pro-inflammatory state in the vascular milieu [19,20]. In turn, macro- and micro- vascular endothelial dysfunction limits the action of insulin on the peripheral extraction of nutrients by limiting the perfusion of insulin-sensitive tissues [21,22]. Endothelial dysfunction and insulin resistance would thus interact in a reciprocal relationship [20,23–25]. Abnormal nitric oxide (NO) production or signaling and endothelial dysfunction, triggered by excessive exposure to high-fat and high-sucrose foods, may be one important mediator of diet-induced insulin resistance and cardiometabolic risk [26,27].

2. The Postprandial Period as a Metabolic Challenge Eliciting Pathophysiological Features Related to Cardiometabolic Risk

A very large body of evidence has demonstrated that a metabolic challenge with a high saturated fat and high sucrose meal results in the transient appearance of low-grade inflammation and endothelial dysfunction [28–40].

The level and chronology of these phenomena are closely associated with the postprandial rise in plasma glucose and lipids [35,41–44]. Postprandial inflammation has been characterized at a systemic level [38,45], in blood leukocytes [42,46,47], in the visceral adipose tissue [48,49], and at the vascular level as an increase in intercellular or vascular adhesion molecules and proteins measured in the plasma ICAM-1 et VCAM-1 [32,50]. Other postprandial changes associated with inflammation have been reported after a high fat meal (HFM), such as changes to markers of angiogenesis (vascular endothelial growth factor-VEGF) [51]. Postprandial vascular endothelial dysfunction has also been repeatedly documented using integrative physiological endpoints such as macrovascular reactivity to acute changes in shear stress (particularly using flow-mediated dilation of the brachial artery-FMD) [52,53].

Although the underlying mechanisms are not fully elucidated, the dramatic rise in plasma glucose and triglycerides (and more precisely chylomicrons and their remnants) are considered to be the trigger factors for the activation of inflammatory signaling pathways in leukocytes, endothelial cells, and possibly other cells or tissues [35,48,54–56]. Postprandial oxidative stress is one mediator of the effect of metabolic stress on inflammation and vascular dysfunction [57,58]. Early evidence for the contribution of oxidative stress was provided by the finding that pre-treatment with high doses of vitamin C and/or vitamin E blunted postprandial endothelial dysfunction and inflammation [32,59]. As we also discuss further below, the initiation of low-grade endotoxemia is considered to be an important mechanism [47,60]. Lastly, of importance to our understanding of cardiometabolic pathophysiology is the fact that postprandial inflammation and macro/micro- vascular endothelial dysfunction are all the more important if individuals present at baseline with markers of dysregulation or cardiometabolic risk factors [21,61], and dysfunction increases when the meal challenge is repeated [62].

At the molecular level, considerable importance has been given to NO, primarily because it is well-known as the pivotal molecule of vascular health, and endothelial dysfunction can be explained by alterations to NO synthesis and/or bioactivity. More specifically, regarding postprandial deregulation the role of NO in the insulin-mediated peripheral extraction of nutrients is becoming increasingly well-established [19,22,63–68]. Furthermore, high fat and high sucrose meals impact NO synthesis and/or NO downstream signaling [26,69,70], and studies have confirmed that impairment of the insulin sensitivity of the vascular NO production pathway may explain the impairment of glucose extraction in the muscle [20,23,24,71]. Finally, because the NO pathway is more sensitive to the oxidative/redox state at many different levels, this pathway may mediate the effect of a postprandial increase in oxidative stress on impairment of endothelial function and the initiation of vascular and systemic inflammation.

The final picture is that the postprandial occurrence of low-grade inflammation and endothelial dysfunction is extremely relevant to the pathophysiological influence of nutrition on cardiometabolic risk for the following reasons: (i) low-grade inflammation and endothelial dysfunction are well known to be pivotal to the initiation and progression of cardiometabolic dysregulations, as discussed previously; (ii) their postprandial appearance is directly related to the degree to which energy nutrients challenge homeostasis and are concurrent with deregulations at the cellular and molecular levels; (iii) their postprandial appearance is graduated according to the basal level of metabolic regulation and in line with the existence of risk factors for CVD and type-II diabetes; and (iv), the level of the postprandial rise in plasma triglycerides, and glucose after a meal challenge is considered to be a potent risk factor for CVD and type-II diabetes [72–74]. Finally, the current paradigm is that repetition of these adverse, silent postprandial events is a mechanism for the initiation and progression of metabolic dysregulation, CVD, and type-II diabetes [36,73].

Accordingly, the postprandial state following a challenge meal offers an interesting, practical, and relevant model for studying the impact of nutrients on metabolic dysregulation, and the initiation of cardiometabolic risk factors such as MS.

3. Fatty Acids, Carbohydrates, and Postprandial Adverse Effects

As mentioned above, there is very convincing evidence that a challenge meal containing both saturated fatty acids and sucrose triggers a vast corpus of inflammatory phenomena and endothelial dysfunction features during the postprandial period. A smaller, yet still high, number of studies have also reported similar findings when the challenge meal only contained saturated fat or simple sugars [75–78], although some studies using a single macronutrient were negative [75]. It should be noted that these studies differed markedly in terms of the methods used to study postprandial metabolism [74].

3.1. Fatty Acids in Challenge Meals

In contrast, the literature is less conclusive regarding the role of the type of fatty acids in the challenge meal [79]. It should, however, be noted that olive oil (as compared to oils rich in palmitic acid, or to milk fat) induces a smaller increase in plasma inflammatory markers, does not result in activation of the NF-κB inflammatory pathway in peripheral blood mononuclear cells, and generates less postprandial endothelial dysfunction in healthy individuals and/or those with risk factors [80–82].

When supplementing a high fat meal, fish oils have also been shown to be beneficial to postprandial vascular function. In a postprandial model combining a high-fat meal and a heparin infusion to increase postprandial non-esterified fatty acids (NEFA), the standard high-fat meal with saturated fatty acids (SFA) impaired flow-mediated dilation (FMD) whereas the addition of fish oil to this meal conversely improved FMD 4 h after ingestion [83]. In another study, the introduction of fish oil as part of a high-fat meal improved (endothelium-independent) microvascular reactivity and increased postprandial plasma nitrite concentration (a marker of nitric oxide synthase activity) [84]. Fish oil enhanced eNOS expression in cultured endothelial cells exposed to triglyceride-rich lipoprotein isolated after the meal. When associated with fibers, unsaturated fatty acids (unSFA) blunted the postprandial expression of the inflammatory genes usually found after a high SFA meal; that is, the postprandial circulation levels of IL-1β, IL-6, MCP-1, and IFN-γ did not rise after an unSFA and fiber-rich meal when compared with an SFA meal [85].

An antioxidant and anti-inflammatory effect of olive oil or monounsaturated fatty acids (versus saturated fatty acids and low-fat meals) during the postprandial state has also been reported when the individuals had been receiving diets of a similar composition before the postprandial challenge [86,87].

The underlying mechanism for the effect of SFA on systemic inflammation has been documented. Studies have suggested that SFA increase the intestinal absorption of lipopolysaccharide (LPS), which in turn increases postprandial endotoxemia and the postprandial inflammatory response. For instance, in individuals with metabolic syndrome, a meal rich in SFA raises plasma LPS concentrations when compared to other meals rich in monounsaturated fatty acids (MUFA) or low in fat, and high in complex carbohydrates and n-3 fatty acids. After the SFA meal, the increase in LPS was correlated with the gene expression of IkBα (an NF-kB inhibitor) and MIF1 (a pro-inflammatory cytokine) in peripheral blood mononuclear cells, suggesting partial mediation by these pro-inflammatory pathways [88,89]. Finally, a high SFA meal could be involved in causing postprandial endotoxemia and also affect other mechanisms, including intestinal absorption and clearance rates of LPS, changes to intestinal microbiota, and intestinal barrier function [88]. However, it remains difficult to assess the significance of endotoxins in plasma, as LPSs are highly heterogeneous. Indeed, stimulatory, non-stimulatory, and inhibitory LPS molecules coexist in plasma, and assays cannot distinguish or quantify them separately [90].

In contrast, the literature remains scarce and still inconclusive regarding the effect of different types of saturated fatty acids, or the role of various unsaturated fatty acids [91–95].

3.2. Carbohydrates in Challenge Meals

There is quite a large body of evidence to suggest that sucrose and glucose loads induce postprandial inflammation and endothelial dysfunction, related to the postprandial increase in plasma glucose [75,96], although there have been some negative reports when these loads were given alone (i.e., without saturated fatty acids). To our knowledge, there are no data regarding the effect of other simple carbohydrates. Given the relationship between postprandial plasma glucose and postprandial dysfunctions, the glycemic index (GI) is expected to be an important factor in the adverse effect of carbohydrates, however, findings are scarce and conflicting [56,97,98]. For instance, nuts have shown potential to manage post-meal glucose when consumed with high GI food content [99] but not with low GI foods [100]. Also, the acute ingestion of low-fat milk has been shown to protect adults with metabolic syndrome from endothelial dysfunction when compared to rice milk (high GI). The postprandial serum glucose peak was higher after rice milk and correlated positively with an increase in malondialdehyde (MDA, a biomarker of oxidative stress mostly related to lipid peroxidation) and a drop in plasma arginine, suggesting that cow's milk may limit postprandial hyperglycemia, which in turn may decrease lipid peroxidation and enhance NO bioavailability [101].

Although most studies have resorted to using experimental artificial meals containing high amounts of simple ingredients such as milk cream and sucrose, postprandial inflammation and dysfunction are not the result of an experimental artefact because they have also been evidenced following the consumption of "real" energy-dense meals, such as those supplied by fast-food outlets [46,102–105]. In contrast, some foods, such as orange juice and certain meals considered to form part of a prudent diet (e.g., meals rich in fibers and fruit, or light regular meals), do not induce adverse postprandial effects [106–110].

4. Relevance to the Effect of the Type of Dietary Protein

As mentioned before, some carbohydrates and fat sources do not appear to elicit any adverse effects during the postprandial period. Although dietary proteins are the third most important energy macronutrient, their effects have been little studied.

Indeed, we previously reported that a mixture of 50 g amino acids (based on the total milk protein composition, and with or without a supplement of l-arginine) did not increase plasma markers of inflammation or induce endothelial dysfunction [111].

In a pioneering work, Westphal and colleagues showed that adding dietary protein (milk or soy protein) to a high-fat meal prevented postprandial endothelial dysfunction [112]. This effect could, however, be explained by a quantitative effect of protein, because a high intake of protein (as compared to fat), (i) slowed down gastric emptying and decreased postprandial exposure to fatty acids in the meal [113], and (ii), raised postprandial insulin, which in this context could have anti-inflammatory and anti-atherogenic properties [114]. However, specific effects of protein quality or specific amino acids have also been documented [115]. The same authors reported that a "dietary" amount (2.5 g) of l-arginine alone (and not phenylalanine or leucine) prevented postprandial endothelial dysfunction [78], confirming the results of a study that used a massive dose of l-arginine [116]. The issue of the dose was raised in one of our studies which consisted of supplementing overweight adults with a low dose of l-arginine. After a high fat meal, reductions in the FMD and fRHI (a reactive hyperemia index that is another measure of endothelial function) compared to baseline were attenuated by arginine supplementation in individuals whose plasma arginine concentration was below the median [117]. Likewise, in a validated rat model [70], we showed that rapeseed protein (an arginine- and cysteine-rich protein when compared to milk protein), and the supplementation of milk protein with l-arginine and l-cysteine, prevented postprandial endothelial dysfunction [118]. Using this model, we were also able to show that rapeseed protein markedly reduced a postprandial increase in the production of reactive oxygen species (ROS) in the aorta [70]. Indeed, dietary arginine and cysteine are known to impact critical metabolic pathways (notably glutathione and nitric oxide) and may exert favorable effects on the initiation of cardiometabolic risk factors such as insulin sensitivity and endothelial function [119,120].

It has also been reported in overweight/obese individuals that neither a palmolein nor an olive oil diet impaired postprandial FMD when consumed in a high-fat, high-protein meal rich in l-arginine [121]. These results were not in line with the findings of a study that could not find a protective effect of proteins on postprandial endothelial dysfunction and low-grade inflammation, apart from a decrease in sVCAM after a protein mix compared to maltodextrin. However, the protein mix that was used during that study was not high in arginine, and this might have been the reason for the discrepancy [122].

Other plausible mechanisms (other than the arginine content) could explain the protective effect of milk on cardiometabolic health and endothelial function [123–125]. For example, acute dairy cheese consumption has been demonstrated to improve NO-dependent vasodilation compared to non-dairy products (soy cheese and pretzels) when eaten with non-dairy sodium. This suggests that dairy proteins may protect against Na-induced reductions in NO-dependent dilation [126].

5. Foods, Nutrients, and Other Dietary Substances That May Protect against Adverse Postprandial Effects

The adverse postprandial effects of a high-saturated fat/high-sucrose meal have been used to determine whether adding a nutrient or dietary substance to that meal might lower or prevent the postprandial inflammatory reaction and endothelial dysfunction. Because high exposure to triglycerides and glucose have been convincingly proposed as trigger factors for adverse postprandial effects, numerous studies have addressed the effects of dietary factors on postprandial increases in glucose and triglycerides. As with the addition of protein, some foods or ingredients may basically act through their added weight/energy, slowing down gastric emptying and modulating plasma insulin. Furthermore, the kinetics of digestion and the availability of carbohydrates and fats differ depending on the type of food or the structure of the meal. For instance, the unique physical structure of nuts may explain their role in postprandial regulation. Indeed, the effects of processing on nuts have been shown to affect the postprandial glycemic response [127] by breaking down the nut cell walls and increasing the bioaccessibility of intracellular lipids [128,129], leading to prolonged gastric emptying. Likewise, we have shown that interactions between macronutrients within a meal may modify the kinetics of the absorption of meal fat and result in a different challenge for postprandial metabolism [130,131].

Several nutrients, micronutrients, and phytochemicals may affect postprandial blood lipid concentrations after both acute and chronic consumption, as recently reviewed in detail by Desmarchelier et al. [132]. Among many examples [133], a blend of antioxidant spices added to a high-fat meal lowered postprandial insulin and triglycerides [134]. Nuts have also been described as improving postprandial FMD [135,136], glycemia [137,138], and triglyceridemia [139]. In contrast, in many cases, certain nutrients and other dietary substances that have been shown to reduce the adverse postprandial effects of a challenge meal, did not affect postprandial plasma lipids [140].

5.1. Adding Nuts to a High-Fat/Carbohydrate Meal Prevents Postprandial Endothelial Dysfunction and Oxidative Stress

Glucose fluctuations have been shown to alter endothelial cells by inducing markers of oxidative stress and DNA damage and the onset of a metabolic memory [141,142]. However, it appears that glucose fluctuations do not impact FMD shortly after intake (within 2 h) [143]. Beyond fluctuations in glucose concentrations, evidence has shown that it is the acute consumption of whole macronutrient meals that has the most influence on FMD within 6 h of intake [144].

Nuts have also been involved in improving endothelial function when combined with a meal. In healthy overweight or obese men, the acute consumption of a control shake significantly reduced FMD whereas a peanut shake, matched for nutrient content, did not significantly decrease FMD 4 h after the meal, regardless of the patients' baseline cholesterol concentrations (total cholesterol -TC or low density lipoprotein-LDL) [139]. The peanut shake reduced the triglycerides area under the curve (TG AUC) by 32%. The impact of nuts on postprandial lipemia still needs to be clarified, as the results regarding improvements to postprandial VLDL, HDL, cholesterol efflux [145], and TG [139] are not always consistent [146].

There is some evidence that consuming walnuts improves postprandial endothelial function after a meal challenge in overweight or obese and hypercholesterolemic populations [135,136,147]. When measured with FMD, endothelial function improved over baseline by 64% following daily consumption for four weeks [147] or 24% after acute consumption [135]. In normocholesterolemic [135] or moderately hypercholesterolemic [136] populations only, a walnut meal has been shown to prevent postprandial endothelial dysfunction as assessed using both FMD and RHI measurements.

To determine the walnut component to which the effect on endothelial function could be ascribed, Berryman et al. [136] studied the effects of separated nut skins, de-fatted nutmeat, and nut oil derived from 85 g of whole walnut in mildly hypercholesterolemic individuals. The effect of walnut oil on fRHI differed from those of the skin and whole nut, and this might be related to its fatty acid composition. This is in line with the results of a study that compared two types of walnuts which differed in terms of their polyunsaturated fatty acid contents [148]. Finally, when compared with olive oil, which is quite low in polyunsaturated fatty acids (PUFA), the acute consumption of walnut with a high-fat meal improved endothelial function [135]. Taken together, these findings suggest a beneficial effect of plant PUFA, or in fact α-linolenic acid (ALA), on endothelial function.

Nuts have favorable effects on certain inflammation and oxidative status indices [149]. English walnuts contain the highest antioxidant content [150], and in healthy young adults the acute consumption of a walnut meal increased postprandial γ-tocopherol, catechins, and hydrophilic and lipophilic oxygen radical absorbance capacity (ORAC, a measure of the antioxidant capacity), while decreasing some markers of oxidative stress, such as MDA, when compared with a refined meal matched for energy nutrients [151]. These results suggest that walnuts exert antioxidant activities in both the lipid and aqueous plasma fractions. However, when comparing the antioxidant capacity of plasma regarding different walnut components in individuals with mild hypercholesterolemia, this antioxidant capacity (as assessed by the ferric reducing antioxidant potential, FRAP) was higher after the intake of walnut oil and skin compared with intake of the nutmeat [136].

Phenolic antioxidants may be more effective in MUFA-rich nuts, such as almonds and pistachios, than in PUFA-rich nuts [152]. One study reported that, in healthy individuals, the acute intake of almonds induced less protein damage during the postprandial period than parboiled rice/mashed potato, cheese, and butter meals, whereas the total antioxidant capacity did not differ between the groups [153].

As for the effects of pistachios on inflammation and oxidative stress, data are scarce in the acute setting. Nonetheless, several studies have shown chronic effects on various markers of oxidative stress in individuals with metabolic syndrome [154], hypercholesterolemia [155], and prediabetes [156], or in healthy populations [157,158]. By contrast, in obese people with metabolic syndrome, the acute consumption of pistachio meals had no significant postprandial effect on RHI [159]. It is still difficult to interpret the overall effects of pistachio nuts on postprandial inflammation and oxidative stress based on the results for various markers in isolation because many antioxidant components have been studied in plasma and tissues and there are few data to infer their final possible combined action. However, one study found a significant increase in blood antioxidant potential and lowering of MDA concentration (an indicator of lipid peroxidation) after substituting pistachio nuts for 20% of daily caloric intake for three weeks in a healthy population [158].

A review concluded that pistachios are singularly rich in nutrients and substances that exert antioxidant and anti-inflammatory effects that may be beneficial to cardiovascular health. There is evidence that three key nutrients/phytochemicals in pistachios could mediate these effects: carotenoids, γ-Tocopherol, and phenolic compounds [152].

5.2. Adding Fruit to a High Fat/Carbohydrate Meal Prevents Postprandial Endothelial Dysfunction and Oxidative Stress

The protective effects of extra virgin olive oil on postprandial oxidative stress have frequently been described during the past decade [160–162] and these effects appear to be comparable to those reported with walnuts. Indeed, the acute consumption of walnuts and olive oil in a high-fat meal by patients with hypercholesterolemia caused similar reductions in postprandial plasma concentrations of soluble inflammatory cytokines, adhesion molecules, and oxidized low-density lipoproteins. Only E-selectin levels fell more after the walnut meal than the olive oil meal. The authors concluded that both walnuts and olive oil preserve the protective phenotype of endothelial cells [31].

As are olives, avocados are a fruit that is specifically rich in MUFA (oleic acid) and n-6 PUFA (linoleic acid), and in this respect have also been studied recently in terms of their potential postprandial metabolic and vascular impacts. In overweight/obese individuals with elevated fasting glucose and insulin, the partial substitution of meal carbohydrates with avocado increased postprandial FMD [163]. However, the control breakfast did not result in a significant reduction in postprandial endothelial function as might have been expected. However, this result is important on practical grounds because the introduction of avocado in the meal represented only ~15% of the meal energy. The effect on FMD might, in part, have resulted from the effect on postprandial lipoprotein profiles, such as lower post-meal VLDL with avocado, which could be ascribed to the exchange of carbohydrates for MUFA. The avocado meal also caused a smaller increase in postprandial plasma insulin [163].

Because of their particular composition of nutrients and other substances, berries have also been studied in terms of their benefits on cardiovascular health [164]. In a well-designed study, Alqurashi et al. showed that in healthy overweight males in an acute setting, an acai-based shake (vs. a control shake) consumed alongside a high-fat breakfast significantly improved postprandial FMD [165]. Acai is well known for its high flavonoids content; however, the mechanism underlying the reported benefits of Acai still needs to be elucidated and further research is required to understand the degree to which this effect could be extended to other berries. Additional positive findings have been reported for other berries such as blueberry and raspberry, and possible mediation by polyphenols has been considered. In healthy males, the acute consumption of processed or unprocessed blueberries caused changes to the profile of polyphenols but not the amount, resulting in different patterns of increase in polyphenol metabolites in the plasma but similar improvements in postprandial FMD [166]. After the consumption of raspberries, increases in plasma urolithin metabolites were found to be associated with improvements to endothelial function [167]. Interestingly, plasma total nitrite concentrations have been reported to rise significantly during the 2 h following intake of cranberries, suggesting that polyphenols increase postprandial circulating nitric oxide and mediate the maintenance of postprandial endothelial function [168].

Berries are rich in phytochemicals, and particularly phenolic compounds (2/3 flavonoids such as anthocyanins, catechins, quercetin, and kaempferol, and 1/3 phenolic acids such as ellagic acid), which are considered to be potent antioxidants inasmuch as they are able to scavenge ROS, chelate metal ions in vitro, and act synergistically between themselves and with micronutrients such as ascorbate and tocopherol [165,169,170]. The effects of berries on post-meal oxidative stress have been described in both the acute [171,172] and chronic settings [172]. In a chronic context, berries may exert anti-oxidative and anti-inflammatory effects by modulating mRNA expression in overweight and hypercholesterolemic individuals. Indeed, a study showed that the intake of an aqueous extract of wolfberry fruit (goji) once a day after a meal for eight weeks significantly decreased erythrocyte superoxide dismutase activity, DNA damage in lymphocytes, and the expression of TNF, IL-6, and other mRNAs related to oxidative or inflammatory stress. In addition, superoxide dismutase (SOD) expression in whole-cell extracts was down-regulated [173].

The effects of strawberries on postprandial hyperlipidemia and oxidized low-density lipoprotein cholesterol (LDL) have previously been studied in hyperlipidemic and overweight individuals using a control beverage supplemented with strawberry powder at a dietary dose (equivalent to 110 g fresh strawberries) or a placebo beverage (matched for energy, macronutrient, micronutrient, and fiber contents) given with a high-fat test meal. In the acute setting, the strawberry beverage (vs. the control) lowered postprandial increases in TG, HDL, and OxLDL at 3, 4, and 6 h after the meal. In the chronic setting, after a 6-week period, the strawberry beverage lowered mean cholesterol, LDL, TG, and OxLDL concentrations (when adjusted for fasting values) following the intake of a high-fat meal [172].

Berries reduce the lowering of ORAC that is usually reported after carbohydrate meals. Furthermore, in healthy women, when adding grape and blueberry powder to a carbohydrate meal, ORAC increases within 2 h of intake. When comparing the AUC for the change in plasma hydrophilic ORAC-FL over 4–5 h after a meal, Burton-Freeman et al. found that the decrease was halved after a grape and blueberry supplemented meal as compared to the control meal [171]. The ultimate health impacts of such postprandial changes to ORAC still need to be determined.

Alternatively, in adults with type 2 diabetes, the addition of cranberries (40 g dried) to a high-fat fast-food-style breakfast lowered some biomarkers of inflammation and lipid oxidation, such as serum IL-18 and MDA, 4 h after the meal, although no significant differences in postprandial concentrations of CRP and IL-6 were observed [168]. Postprandially, a meal composed of an antioxidant-rich concentrate of berry added to a turkey burger and in the water consumed during the meal blunted the postprandial increase in MDA, decreased protein carbonyls (a marker of oxidative stress on protein), and increased plasma antioxidant activity [174].

A similar series of protective effects on postprandial inflammation in mononuclear cells has also been reported regarding the consumption of orange juice with a high-fat meal [104]. Consuming orange juice (300 kcal, i.e., ~600 mL, versus water or a glucose solution) with the meal, lowered the postprandial production of ROS by blood polymorphonuclear cells and resulted in less activation of inflammatory pathways such as mitogen-activated protein kinase (MAPK) and suppressor of cytokine signaling 3 (SOCS-3) in mononuclear cells. As with the aforementioned study, orange juice also lessened postprandial low-grade endotoxemia and the expression of toll-like receptor 4 (TLR-4) [104].

Regarding oxidative stress and the effects of fruit juice, the results should be interpreted with caution as most studies assessing the effects of fruit-based beverages on postprandial stress used as a control a drink matched for macro- and micronutrients and not simple water. Therefore, what was being tested was not the fruit-based juice itself but rather the phytochemicals it contained in the context of a drink and a high-fat meal [175].

By contrast, acute avocado consumption was not associated with postprandial changes to biomarkers of inflammation or oxidative stress/damage to MCP-1, tumor necrosis factor alpha (TNF-α), or Ox-LDL [163].

6. Key Phytochemicals Identified as Mediating Postprandial Antioxidant and Anti-Inflammatory Effects

According to the same type of study design, it has been reported that red wine (but not vodka) consumed with a high-fat meal prevented the postprandial activation of NF-κB in mononuclear cells [176]. Indeed, the ingestion of wine with a meal has been reported to reduce postprandial oxidative stress, although the markers chosen for most studies were of limited value [177]. It has also been shown that the consumption of other foods and nutrients does not result in the postprandial inflammation and dysfunction that are induced by high saturated fat and high sucrose loads.

In the context of elucidating the complex effects of wine on endothelial function and postprandial inflammation, it was reported that combining muscadine grape polyphenols with resveratrol—a phenolic compound in red wine that has been long largely studied for various anti-inflammatory effects [178]—reduced postprandial increases in a set of pro-inflammatory and inflammatory markers in mononuclear cells, such as the expression of IL1-β and SOCS-3 [179]. Because the dose of resveratrol (100 mg) used in this study was very high when compared to the amounts found in wine [180–183], the results cannot be used to conclude that wine polyphenols and resveratrol are candidates for a potentially favorable effect of wine on postprandial inflammation [184,185]. However, they offer a good example of the potential effects of combining different chemicals at neutraceutical doses on postprandial dysfunctions.

The effects of the resveratrol and polyphenols combination were considered in detail in the same study, and this work also provided some interesting insights into the possible mechanisms underlying prevention of the initiation of inflammation in mononuclear cells in the postprandial setting. The combination of resveratrol and polyphenols largely reduced the increase in the expression of the p47 NADPH subunit, which is known to be associated with a postprandial increase in oxidative stress in mononuclear cells. Furthermore, the supplement increased the binding activity of Nrf-2 and the expression of some target genes. Because Nrf-2 is a transcription factor that mediates the physiological antioxidant response to oxidative stress, this supplementation may have limited the production of ROS yet evoked a higher protective antioxidant response, and the nrf-2 pathway might be important in mediating the adverse effect of triglyceride-rich lipoprotein on vascular health [186]. However, in our view, because of the delay required for this antioxidant response to take effect, it might not generally account for the series of protective effects that appear acutely in the postprandial phase. Other authors have confirmed that grape powder (in quantities compatible with dietary modulation) increases the expression of Nrf-2 acutely during a high fat carbohydrate meal [187]. Another result of importance to our understanding of the pathogenesis of postprandial adverse effects and that of the effect of the supplement in the study by Ghanim et al. is that the supplement also reduced or prevented postprandial low-grade endotoxemia, plasma lipoprotein binding protein, and TLR-4 expression in mononuclear cells. As discussed above, an increase in the translocation of endotoxins from the gut has been proposed as a mechanism for the adverse postprandial effect of high-fat meals on low-grade inflammation and endothelial dysfunction [75,188,189]. However, such a mechanism may not strictly require TLR-4 mediation, but rather may act through a combined interplay between TLRs [190]. Therefore, the protective effect of the supplement may be mediated, at least in part, by a reduction in postprandial low-grade endotoxemia and downstream pro-inflammatory signaling [179], although the potential underlying mechanisms for an acute reduction in endotoxemia still need to be fully elucidated.

With respect to berries, urolithins and ellagic acid appear to be the best candidates for their anti-atherogenic effects. In quantities compatible with dietary modulation, these phytochemicals have displayed their potential to affect key processes in the development and progression of atherosclerosis in vitro, such as endothelial activation and resulting monocyte recruitment, cholesterol transport, and foam cell formation [191].

Some polyphenolic compounds are attractive candidates to explain the effects of orange juice. In this regard, it was shown that the consumption for four weeks of 500 mL orange juice, or hesperidin (the major flavonoid in orange juice), increased microvascular endothelium-dependent function during the peak of hesperidin absorption [192]. This result cannot directly be extrapolated to the postprandial phase. However, because hesperidin has a short half-life in plasma, its effect may be mostly transient so it may operate acutely during the postprandial period, with favorable effects on macro- or micro-vascular endothelial function or other related pro-inflammatory postprandial features. More recently, in adults with hypertriglyceridemia or who were overweight/obese and subjected to a double high-fat meal challenge, a study reported that various orange-based drinks containing flavanone (vs. an isoenergetic control) alleviated the postprandial decrease in FMD 7 h after a high-fat meal. The effects were similar despite variations by a factor of four in the amount of flavanone in the drinks. However, the effect on FMD at 7 h coincided with the peak of naringenin and hesperidin metabolites being found in the plasma, and the fraction of hesperidin metabolites assayed in plasma predicted, in part, the magnitude of the changes to FMD [193]. Salden et al. did not evidence any effect of supplementation with hesperidin 2S in their study sample as a whole, either with acute postprandial testing, or after six weeks of supplementation. In individuals with a normal or high baseline FMD (60% of the total sample), hesperidin 2S improved FMD and reduced adhesion molecules after a HFM, when the latter was given after six weeks of supplementation [194].

It is difficult to draw any firm conclusions from the literature on the postprandial effects of polyphenols. First, many studies have investigated the effects of polyphenol-rich foods (such as cocoa, grape, or berries) or food preparations (e.g., juices) rather than purified and well-characterized extracts. Again, adding a food/ingredient with a significant mass and energy (e.g., 500 mL juice) to a challenge meal tends to affect the kinetics of postprandial metabolism directly, and the results are therefore difficult to analyze. More recently, studies have used ingredients such as powders, and control treatments matched for macronutrient content, which is more useful when trying to ascribe the effects to polyphenolic fractions [172,187,195]. Second, studies have resorted to different postprandial endpoints and markers, in limited numbers, giving rise to highly fragmented findings. For instance, the consumption of a juice rich in blackcurrant polyphenols (as compared to a well-made placebo drink) was shown to improve postprandial oxidative status, but in vivo evidence for postprandial anti-inflammatory effects was lacking [196] as was evidence for a beneficial effect on vascular reactivity [197]. Likewise, an anthocyanin-rich blackcurrant extract lowered postprandial glucose and insulin after a high-carbohydrate meal but did not affect 8-isoprostane F2α (a stable and reliable marker of overall lipid peroxidation [164]) or arterial stiffness; endothelial function was not measured [198].

Many phenolic compounds have shown that they can reduce postprandial oxidative stress or acutely affect parameters for oxidative stress [164,199]. This has been largely documented for cocoa flavanols and grape polyphenols [200]. Potential mechanisms of action have also been reported. In vitro, a grape seed extract and a strawberry powder activated NO synthesis pathways in endothelial cells [201,202]. Cocoa flavanols acutely increased plasma concentrations of nitroso compounds, reduced arginase activity [203,204], affected pro-inflammatory pathways in vitro [205], and acutely improved endothelial dysfunction [206–208]. However, for many polyphenolic compounds with interesting in vivo effects after ingestion, the evidence remains limited regarding their effects on important endpoints of cardiometabolic health, such as endothelial function in humans [196,209].

7. Conclusions

As we have shown in this review, acute supplementation with certain whole foods, ingredients, nutrients, and phytochemicals can prevent postprandial endothelial dysfunction and inflammation. Because many studies have lent credence to the current paradigm that oxidative stress mediates adverse postprandial effects [58], further efforts are necessary to determine whether nutrients and substances that display postprandial antioxidant effects also reduce postprandial low-grade inflammation and vascular endothelial dysfunction. Future studies also need to investigate other mechanisms that are good candidates for the acute effects of nutrients during a high-fat meal, such the induction of low-grade endotoxemia. These studies could take advantage of simultaneously analyzing the effects on different endpoints and using various markers. It would be interesting to further clarify the degree to which certain nutrients (e.g., some amino acids) and other substances (in particular various polyphenols) affect potential underlying mechanisms that are directly or indirectly related to oxidative stress, including NO and nitroso-compound metabolism, induction of the antioxidant defense system, the delicate redox status in tissues, and insulin-related signaling pathways [203,210–213]. Further studies could also profitably investigate acute variations in the metabolism of arginine and related compounds (such as homoarginine and methylated arginine) during the postprandial period, and their potential modulation by the nature of the meal [214,215].

The metabolic utilization and effects of any nutrients added to a high-fat high sucrose meal are basically postprandial (e.g., amino acids). Many other dietary substances also exhibit an acute postprandial metabolism; in particular, despite huge heterogeneity, many polyphenols (e.g., flavonoids) and their metabolites display early plasma peaks (e.g., at 2 h) and a short half-life in plasma after ingestion [216]. We can therefore expect that many nutrients and other substances exert most of their biological effects during the postprandial period. This factor warrants dedicated investigations of their specific effects under adverse postprandial conditions.

Finally, we can conclude that based on a very large set of data, the present paradigm is that the postprandial occurrence of cardiometabolic-related dysfunctions, including postprandial inflammation and endothelial dysfunction, are pathogenic to the initiation and progression of MS. The high-saturated-fat/high-sucrose model is therefore highly relevant to preventive nutrition, and as it is also practical for the conduct of human trials, it can be used to study the benefit of nutrients and other dietary substances when added to a challenge meal or consumed immediately beforehand. Although many studies have addressed the postprandial effects of nutrients and other substances, the literature remains largely fragmented. In particular, some nutrients and substances have been shown to lower postprandial oxidative stress and impact inflammatory-related pathways, but further studies are needed and should involve final critical endpoints such as endothelial dysfunction. Nevertheless, we found that nuts, l-arginine, polyphenols from berries, and citrus are good candidates for acute and multiple protective effects during the postprandial phase, and the data so far warrant further investigations involving multiple clear endpoints and valid, sensitive markers to ascertain the global picture.

Author Contributions: Conceptualization, F.M.; Review process, L.D. and F.M.; Writing—original draft preparation, L.D. and F.M.; Writing—review and editing, L.D. and F.M.

Funding: This research received no external funding.

Conflicts of Interest: L.D. declares no conflict of interest. F.M. has been the principal investigator for studies on l-arginine supplementation that have received grants from the Institut de Recherche Pierre Fabre.

References

1. Ford, E.S.; Ajani, U.A.; Mokdad, A.H. The Metabolic Syndrome and Concentrations of C-Reactive Protein Among U.S. Youth. *Diabetes Care* **2005**, *28*, 878–881. [CrossRef] [PubMed]
2. Grundy, S.M.; Brewer, H.B., Jr.; Cleeman, J.I.; Smith, S.C., Jr.; Lenfant, C. Definition of metabolic syndrome: Report of the National Heart, Lung, and Blood Institute/American Heart Association conference on scientific issues related to definition. *Arterioscler. Thromb. Vasc. Biol.* **2004**, *24*, e13–e18. [PubMed]

3. Grundy, S.M. Metabolic syndrome pandemic. *Arterioscler. Thromb. Vasc. Biol.* **2008**, *28*, 629–636. [CrossRef] [PubMed]

4. Saklayen, M.G. The Global Epidemic of the Metabolic Syndrome. *Curr. Hypertens. Rep.* **2018**, *20*, 12. [CrossRef] [PubMed]

5. Alberti, K.G.M.; Zimmet, P.; Shaw, J. The metabolic syndrome—A new worldwide definition. *Lancet* **2005**, *366*, 1059–1062. [CrossRef]

6. Alberti, K.G.; Eckel, R.H.; Grundy, S.M.; Zimmet, P.Z.; Cleeman, J.I.; Donato, K.A.; Fruchart, J.C.; James, W.P.; Loria, C.M.; Smith, S.C., Jr. Harmonizing the metabolic syndrome: A joint interim statement of the International Diabetes Federation Task Force on Epidemiology and Prevention; National Heart, Lung, and Blood Institute; American Heart Association; World Heart Federation; International Atherosclerosis Society; and International Association for the Study of Obesity. *Circulation* **2009**, *120*, 1640–1645. [PubMed]

7. Huang, P.L. A comprehensive definition for metabolic syndrome. *Dis. Models Mech.* **2009**, *2*, 231–237. [CrossRef] [PubMed]

8. Sattar, N. Inflammation and endothelial dysfunction: Intimate companions in the pathogenesis of vascular disease? *Clin. Sci.* **2004**, *106*, 443–445. [CrossRef]

9. Roberts, C.K.; Hevener, A.L.; Barnard, R.J. Metabolic Syndrome and Insulin Resistance: Underlying Causes and Modification by Exercise Training. *Compr. Physiol.* **2013**, *3*, 1–58.

10. Carey, D.G.; Jenkins, A.B.; Campbell, L.V.; Freund, J.; Chisholm, D.J. Abdominal Fat and Insulin Resistance in Normal and Overweight Women: Direct Measurements Reveal a Strong Relationship in Subjects at Both Low and High Risk of NIDDM. *Diabetes* **1996**, *45*, 633–638. [CrossRef]

11. Godo, S.; Shimokawa, H. Endothelial Functions. *Arterioscler. Thromb. Vasc. Biol.* **2017**, *37*, e108–e114. [CrossRef]

12. Reaven, G.M. The Insulin Resistance Syndrome: Definition and Dietary Approaches to Treatment. *Annu. Rev. Nutr.* **2005**, *25*, 391–406. [CrossRef]

13. Rodriguez-Monforte, M.; Sanchez, E.; Barrio, F.; Costa, B.; Flores-Mateo, G. Metabolic syndrome and dietary patterns: A systematic review and meta-analysis of observational studies. *Eur. J. Nutr.* **2017**, *56*, 925–947. [CrossRef]

14. Keane, D.; Kelly, S.; Healy, N.P.; McArdle, M.A.; Holohan, K.; Roche, H.M. Diet and metabolic syndrome: An overview. *Curr. Vasc. Pharmacol.* **2013**, *11*, 842–857. [CrossRef]

15. Kim, F.; Pham, M.; Maloney, E.; Rizzo, N.O.; Morton, G.J.; Wisse, B.E.; Kirk, E.A.; Chait, A.; Schwartz, M.W. Vascular Inflammation, Insulin Resistance, and Reduced Nitric Oxide Production Precede the Onset of Peripheral Insulin Resistance. *Arterioscler. Thromb. Vasc. Biol.* **2008**, *28*, 1982–1988. [CrossRef] [PubMed]

16. Hsueh, W.A.; Lyon, C.J.; Quiñones, M.J. Insulin resistance and the endothelium. *Am. J. Med.* **2004**, *117*, 109–117. [CrossRef]

17. Grandl, G.; Wolfrum, C. Hemostasis, endothelial stress, inflammation, and the metabolic syndrome. *Semin. Immunopathol.* **2018**, *40*, 215–224. [CrossRef] [PubMed]

18. Olver, T.D.; Grunewald, Z.I.; Jurrissen, T.J.; MacPherson, R.E.K.; LeBlanc, P.J.; Schnurbusch, T.R.; Czajkowski, A.M.; Laughlin, M.H.; Rector, R.S.; Bender, S.B.; et al. Microvascular insulin resistance in skeletal muscle and brain occurs early in the development of juvenile obesity in pigs. *Am. J. Physiol. Regul. Integr. Comp. Physiol.* **2018**, *314*, R252–R264. [CrossRef]

19. Muniyappa, R.; Quon, M.J. Insulin action and insulin resistance in vascular endothelium. *Curr. Opin. Clin. Nutr. Metab. Care* **2007**, *10*, 523–530. [CrossRef]

20. Kim, J.A.; Montagnani, M.; Koh, K.K.; Quon, M.J. Reciprocal relationships between insulin resistance and endothelial dysfunction: Molecular and pathophysiological mechanisms. *Circulation* **2006**, *113*, 1888–1904. [CrossRef]

21. Jonk, A.M.; Houben, A.J.; Schaper, N.C.; de Leeuw, P.W.; Serne, E.H.; Smulders, Y.M.; Stehouwer, C.D. Obesity is associated with impaired endothelial function in the postprandial state. *Microvasc. Res.* **2011**, *82*, 423–429. [CrossRef]

22. Sorop, O.; Van De Wouw, J.; Heinonen, I.; Merkus, D.; Olver, T.D.; Van Duin, R.W.; Duncker, D.J. The microcirculation: A key player in obesity-associated cardiovascular disease. *Cardiovasc. Res.* **2017**, *113*, 1035–1045. [CrossRef]

23. Tousoulis, D.; Tsarpalis, K.; Cokkinos, D.; Stefanadis, C. Effects of insulin resistance on endothelial function: Possible mechanisms and clinical implications. *Diabetes Obes. Metab.* **2008**, *10*, 834–842. [CrossRef]

24. Kubota, T.; Kubota, N.; Kumagai, H.; Yamaguchi, S.; Kozono, H.; Takahashi, T.; Inoue, M.; Itoh, S.; Takamoto, I.; Sasako, T.; et al. Impaired Insulin Signaling in Endothelial Cells Reduces Insulin-Induced Glucose Uptake by Skeletal Muscle. *Cell Metab.* **2011**, *13*, 294–307. [CrossRef]

25. Loader, J.; Khouri, C.; Taylor, F.; Stewart, S.; Lorenzen, C.; Cracowski, J.; Walther, G.; Roustit, M. The continuums of impairment in vascular reactivity across the spectrum of cardiometabolic health: A systematic review and network meta-analysis. *Obes. Rev.* **2019**, *20*, 906–920. [CrossRef]

26. Rizzo, N.O.; Maloney, E.; Pham, M.; Luttrell, I.; Wessell, H.; Tateya, S.; Daum, G.; Handa, P.; Schwartz, M.W.; Kim, F. Reduced NO-cGMP Signaling Contributes to Vascular Inflammation and Insulin Resistance Induced by High-Fat Feeding. *Arterioscler. Thromb. Vasc. Biol.* **2010**, *30*, 758–765. [CrossRef]

27. Sweazea, K.L.; Lekic, M.; Walker, B.R. Comparison of mechanisms involved in impaired vascular reactivity between high sucrose and high fat diets in rats. *Nutr. Metab.* **2010**, *7*, 48. [CrossRef]

28. Vogel, R.A.; Corretti, M.C.; Plotnick, G.D. Effect of a Single High-Fat Meal on Endothelial Function in Healthy Subjects. *Am. J. Cardiol.* **1997**, *79*, 350–354. [CrossRef]

29. Ceriello, A. The post-prandial state and cardiovascular disease: Relevance to diabetes mellitus. *Diabetes Metab. Res. Rev.* **2000**, *16*, 125–132. [CrossRef]

30. De Koning, E.J.; Rabelink, T.J. Endothelial function in the post-prandial state. *Atheroscler. Suppl.* **2002**, *3*, 11–16. [CrossRef]

31. Lee, I.K.; Kim, H.S.; Bae, J.H. Endothelial dysfunction: Its relationship with acute hyperglycaemia and hyperlipidemia. *Int. J. Clin. Pract. Suppl.* **2002**, *129*, 59–64.

32. Nappo, F.; Esposito, K.; Cioffi, M.; Giugliano, G.; Molinari, A.M.; Paolisso, G.; Marfella, R.; Giugliano, D. Postprandial endothelial activation in healthy subjects and in type 2 diabetic patients: Role of fat and carbohydrate meals. *J. Am. Coll. Cardiol.* **2002**, *39*, 1145–1150. [CrossRef]

33. Bae, J.H.; Schwemmer, M.; Lee, I.K.; Lee, H.J.; Park, K.R.; Kim, K.Y.; Bassenge, E. Postprandial hypertriglyceridemia-induced endothelial dysfunction in healthy subjects is independent of lipid oxidation. *Int. J. Cardiol.* **2003**, *87*, 259–267. [CrossRef]

34. Alipour, A.; Elte, J.; Van Zaanen, H.; Rietveld, A.; Cabezas, M.C. Postprandial inflammation and endothelial dysfuction: Figure. *Biochem. Soc. Trans.* **2007**, *35*, 466–469. [CrossRef]

35. Wautier, J.L.; Boulanger, E.; Wautier, M.P. Postprandial hyperglycemia alters inflammatory and hemostatic parameters. *Diabetes Metab.* **2006**, *32*, 34–36. [CrossRef]

36. Burdge, G.C.; Calder, P.C. Plasma cytokine response during the postprandial period: A potential causal process in vascular disease? *Br. J. Nutr.* **2005**, *93*, 3–9. [CrossRef]

37. van Oostrom, A.J.; Sijmonsma, T.P.; Verseyden, C.; Jansen, E.H.; de Koning, E.J.; Rabelink, T.J.; Castro Cabezas, M. Postprandial recruitment of neutrophils may contribute to endothelial dysfunction. *J. Lipid Res.* **2003**, *44*, 576–583. [CrossRef]

38. Blackburn, P.; Després, J.P.; Lamarche, B.; Tremblay, A.; Bergeron, J.; Lemieux, I.; Couillard, C. Postprandial Variations of Plasma Inflammatory Markers in Abdominally Obese Men. *Obesity* **2006**, *14*, 1747–1754. [CrossRef]

39. Poppitt, S.D. Postprandial Lipaemia, Haemostasis, Inflammatory Response and other Emerging Risk Factors for Cardiovascular Disease: The Influence of Fatty Meals. *Curr. Nutr. Food Sci.* **2005**, *1*, 23–34. [CrossRef]

40. Esposito, K.; Ciotola, M.; Sasso, F.C.; Cozzolino, D.; Saccomanno, F.; Assaloni, R.; Ceriello, A.; Giugliano, D. Effect of a single high-fat meal on endothelial function in patients with the metabolic syndrome: Role of tumor necrosis factor-alpha. *Nutr. Metab. Cardiovasc. Dis.* **2007**, *17*, 274–279. [CrossRef]

41. Ceriello, A. Impaired glucose tolerance and cardiovascular disease: The possible role of post-prandial hyperglycemia. *Am. Heart J.* **2004**, *147*, 803–807. [CrossRef]

42. Van Oostrom, A.; Rabelink, T.; Verseyden, C.; Sijmonsma, T.; Plokker, H.; De Jaegere, P.; Cabezas, M.C. Activation of leukocytes by postprandial lipemia in healthy volunteers. *Atherosclerosis* **2004**, *177*, 175–182. [CrossRef]

43. Båvenholm, P.N.; Efendic, S. Postprandial hyperglycaemia and vascular damage—The benefits of acarbose. *Diabetes Vasc. Dis. Res.* **2006**, *3*, 72–79. [CrossRef]

44. Alipour, A.; Elte, J.; Van Zaanen, H.; Rietveld, A.; Cabezas, M.C. Novel aspects of postprandial lipemia in relation to atherosclerosis. *Atheroscler. Suppl.* **2008**, *9*, 39–44. [CrossRef]

45. Ceriello, A.; Quagliaro, L.; Piconi, L.; Assaloni, R.; Da Ros, R.; Maier, A.; Esposito, K.; Giugliano, D. Effect of Postprandial Hypertriglyceridemia and Hyperglycemia on Circulating Adhesion Molecules and Oxidative Stress Generation and the Possible Role of Simvastatin Treatment. *Diabetes* **2004**, *53*, 701–710. [CrossRef]

46. Aljada, A.; Mohanty, P.; Ghanim, H.; Abdo, T.; Tripathy, D.; Chaudhuri, A.; Dandona, P. Increase in intranuclear nuclear factor kappaB and decrease in inhibitor kappaB in mononuclear cells after a mixed meal: Evidence for a proinflammatory effect. *Am. J. Clin. Nutr.* **2004**, *79*, 682–690. [CrossRef]

47. Herieka, M.; Erridge, C. High-fat meal induced postprandial inflammation. *Mol. Nutr. Food Res.* **2014**, *58*, 136–146. [CrossRef]

48. Magné, J.; Mariotti, F.; Fischer, R.; Mathé, V.; Tomé, D.; Huneau, J.F. Early postprandial low-grade inflammation after high-fat meal in healthy rats: Possible involvement of visceral adipose tissue. *J. Nutr. Biochem.* **2010**, *21*, 550–555. [CrossRef]

49. Meneses, M.E.; Camargo, A.; Jimenez-Gomez, Y.; Paniagua, J.A.; Tinahones, F.J.; Roche, H.M.; Perez-Jimenez, F.; Malagón, M.M.; Perez-Martinez, P.; Delgado-Lista, J.; et al. Postprandial inflammatory response in adipose tissue of patients with metabolic syndrome after the intake of different dietary models. *Mol. Nutr. Food Res.* **2011**, *55*, 1759–1770. [CrossRef]

50. Rubin, D.; Claas, S.; Pfeuffer, M.; Nothnagel, M.; Foelsch, U.R.; Schrezenmeir, J. s-ICAM-1 and s-VCAM-1 in healthy men are strongly associated with traits of the metabolic syndrome, becoming evident in the postprandial response to a lipid-rich meal. *Lipids Health Dis.* **2008**, *7*, 32. [CrossRef]

51. Emerson, S.R.; Sciarrillo, C.M.; Kurti, S.P.; Emerson, E.M.; Rosenkranz, S.K. High-Fat Meal-Induced Changes in Markers of Inflammation and Angiogenesis in Healthy Adults Who Differ by Age and Physical Activity Level. *Curr. Dev. Nutr.* **2019**, *3*. [CrossRef]

52. Korkmaz, H.; Onalan, O. Evaluation of Endothelial Dysfunction: Flow-Mediated Dilation. *Endothelium* **2008**, *15*, 157–163. [CrossRef]

53. Thijssen, D.H.J.; Black, M.A.; Pyke, K.E.; Padilla, J.; Atkinson, G.; Harris, R.A.; Parker, B.; Widlansky, M.E.; Tschakovsky, M.E.; Green, D.J. Assessment of flow-mediated dilation in humans: A methodological and physiological guideline. American journal of physiology. *Heart Circ. Physiol.* **2011**, *300*, H2–H12. [CrossRef]

54. Alipour, A.; Van Oostrom, A.J.H.; Izraeljan, A.; Verseyden, C.; Collins, J.M.; Frayn, K.N.; Plokker, T.W.; Elte, J.W.F.; Cabezas, M.C. Leukocyte Activation by Triglyceride-Rich Lipoproteins. *Arter. Thromb. Vasc. Biol.* **2008**, *28*, 792–797. [CrossRef]

55. Norata, G.D.; Grigore, L.; Raselli, S.; Redaelli, L.; Hamsten, A.; Maggi, F.; Eriksson, P.; Catapano, A.L. Post-prandial endothelial dysfunction in hypertriglyceridemic subjects: Molecular mechanisms and gene expression studies. *Atherosclerosis* **2007**, *193*, 321–327. [CrossRef]

56. Dickinson, S.; Hancock, D.P.; Petocz, P.; Ceriello, A.; Brand-Miller, J. High-glycemic index carbohydrate increases nuclear factor-kappaB activation in mononuclear cells of young, lean healthy subjects. *Am. J. Clin. Nutr.* **2008**, *87*, 1188–1193.

57. Bowen, P.E.; Borthakur, G. Postprandial lipid oxidation and cardiovascular disease risk. *Curr. Atheroscler. Rep.* **2004**, *6*, 477–484. [CrossRef]

58. Sies, H.; Stahl, W.; Sevanian, A. Nutritional, Dietary and Postprandial Oxidative Stress. *J. Nutr.* **2005**, *135*, 969–972. [CrossRef]

59. Plotnick, G.D. Effect of antioxidant vitamins on the transient impairment of endothelium-dependent brachial artery vasoactivity following a single high-fat meal. *JAMA* **1997**, *278*, 1682–1686. [CrossRef]

60. Schwander, F.; Kopf-Bolanz, K.A.; Buri, C.; Portmann, R.; Egger, L.; Chollet, M.; McTernan, P.G.; Piya, M.K.; Gijs, M.A.M.; Vionnet, N.; et al. A Dose-Response Strategy Reveals Differences between Normal-Weight and Obese Men in Their Metabolic and Inflammatory Responses to a High-Fat Meal. *J. Nutr.* **2014**, *144*, 1517–1523. [CrossRef]

61. Patel, C.; Ghanim, H.; Ravishankar, S.; Sia, C.L.; Viswanathan, P.; Mohanty, P.; Dandona, P. Prolonged reactive oxygen species generation and nuclear factor-kappaB activation after a high-fat, high-carbohydrate meal in the obese. *J. Clin. Endocrinol. Metab.* **2007**, *92*, 4476–4479. [CrossRef]

62. Tushuizen, M.E.; Nieuwland, R.; Scheffer, P.G.; Sturk, A.; Heine, R.J.; Diamant, M. Two consecutive high-fat meals affect endothelial-dependent vasodilation, oxidative stress and cellular microparticles in healthy men. *J. Thromb. Haemost.* **2006**, *4*, 1003–1010. [CrossRef]

63. Vincent, M.A.; Clerk, L.H.; Lindner, J.R.; Klibanov, A.L.; Clark, M.G.; Rattigan, S.; Barrett, E.J. Microvascular Recruitment Is an Early Insulin Effect That Regulates Skeletal Muscle Glucose Uptake In Vivo. *Diabetes* **2004**, *53*, 1418–1423. [CrossRef]

64. Vincent, M.A.; Montagnani, M.; Quon, M.J. Molecular and physiologic actions of insulin related to production of nitric oxide in vascular endothelium. *Curr. Diabetes Rep.* **2003**, *3*, 279–288. [CrossRef]

65. Vincent, M.A.; Barrett, E.J.; Lindner, J.R.; Clark, M.G.; Rattigan, S. Inhibiting NOS blocks microvascular recruitment and blunts muscle glucose uptake in response to insulin. *Am. J. Physiol. Metab.* **2003**, *285*, 123–129. [CrossRef]

66. Tessari, P.; Coracina, A.; Puricelli, L.; Vettore, M.; Cosma, A.; Millioni, R.; Cecchet, D.; Avogaro, A.; Tiengo, A.; Kiwanuka, E. Acute effect of insulin on nitric oxide synthesis in humans: A precursor-product isotopic study. *Am. J. Physiol. Metab.* **2007**, *293*, 776–782. [CrossRef]

67. Delgado-Lista, J.; Garcia-Rios, A.; Perez-Martinez, P.; Fuentes, F.; Jimenez-Gomez, Y.; Gomez-Luna, M.J.; Parnell, L.D.; Marin, C.; Lai, C.Q.; Perez-Jimenez, F.; et al. Gene variations of nitric oxide synthase regulate the effects of a saturated fat rich meal on endothelial function. *Clin. Nutr.* **2011**, *30*, 234–238. [CrossRef]

68. Barrett, E.J.; Eggleston, E.M.; Inyard, A.C.; Wang, H.; Li, G.; Chai, W.; Liu, Z. The vascular actions of insulin control its delivery to muscle and regulate the rate-limiting step in skeletal muscle insulin action. *Diabetologia* **2009**, *52*, 752–764. [CrossRef]

69. Giugliano, D.; Marfella, R.; Coppola, L.; Verrazzo, G.; Acampora, R.; Giunta, R.; Nappo, F.; Lucarelli, C.; D'Onofrio, F. Vascular effects of acute hyperglycemia in humans are reversed by L-arginine. Evidence for reduced availability of nitric oxide during hyperglycemia. *Circulation* **1997**, *95*, 1783–1790. [CrossRef]

70. Magné, J.; Huneau, J.F.; Delemasure, S.; Rochette, L.; Tomé, D.; Mariotti, F. Whole-body basal nitric oxide production is impaired in postprandial endothelial dysfunction in healthy rats. *Nitric Oxide* **2009**, *21*, 37–43. [CrossRef]

71. Anfossi, G.; Russo, I.; Doronzo, G.; Trovati, M. Contribution of insulin resistance to vascular dysfunction. *Arch. Physiol. Biochem.* **2009**, *115*, 199–217. [CrossRef]

72. Cohn, J.S. Are we ready for a prospective study to investigate the role of chylomicrons in cardiovascular disease? *Atheroscler. Suppl.* **2008**, *9*, 15–18. [CrossRef]

73. O'Keefe, J.H.; Bell, D.S. Postprandial Hyperglycemia/Hyperlipidemia (Postprandial Dysmetabolism) Is a Cardiovascular Risk Factor. *Am. J. Cardiol.* **2007**, *100*, 899–904. [CrossRef]

74. Lairon, D.; Lopez-Miranda, J.; Williams, C. Methodology for studying postprandial lipid metabolism. *Eur. J. Clin. Nutr.* **2007**, *61*, 1145–1161. [CrossRef]

75. Deopurkar, R.; Ghanim, H.; Friedman, J.; Abuaysheh, S.; Sia, C.L.; Mohanty, P.; Viswanathan, P.; Chaudhuri, A.; Dandona, P. Differential effects of cream, glucose, and orange juice on inflammation, endotoxin, and the expression of Toll-like receptor-4 and suppressor of cytokine signaling-3. *Diabetes Care* **2010**, *33*, 991–997. [CrossRef]

76. Nicholls, S.J.; Lundman, P.; Harmer, J.A.; Cutri, B.; Griffiths, K.A.; Rye, K.A.; Barter, P.J.; Celermajer, D.S. Consumption of Saturated Fat Impairs the Anti-Inflammatory Properties of High-Density Lipoproteins and Endothelial Function. *J. Am. Coll. Cardiol.* **2006**, *48*, 715–720. [CrossRef]

77. Spallarossa, P.; Garibaldi, S.; Barisione, C.; Ghigliotti, G.; Altieri, P.; Tracchi, I.; Fabbi, P.; Barsotti, A.; Brunelli, C. Postprandial serum induces apoptosis in endothelial cells: Role of polymorphonuclear-derived myeloperoxidase and metalloproteinase-9 activity. *Atherosclerosis* **2008**, *198*, 458–467. [CrossRef]

78. Borucki, K.; Aronica, S.; Starke, I.; Luley, C.; Westphal, S. Addition of 2.5 g l-arginine in a fatty meal prevents the lipemia-induced endothelial dysfunction in healthy volunteers. *Atherosclerosis* **2009**, *205*, 251–254. [CrossRef]

79. Rathnayake, K.M.; Weech, M.; Jackson, K.G.; Lovegrove, J.A. Impact of meal fatty acid composition on postprandial lipaemia, vascular function and blood pressure in postmenopausal women. *Nutr. Res. Rev.* **2018**, *31*, 193–203. [CrossRef]

80. Jiménez-Gómez, Y.; López-Miranda, J.; Blanco-Colio, L.M.; Marín, C.; Perez-Martinez, P.; Ruano, J.; Paniagua, J.A.; Rodríguez, F.; Egido, J.; Pérez-Jiménez, F. Olive oil and walnut breakfasts reduce the postprandial inflammatory response in mononuclear cells compared with a butter breakfast in healthy men. *Atherosclerosis* **2009**, *204*, e70–e76. [CrossRef]

81. Pacheco, Y.M.; López, S.; Bermudez, B.; Abia, R.; Villar, J.; Muriana, F.J. A meal rich in oleic acid beneficially modulates postprandial sICAM-1 and sVCAM-1 in normotensive and hypertensive hypertriglyceridemic subjects. *J. Nutr. Biochem.* **2008**, *19*, 200–205. [CrossRef]

82. Tentolouris, N.; Arapostathi, C.; Perrea, D.; Kyriaki, D.; Revenas, C.; Katsilambros, N. Differential Effects of Two Isoenergetic Meals Rich in Saturated or Monounsaturated Fat on Endothelial Function in Subjects With Type 2 Diabetes. *Diabetes Care* **2008**, *31*, 2276–2278. [CrossRef]

83. Newens, K.J.; Thompson, A.K.; Jackson, K.G.; Wright, J.; Williams, C.M. DHA-rich fish oil reverses the detrimental effects of saturated fatty acids on postprandial vascular reactivity. *Am. J. Clin. Nutr.* **2011**, *94*, 742–748. [CrossRef]

84. Armah, C.K.; Jackson, K.G.; Doman, I.; James, L.; Cheghani, F.; Minihane, A.M. Fish oil fatty acids improve postprandial vascular reactivity in healthy men. *Clin. Sci.* **2008**, *114*, 679–686. [CrossRef]

85. Monfort-Pires, M.; Crisma, A.R.; Bordin, S.; Ferreira, S.R.G. Greater expression of postprandial inflammatory genes in humans after intervention with saturated when compared to unsaturated fatty acids. *Eur. J. Nutr.* **2018**, *57*, 2887–2895. [CrossRef]

86. Perez-Martinez, P.; Garcia-Quintana, J.M.; Yubero-Serrano, E.M.; Tasset-Cuevas, I.; Tunez, I.; Garcia-Rios, A.; Delgado-Lista, J.; Marín, C.; Perez-Jimenez, F.; Roche, H.M.; et al. Postprandial oxidative stress is modified by dietary fat: Evidence from a human intervention study. *Clin. Sci.* **2010**, *119*, 251–261. [CrossRef]

87. Perez-Martinez, P.; Moreno-Conde, M.; Cruz-Teno, C.; Ruano, J.; Fuentes, F.; Delgado-Lista, J.; Garcia-Rios, A.; Marín, C.; Gómez-Luna, M.J.; Pérez-Jiménez, F.; et al. Dietary fat differentially influences regulatory endothelial function during the postprandial state in patients with metabolic syndrome: From the LIPGENE study. *Atherosclerosis* **2010**, *209*, 533–538. [CrossRef]

88. López-Moreno, J.; García-Carpintero, S.; Jimenez-Lucena, R.; Haro, C.; Rangel-Zúñiga, O.A.; Blanco-Rojo, R.; Yubero-Serrano, E.M.; Tinahones, F.J.; Delgado-Lista, J.; Pérez-Martínez, P.; et al. Effect of Dietary Lipids on Endotoxemia Influences Postprandial Inflammatory Response. *J. Agric. Food Chem.* **2017**, *65*, 7756–7763. [CrossRef]

89. Michalski, M.C.; Vors, C.; LeComte, M.; Laugerette, F. Dietary lipid emulsions and endotoxemia. *OCL Oilseeds Fats Crops Lipids* **2016**, *23*. [CrossRef]

90. Munford, R.S. Endotoxemia—Menace, marker, or mistake? *J. Leukoc. Biol.* **2016**, *100*, 687–698. [CrossRef]

91. West, S.G.; Hecker, K.D.; Mustad, V.A.; Nicholson, S.; Schoemer, S.L.; Wagner, P.; Hinderliter, A.L.; Ulbrecht, J.; Ruey, P.; Kris-Etherton, P.M. Acute effects of monounsaturated fatty acids with and without omega-3 fatty acids on vascular reactivity in individuals with type 2 diabetes. *Diabetologia* **2005**, *48*, 113–122. [CrossRef]

92. Tulk, H.M.; Robinson, L.E. Modifying the n-6/n-3 polyunsaturated fatty acid ratio of a high–saturated fat challenge does not acutely attenuate postprandial changes in inflammatory markers in men with metabolic syndrome. *Metabolism* **2009**, *58*, 1709–1716. [CrossRef]

93. Egert, S.; Stehle, P. Impact of n − 3 fatty acids on endothelial function: Results from human interventions studies. *Curr. Opin. Clin. Nutr. Metab. Care* **2011**, *14*, 121–131. [CrossRef]

94. Jackson, K.; Armah, C.; Minihane, A. Meal fatty acids and postprandial vascular reactivity: Table. *Biochem. Soc. Trans.* **2007**, *35*, 451–453. [CrossRef]

95. Moreira, A.P.B.; Texeira, T.F.S.; Ferreira, A.B.; Peluzio, M.D.C.G.; Alfenas, R.D.C.G. Influence of a high-fat diet on gut microbiota, intestinal permeability and metabolic endotoxaemia. *Br. J. Nutr.* **2012**, *108*, 801–809. [CrossRef]

96. Thazhath, S.S.; Wu, T.; Bound, M.J.; Checklin, H.L.; Jones, K.L.; Willoughby, S.R.; Horowitz, M.; Rayner, C.K. Changes in meal composition and duration affect postprandial endothelial function in healthy humans. *Am. J. Physiol. Liver Physiol.* **2014**, *307*, 1191–1197. [CrossRef]

97. Motton, D.D.; Keim, N.L.; Tenorio, F.A.; Horn, W.F.; Rutledge, J.C. Postprandial monocyte activation in response to meals with high and low glycemic loads in overweight women. *Am. J. Clin. Nutr.* **2007**, *85*, 60–65. [CrossRef]

98. Lee, E.J.; Kim, J.Y.; Kim, D.R.; Kim, K.S.; Kim, M.K.; Kwon, O. Glycemic index of dietary formula may not be predictive of postprandial endothelial inflammation: A double-blinded, randomized, crossover study in non-diabetic subjects. *Nutr. Res. Pract.* **2013**, *7*, 302–308. [CrossRef]

99. Kendall, C.W.C.; Josse, A.R.; Esfahani, A.; Jenkins, D.J.A. The impact of pistachio intake alone or in combination with high-carbohydrate foods on post-prandial glycemia. *Eur. J. Clin. Nutr.* **2011**, *65*, 696–702. [CrossRef]

100. Zhu, R.; Fan, Z.; Dong, Y.; Liu, M.; Wang, L.; Pan, H. Postprandial Glycaemic Responses of Dried Fruit-Containing Meals in Healthy Adults: Results from a Randomised Trial. *Nutrients* **2018**, *10*, 694. [CrossRef]
101. Ballard, K.D.; Mah, E.; Guo, Y.; Pei, R.; Volek, J.S.; Bruno, R.S. Low-Fat Milk Ingestion Prevents Postprandial Hyperglycemia-Mediated Impairments in Vascular Endothelial Function in Obese Individuals with Metabolic Syndrome. *J. Nutr.* **2013**, *143*, 1602–1610. [CrossRef]
102. Carroll, M.F.; Schade, D.S. Timing of Antioxidant Vitamin Ingestion Alters Postprandial Proatherogenic Serum Markers. *Circulation* **2003**, *108*, 24–31. [CrossRef]
103. Devaraj, S.; Wang-Polagruto, J.; Polagruto, J.; Keen, C.L.; Jialal, I. High-fat, energy-dense, fast-food–style breakfast results in an increase in oxidative stress in metabolic syndrome. *Metabolism* **2008**, *57*, 867–870. [CrossRef]
104. Ghanim, H.; Sia, C.L.; Upadhyay, M.; Korzeniewski, K.; Viswanathan, P.; Abuaysheh, S.; Mohanty, P.; Dandona, P. Orange juice neutralizes the proinflammatory effect of a high-fat, high-carbohydrate meal and prevents endotoxin increase and Toll-like receptor expression. *Am. J. Clin. Nutr.* **2010**, *91*, 940–949. [CrossRef]
105. Sawyer, B.J.; Jarrett, C.L.; Bhammar, D.M.; Ryder, J.R.; Angadi, S.S.; Gaesser, G.A.; Tucker, W.J. High-intensity interval exercise attenuates but does not eliminate endothelial dysfunction after a fast food meal. *Am. J. Physiol. Circ. Physiol.* **2018**, *314*, H188–H194.
106. Ghanim, H.; Abuaysheh, S.; Sia, C.L.; Korzeniewski, K.; Chaudhuri, A.; Fernandez-Real, J.M.; Dandona, P. Increase in plasma endotoxin concentrations and the expression of Toll-like receptors and suppressor of cytokine signaling-3 in mononuclear cells after a high-fat, high-carbohydrate meal: Implications for insulin resistance. *Diabetes Care* **2009**, *32*, 2281–2287. [CrossRef]
107. Järvisalo, M.J.; Jartti, L.; Marniemi, J.; Rönnemaa, T.; Viikari, J.S.A.; Lehtimäki, T.; Raitakari, O.T. Determinants of short-term variation in arterial flow-mediated dilatation in healthy young men. *Clin. Sci.* **2006**, *110*, 475–482. [CrossRef]
108. Dandona, P.; Ghanim, H.; Chaudhuri, A.; Dhindsa, S.; Kim, S.S. Macronutrient intake induces oxidative and inflammatory stress: Potential relevance to atherosclerosis and insulin resistance. *Exp. Mol. Med.* **2010**, *42*, 245–253. [CrossRef]
109. George, T.W.; Waroonphan, S.; Niwat, C.; Gordon, M.H.; Lovegrove, J.A. Effects of acute consumption of a fruit and vegetable puree-based drink on vasodilation and oxidative status. *Br. J. Nutr.* **2013**, *109*, 1442–1452. [CrossRef]
110. Lacroix, S.; Des Rosiers, C.; Gayda, M.; Nozza, A.; Thorin, E.; Tardif, J.C.; Nigam, A. A single Mediterranean meal does not impair postprandial flow-mediated dilatation in healthy men with subclinical metabolic dysregulations. *Appl. Physiol. Nutr. Metab.* **2016**, *41*, 888–894. [CrossRef]
111. Mariotti, F.; Huneau, J.F.; Szezepanski, I.; Petzke, K.J.; Aggoun, Y.; Tomé, D.; Bonnet, D. Meal amino acids with varied levels of arginine do not affect postprandial vascular endothelial function in healthy young men. *J. Nutr.* **2007**, *137*, 1383–1389. [CrossRef]
112. Westphal, S.; Taneva, E.; Kästner, S.; Martens-Lobenhoffer, J.; Bode-Böger, S.; Kropf, S.; Dierkes, J.; Luley, C. Endothelial dysfunction induced by postprandial lipemia is neutralized by addition of proteins to the fatty meal. *Atherosclerosis* **2006**, *185*, 313–319. [CrossRef]
113. Westphal, S.; Kastner, S.; Taneva, E.; Leodolter, A.; Dierkes, J.; Luley, C. Postprandial lipid and carbohydrate responses after the ingestion of a casein-enriched mixed meal. *Am. J. Clin. Nutr.* **2004**, *80*, 284–290. [CrossRef]
114. Dandona, P.; Chaudhuri, A.; Ghanim, H.; Mohanty, P. Insulin as an Anti-Inflammatory and Antiatherogenic Modulator. *J. Am. Coll. Cardiol.* **2009**, *53*, S14–S20. [CrossRef]
115. Mariotti, F. Postprandial low-grade inflammation and the recent data suggesting a protective impact of dietary protein quantity and sources [L'inflammation postprandiale: Les données récentes suggèrent un rôle préventif des protéines alimentaires et de leur nature]. *OCL Ol. Corps Gras Lipides* **2011**, *18*, 14–20. [CrossRef]
116. Lin, C.C.; Tsai, W.C.; Chen, J.Y.; Li, Y.H.; Lin, L.J.; Chen, J.H. Supplements of l-arginine attenuate the effects of high-fat meal on endothelial function and oxidative stress. *Int. J. Cardiol.* **2008**, *127*, 337–341. [CrossRef]
117. Deveaux, A.; Pham, I.; West, S.G.; Andre, E.; Lantoine-Adam, F.; Bunouf, P.; Sadi, S.; Hermier, D.; Mathé, V.; Fouillet, H.; et al. L-Arginine Supplementation Alleviates Postprandial Endothelial Dysfunction When Baseline Fasting Plasma Arginine Concentration Is Low: A Randomized Controlled Trial in Healthy Overweight Adults with Cardiometabolic Risk Factors. *J. Nutr.* **2016**, *146*, 1330–1340. [CrossRef]

118. Magné, J.; Huneau, J.F.; Tsikas, D.; Delemasure, S.; Rochette, L.; Tomé, D.; Mariotti, F. Rapeseed Protein in a High-Fat Mixed Meal Alleviates Postprandial Systemic and Vascular Oxidative Stress and Prevents Vascular Endothelial Dysfunction in Healthy Rats. *J. Nutr.* **2009**, *139*, 1660–1666. [CrossRef]

119. Blouet, C.; Mariotti, F.; Azzout-Marniche, D.; Mathé, V.; Mikogami, T.; Tomé, D.; Huneau, J.F. Dietary cysteine alleviates sucrose-induced oxidative stress and insulin resistance. *Free Radic. Biol. Med.* **2007**, *42*, 1089–1097. [CrossRef]

120. Luiking, Y.C.; Engelen, M.P.; Deutz, N.E. Regulation of nitric oxide production in health and disease. *Curr. Opin. Clin. Nutr. Metab. Care* **2010**, *13*, 97–104. [CrossRef]

121. Stonehouse, W.; Brinkworth, G.D.; Noakes, M. Palmolein and olive oil consumed within a high protein test meal have similar effects on postprandial endothelial function in overweight and obese men: A randomized controlled trial. *Atherosclerosis* **2015**, *239*, 178–185. [CrossRef]

122. Teunissen-Beekman, K.F.M.; Dopheide, J.; Geleijnse, J.M.; Bakker, S.J.L.; Brink, E.J.; De Leeuw, P.W.; Schalkwijk, C.G.; Van Baak, M.A. Dietary proteins improve endothelial function under fasting conditions but not in the postprandial state, with no effects on markers of low-grade inflammation. *Br. J. Nutr.* **2015**, *114*, 1819–1828. [CrossRef]

123. Fekete, Á.A.; Givens, D.I.; Lovegrove, J.A. Can milk proteins be a useful tool in the management of cardiometabolic health? An updated review of human intervention trials. *Proc. Nutr. Soc.* **2016**, *75*, 328–341. [CrossRef]

124. Fekete, Á.A.; Giromini, C.; Chatzidiakou, Y.; Givens, D.I.; Lovegrove, J.A. Whey protein lowers systolic blood pressure and Ca-caseinate reduces serum TAG after a high-fat meal in mildly hypertensive adults. *Sci. Rep.* **2018**, *8*, 5026. [CrossRef]

125. Lovegrove, J.A.; Givens, D.I. Dairy food products: Good or bad for cardiometabolic disease? *Nutr. Res. Rev.* **2016**, *29*, 249–267. [CrossRef]

126. Stanhewicz, A.E.; Alba, B.K.; Kenney, W.L.; Alexander, L.M. Dairy cheese consumption ameliorates single-meal sodium-induced cutaneous microvascular dysfunction by reducing ascorbate-sensitive oxidants in healthy older adults. *Br. J. Nutr.* **2016**, *116*, 658–665. [CrossRef]

127. Reis, C.E.; Bordalo, L.A.; Rocha, A.L.; Freitas, D.M.; da Silva, M.V.; de Faria, V.C.; Martino, H.S.; Costa, N.M.; Alfenas, R.C. Ground roasted peanuts leads to a lower post-prandial glycemic response than raw peanuts. *Nutr. Hosp.* **2011**, *26*, 745–751. [CrossRef]

128. Ellis, P.R.; Kendall, C.W.; Ren, Y.; Parker, C.; Pacy, J.F.; Waldron, K.W.; Jenkins, D.J. Role of cell walls in the bioaccessibility of lipids in almond seeds. *Am. J. Clin. Nutr.* **2004**, *80*, 604–613. [CrossRef]

129. Grundy, M.M.; Grassby, T.; Mandalari, G.; Waldron, K.W.; Butterworth, P.J.; Berry, S.E.; Ellis, P.R. Effect of mastication on lipid bioaccessibility of almonds in a randomized human study and its implications for digestion kinetics, metabolizable energy, and postprandial lipemia. *Am. J. Clin. Nutr.* **2015**, *101*, 25–33. [CrossRef]

130. Mariotti, F.; Valette, M.; Lopez, C.; Fouillet, H.; Famelart, M.H.; Mathé, V.; Airinei, G.; Benamouzig, R.; Gaudichon, C.; Tome, D.; et al. Casein Compared with Whey Proteins Affects the Organization of Dietary Fat during Digestion and Attenuates the Postprandial Triglyceride Response to a Mixed High-Fat Meal in Healthy, Overweight Men. *J. Nutr.* **2015**, *145*, 2657–2664. [CrossRef]

131. Pujos-Guillot, E.; Brandolini-Bunlon, M.; Fouillet, H.; Joly, C.; Martin, J.F.; Huneau, J.F.; Dardevet, D.; Mariotti, F. Metabolomics Reveals that the Type of Protein in a High-Fat Meal Modulates Postprandial Mitochondrial Overload and Incomplete Substrate Oxidation in Healthy Overweight Men. *J. Nutr.* **2018**, *148*, 876–884. [CrossRef]

132. Desmarchelier, C.; Borel, P.; Lairon, D.; Maraninchi, M.; Valéro, R. Effect of Nutrient and Micronutrient Intake on Chylomicron Production and Postprandial Lipemia. *Nutrients* **2019**, *11*, 1299. [CrossRef]

133. Abubakar, S.M.; Ukeyima, M.T.; Spencer, J.P.E.; Lovegrove, J.A. Acute Effects of Hibiscus sabdariffa Calyces on Postprandial Blood Pressure, Vascular Function, Blood Lipids, Biomarkers of Insulin Resistance and Inflammation in Humans. *Nutrients* **2019**, *11*, 341. [CrossRef]

134. Skulas-Ray, A.C.; Kris-Etherton, P.M.; Teeter, D.L.; Chen, C.Y.; Vanden Heuvel, J.P.; West, S.G. A high antioxidant spice blend attenuates postprandial insulin and triglyceride responses and increases some plasma measures of antioxidant activity in healthy, overweight men. *J. Nutr.* **2011**, *141*, 1451–1457. [CrossRef]

135. Cortés, B.; Núñez, I.; Cofán, M.; Gilabert, R.; Pérez-Heras, A.; Casals, E.; Deulofeu, R.; Ros, E. Acute Effects of High-Fat Meals Enriched With Walnuts or Olive Oil on Postprandial Endothelial Function. *J. Am. Coll. Cardiol.* **2006**, *48*, 1666–1671. [CrossRef]

136. Berryman, C.E.; Grieger, J.A.; West, S.G.; Chen, C.Y.O.; Blumberg, J.B.; Rothblat, G.H.; Sankaranarayanan, S.; Kris-Etherton, P.M. Acute Consumption of Walnuts and Walnut Components Differentially Affect Postprandial Lipemia, Endothelial Function, Oxidative Stress, and Cholesterol Efflux in Humans with Mild Hypercholesterolemia. *J. Nutr.* **2013**, *143*, 788–794. [CrossRef]

137. Kendall, C.W.; Esfahani, A.; Josse, A.R.; Augustin, L.S.; Vidgen, E.; Jenkins, D.J. The glycemic effect of nut-enriched meals in healthy and diabetic subjects. *Nutr. Metab. Cardiovasc. Dis.* **2011**, *21*, S34–S39. [CrossRef]

138. Kendall, C.W.C.; Josse, A.R.; Esfahani, A.; Jenkins, D.J.A. Nuts, metabolic syndrome and diabetes. *Br. J. Nutr.* **2010**, *104*, 465–473. [CrossRef]

139. Liu, X.; Hill, A.M.; West, S.G.; Gabauer, R.M.; McCrea, C.E.; Fleming, J.A.; Kris-Etherton, P.M. Acute Peanut Consumption Alters Postprandial Lipids and Vascular Responses in Healthy Overweight or Obese Men. *J. Nutr.* **2017**, *147*, 835–840. [CrossRef]

140. Naissides, M.; Mamo, J.C.; James, A.P.; Pal, S. The effect of acute red wine polyphenol consumption on postprandial lipaemia in postmenopausal women. *Atherosclerosis* **2004**, *177*, 401–408. [CrossRef]

141. Testa, R.; Bonfigli, A.R.; Prattichizzo, F.; La Sala, L.; De Nigris, V.; Ceriello, A. The "Metabolic Memory" Theory and the Early Treatment of Hyperglycemia in Prevention of Diabetic Complications. *Nutrients* **2017**, *9*, 437. [CrossRef]

142. Schisano, B.; Tripathi, G.; McGee, K.; McTernan, P.G.; Ceriello, A. Glucose oscillations, more than constant high glucose, induce p53 activation and a metabolic memory in human endothelial cells. *Diabetologia* **2011**, *54*, 1219–1226. [CrossRef]

143. Suzuki, K.; Watanabe, K.; Futami-Suda, S.; Yano, H.; Motoyama, M.; Matsumura, N.; Igari, Y.; Suzuki, T.; Nakano, H.; Oba, K. The effects of postprandial glucose and insulin levels on postprandial endothelial function in subjects with normal glucose tolerance. *Cardiovasc. Diabetol.* **2012**, *11*, 98. [CrossRef]

144. Thom, N.J.; Early, A.R.; Hunt, B.E.; Harris, R.A.; Herring, M.P. Eating and arterial endothelial function: A meta-analysis of the acute effects of meal consumption on flow-mediated dilation. *Obes. Rev.* **2016**, *17*, 1080–1090. [CrossRef]

145. Tuccinardi, D.; Farr, O.M.; Upadhyay, J.; Oussaada, S.M.; Klapa, M.I.; Candela, M.; Rampelli, S.; Lehoux, S.; Lazaro, I.; Sala-Vila, A.; et al. Mechanisms Underlying the Cardiometabolic Protective Effect of Walnut Consumption in Obese Subjects: A Cross-Over, Randomized, Double-Blinded, Controlled Inpatient Physiology Study. *Diabetes Obes. Metab.* **2019**. [CrossRef]

146. Wu, L.; Piotrowski, K.; Rau, T.; Waldmann, E.; Broedl, U.C.; Demmelmair, H.; Koletzko, B.; Stark, R.G.; Nagel, J.M.; Mantzoros, C.S.; et al. Walnut-enriched diet reduces fasting non-HDL-cholesterol and apolipoprotein B in healthy Caucasian subjects: A randomized controlled cross-over clinical trial. *Metabolism* **2014**, *63*, 382–391. [CrossRef]

147. Ros, E.; Núñez, I.; Pérez-Heras, A.; Serra, M.; Gilabert, R.; Casals, E.; Deulofeu, R. A walnut diet improves endothelial function in hypercholesterolemic subjects: A randomized crossover trial. *Circulation* **2004**, *109*, 1609–1614. [CrossRef]

148. Fitschen, P.J.; Rolfhus, K.R.; Winfrey, M.R.; Allen, B.K.; Manzy, M.; Maher, M.A. Cardiovascular Effects of Consumption of Black Versus English Walnuts. *J. Med. Food* **2011**, *14*, 890–898. [CrossRef]

149. Casas-Agustench, P.; Bulló, M.; Salas-Salvadó, J. Nuts, inflammation and insulin resistance. *Asia Pac. J. Clin. Nutr.* **2010**, *19*, 124–130.

150. Halvorsen, B.L.; Holte, K.; Barikmo, I.; Remberg, S.F.; Wold, A.B.; Haffner, K.; Baugerød, H.; Andersen, L.F.; Moskaug, O.; Jacobs, D.R.; et al. A Systematic Screening of Total Antioxidants in Dietary Plants. *J. Nutr.* **2002**, *132*, 461–471. [CrossRef]

151. Haddad, E.H.; Gaban-Chong, N.; Oda, K.; Sabaté, J. Effect of a walnut meal on postprandial oxidative stress and antioxidants in healthy individuals. *Nutr. J.* **2014**, *13*, 4. [CrossRef]

152. Dreher, M.L. Pistachio nuts: Composition and potential health benefits. *Nutr. Rev.* **2012**, *70*, 234–240. [CrossRef]

153. Jenkins, D.J.A.; Josse, A.R.; Salvatore, S.; Vidgen, E.; Rao, A.V.; Kendall, C.W.C.; Brighenti, F.; Augustin, L.S.A.; Ellis, P.R. Almonds Decrease Postprandial Glycemia, Insulinemia, and Oxidative Damage in Healthy Individuals. *J. Nutr.* **2006**, *136*, 2987–2992. [CrossRef]

154. Wang, X.; Li, Z.; Liu, Y.; Lv, X.; Yang, W. Effects of pistachios on body weight in Chinese subjects with metabolic syndrome. *Nutr. J.* **2012**, *11*, 20. [CrossRef]

155. Kay, C.D.; Gebauer, S.K.; West, S.G.; Kris-Etherton, P.M. Pistachios Increase Serum Antioxidants and Lower Serum Oxidized-LDL in Hypercholesterolemic Adults. *J. Nutr.* **2010**, *140*, 1093–1098. [CrossRef]

156. Salas-Salvadó, J.; Baldrich-Mora, M.; Juanola-Falgarona, M.; Hernández-Alonso, P.; Bulló, M. Beneficial Effect of Pistachio Consumption on Glucose Metabolism, Insulin Resistance, Inflammation, and Related Metabolic Risk Markers: A Randomized Clinical Trial. *Diabetes Care* **2014**, *37*, 3098–3105.

157. Sari, I.; Baltaci, Y.; Bagci, C.; Davutoglu, V.; Erel, O.; Çelik, H.; Ozer, O.; Aksoy, N.; Aksoy, M. Effect of pistachio diet on lipid parameters, endothelial function, inflammation, and oxidative status: A prospective study. *Nutrition* **2010**, *26*, 399–404. [CrossRef]

158. Kocyigit, A.; Koylu, A.; Keles, H. Effects of pistachio nuts consumption on plasma lipid profile and oxidative status in healthy volunteers. *Nutr. Metab. Cardiovasc. Dis.* **2006**, *16*, 202–209. [CrossRef]

159. Kendall, C.W.C.; West, S.G.; Augustin, L.S.; Esfahani, A.; Vidgen, E.; Bashyam, B.; Sauder, K.A.; Campbell, J.; Chiavaroli, L.; Jenkins, A.L.; et al. Acute effects of pistachio consumption on glucose and insulin, satiety hormones and endothelial function in the metabolic syndrome. *Eur. J. Clin. Nutr.* **2014**, *68*, 370–375. [CrossRef]

160. Rangel-Zuniga, O.A.; Haro, C.; Tormos, C.; Perez-Martinez, P.; Delgado-Lista, J.; Marin, C.; Quintana-Navarro, G.M.; Cerda, C.; Saez, G.T.; Lopez-Segura, F.; et al. Frying oils with high natural or added antioxidants content, which protect against postprandial oxidative stress, also protect against DNA oxidation damage. *Eur. J. Nutr.* **2017**, *56*, 1597–1607. [CrossRef]

161. Lyte, J.M.; Gabler, N.K.; Hollis, J.H. Postprandial serum endotoxin in healthy humans is modulated by dietary fat in a randomized, controlled, cross-over study. *Lipids Health Dis.* **2016**, *15*, 186. [CrossRef]

162. Carnevale, R.; Pignatelli, P.; Nocella, C.; Loffredo, L.; Pastori, D.; Vicario, T.; Petruccioli, A.; Bartimoccia, S.; Violi, F. Extra virgin olive oil blunt post-prandial oxidative stress via NOX2 down-regulation. *Atherosclerosis* **2014**, *235*, 649–658. [CrossRef]

163. Park, E.; Edirisinghe, I.; Burton-Freeman, B. Avocado Fruit on Postprandial Markers of Cardio-Metabolic Risk: A Randomized Controlled Dose Response Trial in Overweight and Obese Men and Women. *Nutrients* **2018**, *10*, 1287. [CrossRef]

164. Burton-Freeman, B. Postprandial metabolic events and fruit-derived phenolics: A review of the science. *Br. J. Nutr.* **2010**, *104* (Suppl. S3), S1–S14. [CrossRef]

165. AlQurashi, R.M.; Galante, L.A.; Rowland, I.R.; Spencer, J.P.; Commane, D.M. Consumption of a flavonoid-rich açai meal is associated with acute improvements in vascular function and a reduction in total oxidative status in healthy overweight men. *Am. J. Clin. Nutr.* **2016**, *104*, 1227–1235. [CrossRef]

166. Rodriguez-Mateos, A.; Del Pino-García, R.; George, T.W.; Vidal-Diez, A.; Heiss, C.; Spencer, J.P.E.; Rodriguez-Mateos, A.; Del Pino-García, R.; Vidal-Diez, A. Impact of processing on the bioavailability and vascular effects of blueberry (poly)phenols. *Mol. Nutr. Food Res.* **2014**, *58*, 1952–1961. [CrossRef]

167. Istas, G.; Feliciano, R.P.; Weber, T.; García-Villalba, R.; Tomás-Barberán, F.; Heiss, C.; Rodriguez-Mateos, A. Plasma urolithin metabolites correlate with improvements in endothelial function after red raspberry consumption: A double-blind randomized controlled trial. *Arch. Biochem. Biophys.* **2018**, *651*, 43–51. [CrossRef]

168. Schell, J.; Betts, N.M.; Foster, M.; Scofield, R.H.; Basu, A. Cranberries improve postprandial glucose excursions in type 2 diabetes. *Food Funct.* **2017**, *8*, 3083–3090. [CrossRef]

169. Hannum, S.M. Potential Impact of Strawberries on Human Health: A Review of the Science. *Crit. Rev. Food Sci. Nutr.* **2004**, *44*, 1–17. [CrossRef]

170. Joseph, S.V.; Edirisinghe, I.; Burton-Freeman, B.M. Fruit Polyphenols: A Review of Anti-inflammatory Effects in Humans. *Crit. Rev. Food Sci. Nutr.* **2016**, *56*, 419–444. [CrossRef]

171. Burton-Freeman, B.; Linares, A.; Hyson, D.; Kappagoda, T. Strawberry modulates LDL oxidation and postprandial lipemia in response to high-fat meal in overweight hyperlipidemic men and women. *J. Am. Coll. Nutr.* **2010**, *29*, 46–54. [CrossRef]

172. Richter, C.K.; Skulas-Ray, A.C.; Gaugler, T.L.; Lambert, J.D.; Proctor, D.N.; Kris-Etherton, P.M. Incorporating freeze-dried strawberry powder into a high-fat meal does not alter postprandial vascular function or blood markers of cardiovascular disease risk: A randomized controlled trial. *Am. J. Clin. Nutr.* **2017**, *105*, 313–322. [CrossRef]

173. Lee, Y.J.; Ahn, Y.; Kwon, O.; Lee, M.Y.; Lee, C.H.; Lee, S.; Park, T.; Kwon, S.W.; Kim, J.Y. Dietary Wolfberry Extract Modifies Oxidative Stress by Controlling the Expression of Inflammatory mRNAs in Overweight and Hypercholesterolemic Subjects: A Randomized, Double-Blind, Placebo-Controlled Trial. *J. Agric. Food Chem.* **2017**, *65*, 309–316. [CrossRef]

174. Urquiaga, I.; Ávila, F.; Echeverria, G.; Perez, D.; Trejo, S.; Leighton, F. A Chilean Berry Concentrate Protects against Postprandial Oxidative Stress and Increases Plasma Antioxidant Activity in Healthy Humans. *Oxidative Med. Cell. Longev.* **2017**, *2017*, 8361493. [CrossRef]

175. Peluso, I.; Palmery, M. Risks of Misinterpretation in the Evaluation of the Effect of Fruit-Based Drinks in Postprandial Studies. *Gastroenterol. Res. Pract.* **2014**, *2014*, 870547. [CrossRef]

176. Blanco-Colio, L.M.; Valderrama, M.; Alvarez-Sala, L.A.; Bustos, C.; Ortego, M.; Hernández-Presa, M.A.; Cancelas, P.; Gómez-Gerique, J.; Millán, J.; Egido, J. Red wine intake prevents nuclear factor-kappaB activation in peripheral blood mononuclear cells of healthy volunteers during postprandial lipemia. *Circulation* **2000**, *102*, 1020–1026. [CrossRef]

177. Covas, M.I.; Gambert, P.; Fito, M.; de la Torre, R. Wine and oxidative stress: Up-to-date evidence of the effects of moderate wine consumption on oxidative damage in humans. *Atherosclerosis* **2010**, *208*, 297–304. [CrossRef]

178. Schmitt, C.A.; Heiss, E.H.; Dirsch, V.M. Effect of resveratrol on endothelial cell function: Molecular mechanisms. *BioFactors* **2010**, *36*, 342–349. [CrossRef]

179. Ghanim, H.; Sia, C.L.; Korzeniewski, K.; Lohano, T.; Abuaysheh, S.; Marumganti, A.; Chaudhuri, A.; Dandona, P. A Resveratrol and Polyphenol Preparation Suppresses Oxidative and Inflammatory Stress Response to a High-Fat, High-Carbohydrate Meal. *J. Clin. Endocrinol. Metab.* **2011**, *96*, 1409–1414. [CrossRef]

180. Hampton, S.M.; Isherwood, C.; Kirkpatrick, V.J.E.; Lynne-Smith, A.C.; Griffin, B.A. The influence of alcohol consumed with a meal on endothelial function in healthy individuals. *J. Hum. Nutr. Diet.* **2010**, *23*, 120–125. [CrossRef]

181. Karatzi, K.; Papamichael, C.; Karatzis, E.; Papaioannou, T.G.; Voidonikola, P.T.; Vamvakou, G.D.; Lekakis, J.; Zampelas, A. Postprandial improvement of endothelial function by red wine and olive oil antioxidants: A synergistic effect of components of the Mediterranean diet. *J. Am. Coll. Nutr.* **2008**, *27*, 448–453. [CrossRef]

182. Dhindsa, S.; Tripathy, D.; Mohanty, P.; Ghanim, H.; Syed, T.; Aljada, A.; Dandona, P. Differential effects of glucose and alcohol on reactive oxygen species generation and intranuclear nuclear factor-kappaB in mononuclear cells. *Metabolism* **2004**, *53*, 330–334. [CrossRef]

183. Bulut, D.; Jelich, U.; Dacanay-Schwarz, R.; Mügge, A. Red Wine Ingestion Prevents Microparticle Formation After a Single High-Fat Meal—A Crossover Study in Healthy Humans. *J. Cardiovasc. Pharmacol.* **2013**, *61*, 489–494. [CrossRef]

184. Fragopoulou, E.; Choleva, M.; Antonopoulou, S.; Demopoulos, C.A. Wine and its metabolic effects. A comprehensive review of clinical trials. *Metabolism* **2018**, *83*, 102–119. [CrossRef]

185. Argyrou, C.; Vlachogianni, I.; Stamatakis, G.; Demopoulos, C.A.; Antonopoulou, S.; Fragopoulou, E. Postprandial effects of wine consumption on Platelet Activating Factor metabolic enzymes. *Prostaglandins Other Lipid Mediat.* **2017**, *130*, 23–29. [CrossRef]

186. Botham, K.M.; Wheeler-Jones, C.P. Postprandial lipoproteins and the molecular regulation of vascular homeostasis. *Prog. Lipid Res.* **2013**, *52*, 446–464. [CrossRef]

187. Bardagjy, A.S.; Hu, Q.; Giebler, K.A.; Ford, A.; Steinberg, F.M. Effects of grape consumption on biomarkers of inflammation, endothelial function, and PBMC gene expression in obese subjects. *Arch. Biochem. Biophys.* **2018**, *646*, 145–152. [CrossRef]

188. Erridge, C.; Attinà, T.; Spickett, C.M.; Webb, D.J. A high-fat meal induces low-grade endotoxemia: Evidence of a novel mechanism of postprandial inflammation. *Am. J. Clin. Nutr.* **2007**, *86*, 1286–1292. [CrossRef]

189. Cani, P.D.; Amar, J.; Iglesias, M.A.; Poggi, M.; Knauf, C.; Bastelica, D.; Neyrinck, A.M.; Fava, F.; Tuohy, K.M.; Chabo, C.; et al. Metabolic Endotoxemia Initiates Obesity and Insulin Resistance. *Diabetes* **2007**, *56*, 1761–1772. [CrossRef]

190. Hermier, D.; Mathé, V.; Lan, A.; Santini, C.; Quignard-Boulangé, A.; Huneau, J.F.; Mariotti, F. Postprandial low-grade inflammation does not specifically require TLR4 activation in the rat. *Nutr. Metab.* **2017**, *14*, 65. [CrossRef]

191. Mele, L.; Mena, P.; Piemontese, A.; Marino, V.; López-Gutiérrez, N.; Bernini, F.; Brighenti, F.; Zanotti, I.; Del Rio, D. Antiatherogenic effects of ellagic acid and urolithins in vitro. *Arch. Biochem. Biophys.* **2016**, *599*, 42–50. [CrossRef]

192. Morand, C.; Dubray, C.; Milenkovic, D.; Lioger, D.; Martin, J.F.; Scalbert, A.; Mazur, A. Hesperidin contributes to the vascular protective effects of orange juice: A randomized crossover study in healthy volunteers. *Am. J. Clin. Nutr.* **2010**, *93*, 73–80. [CrossRef]

193. Rendeiro, C.; Dong, H.; Saunders, C.; Harkness, L.; Blaze, M.; Hou, Y.; Belanger, R.L.; Corona, G.; Lovegrove, J.A.; Spencer, J.P.E. Flavanone-rich citrus beverages counteract the transient decline in postprandial endothelial function in humans: A randomised, controlled, double-masked, cross-over intervention study. *Br. J. Nutr.* **2016**, *116*, 1999–2010. [CrossRef]

194. Salden, B.N.; Troost, F.J.; De Groot, E.; Stevens, Y.R.; Garces-Rimon, M.; Possemiers, S.; Winkens, B.; Masclee, A.A. Randomized clinical trial on the efficacy of hesperidin 2S on validated cardiovascular biomarkers in healthy overweight individuals. *Am. J. Clin. Nutr.* **2016**, *104*, 1523–1533. [CrossRef]

195. Ono-Moore, K.D.; Snodgrass, R.G.; Huang, S.; Singh, S.; Freytag, T.L.; Burnett, D.J.; Bonnel, E.L.; Woodhouse, L.R.; Zunino, S.J.; Peerson, J.M.; et al. Postprandial Inflammatory Responses and Free Fatty Acids in Plasma of Adults Who Consumed a Moderately High-Fat Breakfast with and without Blueberry Powder in a Randomized Placebo-Controlled Trial. *J. Nutr.* **2016**, *146*, 1411–1419. [CrossRef]

196. Huebbe, P.; Giller, K.; de Pascual-Teresa, S.; Arkenau, A.; Adolphi, B.; Portius, S.; Arkenau, C.N.; Rimbach, G. Effects of blackcurrant-based juice on atherosclerosis-related biomarkers in cultured macrophages and in human subjects after consumption of a high-energy meal. *Br. J. Nutr.* **2012**, *108*, 234–244. [CrossRef]

197. Jin, Y.; Alimbetov, D.; George, T.; Gordon, M.H.; Lovegrove, J.A. A randomised trial to investigate the effects of acute consumption of a blackcurrant juice drink on markers of vascular reactivity and bioavailability of anthocyanins in human subjects. *Eur. J. Clin. Nutr.* **2011**, *65*, 849–856. [CrossRef]

198. Castro-Acosta, M.L.; Smith, L.; Miller, R.J.; McCarthy, D.I.; Farrimond, J.A.; Hall, W.L. Drinks containing anthocyanin-rich blackcurrant extract decrease postprandial blood glucose, insulin and incretin concentrations. *J. Nutr. Biochem.* **2016**, *38*, 154–161. [CrossRef]

199. Wiswedel, I.; Hirsch, D.; Kropf, S.; Gruening, M.; Pfister, E.; Schewe, T.; Sies, H. Flavanol-rich cocoa drink lowers plasma F 2 -isoprostane concentrations in humans. *Free Radic. Biol. Med.* **2004**, *37*, 411–421. [CrossRef]

200. Magrone, T.; Russo, M.A.; Jirillo, E. Cocoa and Dark Chocolate Polyphenols: From Biology to Clinical Applications. *Front. Immunol.* **2017**, *8*, 677. [CrossRef]

201. Edirisinghe, I.; Burton-Freeman, B.; Kappagoda, C.T. Mechanism of the endothelium-dependent relaxation evoked by a grape seed extract. *Clin. Sci.* **2008**, *114*, 331–337. [CrossRef]

202. Edirisinghe, I.; Burton-Freeman, B.; Varelis, P.; Kappagoda, T. Strawberry Extract Caused Endothelium-Dependent Relaxation through the Activation of PI3 Kinase/Akt. *J. Agric. Food Chem.* **2008**, *56*, 9383–9390. [CrossRef]

203. Sies, H.; Schewe, T.; Heiss, C.; Kelm, M. Cocoa polyphenols and inflammatory mediators. *Am. J. Clin. Nutr.* **2005**, *81*, 304–312. [CrossRef]

204. Schnorr, O.; Brossette, T.; Momma, T.Y.; Kleinbongard, P.; Keen, C.L.; Schroeter, H.; Sies, H. Cocoa flavanols lower vascular arginase activity in human endothelial cells in vitro and in erythrocytes in vivo. *Arch. Biochem. Biophys.* **2008**, *476*, 211–215. [CrossRef]

205. Selmi, C.; Cocchi, C.A.; Lanfredini, M.; Keen, C.L.; Gershwin, M.E. Chocolate at heart: The anti-inflammatory impact of cocoa flavanols. *Mol. Nutr. Food Res.* **2008**, *52*, 1340–1348. [CrossRef]

206. Grassi, D.; Desideri, G.; Necozione, S.; Ruggieri, F.; Blumberg, J.B.; Stornello, M.; Ferri, C. Protective Effects of Flavanol-Rich Dark Chocolate on Endothelial Function and Wave Reflection During Acute Hyperglycemia. *Hypertension* **2012**, *60*, 827–832. [CrossRef]

207. Westphal, S.; Luley, C. Flavanol-rich cocoa ameliorates lipemia-induced endothelial dysfunction. *Heart Vessel.* **2011**, *26*, 511–515. [CrossRef]

208. Varadhan, K.; Limb, M.C.; Phillips, B.E.; Atherton, P.J.; Williams, J.P.; Smith, K. Acute cocoa flavanol supplementation improves muscle macro- and microvascular but not anabolic responses to amino acids in older men. *Appl. Physiol. Nutr. Metab.* **2016**, *41*, 548–556.

209. García-Conesa, M.T.; Chambers, K.; Combet, E.; Pinto, P.; Garcia-Aloy, M.; Andrés-Lacueva, C.; De Pascual-Teresa, S.; Mena, P.; Ristic, A.K.; Hollands, W.J.; et al. Meta-Analysis of the Effects of Foods and Derived Products Containing Ellagitannins and Anthocyanins on Cardiometabolic Biomarkers: Analysis of Factors Influencing Variability of the Individual Responses. *Int. J. Mol. Sci.* **2018**, *19*, 694. [CrossRef]

210. Ndiaye, M.; Chataigneau, M.; Lobysheva, I.; Schini-Kerth, V.B. Red wine polyphenol-induced, endothelium-dependent NO-mediated relaxation is due to the redox-sensitive PI3-kinase/Akt-dependent phosphorylation of endothelial NO-synthase in the isolated porcine coronary artery. *FASEB J.* **2005**, *19*, 455–457. [CrossRef]

211. Ndiaye, M.; Chataigneau, T.; Chataigneau, M.; Schini-Kerth, V.B. Red wine polyphenols induce EDHF-mediated relaxations in porcine coronary arteries through the redox-sensitive activation of the PI3-kinase/Akt pathway. *Br. J. Pharmacol.* **2004**, *142*, 1131–1136. [CrossRef]

212. Heiss, C.; Kleinbongard, P.; Dejam, A.; Perré, S.; Schroeter, H.; Sies, H.; Kelm, M. Acute Consumption of Flavanol-Rich Cocoa and the Reversal of Endothelial Dysfunction in Smokers. *J. Am. Coll. Cardiol.* **2005**, *46*, 1276–1283. [CrossRef]

213. Schmitt, C.A.; Dirsch, V.M. Modulation of endothelial nitric oxide by plant-derived products. *Nitric Oxide* **2009**, *21*, 77–91. [CrossRef]

214. Bollenbach, A.; Huneau, J.F.; Mariotti, F.; Tsikas, D. Asymmetric and Symmetric Protein Arginine Dimethylation: Concept and Postprandial Effects of High-Fat Protein Meals in Healthy Overweight Men. *Nutrients* **2019**, *11*, 1463. [CrossRef]

215. McDonald, J.D.; Mah, E.; Chitchumroonchokchai, C.; Reverri, E.J.; Li, J.; Volek, J.S.; Villamena, F.A.; Bruno, R.S. Co-ingestion of whole eggs or egg whites with glucose protects against postprandial hyperglycaemia-induced oxidative stress and dysregulated arginine metabolism in association with improved vascular endothelial function in prediabetic men. *Br. J. Nutr.* **2018**, *120*, 901–913. [CrossRef]

216. Manach, C.; Williamson, G.; Morand, C.; Scalbert, A.; Remesy, C. Bioavailability and bioefficacy of polyphenols in humans. I. Review of 97 bioavailability studies. *Am. J. Clin. Nutr.* **2005**, *81*, 230–242. [CrossRef]

Review

Effect of Nutrient and Micronutrient Intake on Chylomicron Production and Postprandial Lipemia

Charles Desmarchelier [1,2,3,4], **Patrick Borel** [1,2,3,4], **Denis Lairon** [1,2,3,4], **Marie Maraninchi** [1,2,3,4,5] **and René Valéro** [1,2,3,4,5,*]

1   Aix-Marseille Université, Faculty of Medicine, 27 Boulevard Jean Moulin, 13385 Marseille, France; Charles.DESMARCHELIER@univ-amu.fr (C.D.); Patrick.BOREL@univ-amu.fr (P.B.); denis.lairon@orange.fr (D.L.); Marie.MARANINCHI@univ-amu.fr (M.M.)
2   C2VN (Center for Cardiovascular and Nutrition Research), Faculty of Medicine, 27 Boulevard Jean Moulin, 13385 Marseille, France
3   INSERM, Faculty of Medicine, 27 Boulevard Jean Moulin, 13385 Marseille, France
4   INRA, Faculty of Medicine, 27 Boulevard Jean Moulin, 13385 Marseille, France
5   APHM (Assistance Publique-Hôpitaux de Marseille), CHU Conception, 147 Boulevard Baille, 13385 Marseille, France
*   Correspondence: RVALERO@mail.ap-hm.fr; Tel.: +33-491-383-650

Received: 30 April 2019; Accepted: 4 June 2019; Published: 8 June 2019

Abstract: Postprandial lipemia, which is one of the main characteristics of the atherogenic dyslipidemia with fasting plasma hypertriglyceridemia, low high-density lipoprotein cholesterol and an increase of small and dense low-density lipoproteins is now considered a causal risk factor for atherosclerotic cardiovascular disease and all-cause mortality. Postprandial lipemia, which is mainly related to the increase in chylomicron production, is frequently elevated in individuals at high cardiovascular risk such as obese or overweight patients, type 2 diabetic patients and subjects with a metabolic syndrome who share an insulin resistant state. It is now well known that chylomicron production and thus postprandial lipemia is highly regulated by many factors such as endogenous factors: circulating factors such as hormones or free fatty acids, genetic variants, circadian rhythms, or exogenous factors: food components, dietary supplements and prescription drugs. In this review, we focused on the effect of nutrients, micronutrients and phytochemicals but also on food structure on chylomicron production and postprandial lipemia.

Keywords: carbohydrates; cardiovascular disease; cholesterol; fibers; food structure; lipids; polyphenols; proteins; triglycerides; vitamins

---

1. Introduction

Cardiovascular diseases (CVD) are the leading cause of death in the world [1]. Atherogenic dyslipidemia (AD), which is mainly characterized by plasma fasting and postprandial hypertriglyceridemia (postprandial hyperlipemia), low high-density lipoprotein cholesterol (HDL-C) and an increase of small and dense low-density lipoproteins (LDL), is frequently seen in individuals at high cardiovascular risk such as obese or overweight patients, type 2 diabetic (T2D) patients and subjects with a metabolic syndrome who share an insulin resistant state [2,3]. The pathophysiology of the AD is widely explained by the blood accumulation of triglyceride-rich lipoproteins (TRL) synthesized by the liver (very low-density lipoproteins (VLDL)) [4] and the intestine (chylomicrons (CM)) [5]. This accumulation has been attributed to the overproduction of both VLDL and CM and to a defective TRL removal process [6,7]. Elevated fasting and postprandial blood TRL concentrations, which are mainly related to the increase in CM production, are now considered a causal risk factor for low-grade inflammation, atherosclerotic CVD and all-cause mortality [8]. It is now well known that

CM production is highly regulated by many factors such as endogenous factors: circulating factors such as hormones or free fatty acids (FFA), genetic variants, circadian rhythms, or exogenous factors: food components, nutraceuticals and therapeutic interventions. In this review, we will focus on the effect of nutrients, micronutrients and phytochemicals but also on food structure on CM production and postprandial lipemia (TRL metabolism only) in humans [9].

2. Methodological Introduction

A large number of studies have assessed the effects of acute or chronic ingestion of meals containing different types of fat or other nutrients on postprandial lipemia but have yielded conflicting results. A number of potentially confounding factors reflecting the lack of standardization among studies could explain this: population, amount of fat, type of fat, amount and types of other nutrients, physicochemical composition of the meal and fatty acid (FA) or other nutrients, composition of habitual chronic food intake. Moreover, the measurement of postprandial plasma triglyceride (TG) response may provide only a limited evaluation of the true impact of meals and nutrients on postprandial lipoprotein metabolism. Studies have analyzed TG or retinyl-palmitate in a variety of sample types, including whole blood, plasma, serum, lipoproteins and their remnants, over a wide range from two up to 12 h postprandially [10]. Finally, qualitative and not only quantitative changes (size beyond the number of lipoproteins, lipidomic changes) have been described in several studies [11,12].

3. Effect of Dietary Lipids

3.1. Lipid Amount

A study performed in healthy men showed an increase in postprandial lipemia (plasma-TG peak concentration) following an 80 g fat meal compared to a 20 g fat meal with an intermediate result following a 40 g fat meal. A parallel elevation of glucose-dependent insulinotropic polypeptide (GIP) concentration and postheparin lipoprotein lipase (LPL) activity with a trend but no significant change in the increase of insulin response were seen following the 80 g fat meal compared to the 20 g fat meal [13]. Another study in healthy men individuals did not report any effect of a 15 g low-fat meal on postprandial lipemia compared to a nonfat meal and confirmed the dose-dependent increased postprandial serum-TG and CM-TG concentrations for moderate doses of fat per meal (30 to 50 g). The serum insulin response was significantly higher only following the 50 g fat meal compared to the nonfat and 15 g fat meal [14]. These results were confirmed in normal-weight and obese individuals with an increase in CM-TG concentration following a 40 g fat meal compared to a 10 g fat meal [15] and in obese boys with an increase in postprandial plasma-TG concentration following a 33 g fat meal compared to a 18 g fat meal. Glucagon-like-peptide-1 (GLP-1) concentration was significantly higher after the high-fat meal in the latter study [16]. Very high doses of fat (80 g and above) exaggerated postprandial serum-TG in healthy men [17]. This stepwise increase in the postprandial lipemia seen with the increase amount of fat intake suggests that the clearance capacity of the individuals is overloaded proportionally to the amount of fat assimilated. Moreover, consecutive meals containing fat appear to enhance postprandial lipemia [7,18].

3.2. Fatty Acid Composition

3.2.1. Test Meal (Acute)

First, it should be remembered that dietary short- or medium-chain FA have limited effect on postprandial lipemia because they directly enter the general circulation via portal route instead of CM secretion. It is probably why studies using dairy fats, which contain significant amount of short- and medium-chain FA, as the only source of saturated fatty acids (SFA), generally report a lower postprandial TG response compared to other sources of SFA or other types of fats [7,19]. In healthy individuals, a recent randomized, cross-over, single-blinded design study, showed that a medium-chain

SFA-rich meal (coconut biscuit) resulted in a significant lower postprandial whole blood TG response (concentration and net area under the curve (AUC)) compared to a short-chain SFA-rich meal (butter biscuit) and a long-chain SFA-rich meal (lard biscuit) despite identical fat and caloric content [20]. In healthy men, a study showed an increase in CM-TG concentration after an SFA-rich meal compared to an *n*-6 polyunsaturated fatty acid (PUFA)-rich meal but no difference with the monounsaturated fatty acid (MUFA)-rich meal [21]. The postprandial lipemic response to a SFA-rich meal was comparable to that of a *n*-6 PUFA-rich meal when consumed with *n*-3 PUFA in one study conducted in healthy individuals [22] whereas, the increase of the *n*-3 PUFA content of a SFA-rich meal fat meal did not acutely change postprandial TG concentration in another study performed in subjects with a metabolic syndrome [23]. In overweight men, a difference in the postprandial serum TG concentrations between an SFA-rich meal and an *n*-6 PUFA-rich meal was only seen in the late postprandial phase [24]. Four studies showed different effects of SFA compared to MUFA consumption on postprandial lipemia showing either an increase [25] or a decrease in healthy individuals [25,26] or in overweight and obese subjects [27], and one no change in healthy individuals [28]. In healthy individuals, another study showed a decrease in postprandial lipemia following a stearic acid-rich fat meal compared to other SFA (palmitic acid) or a high-oleic acid sunflower oil [29] but there was no difference between a high-oleic acid sunflower oil and a stearic acid-rich fat meal using cocoa butter [30]. Consumption of six different test meals rich in stearic, palmitic, palmitic plus myristic, oleic, elaidic or linoleic acids by healthy individuals resulted in a relatively lower postprandial lipemia response with long-chain SFA than did the intake of the unsaturated FA. The significant differences in LPL activities between groups did not explain the postprandial response that could be due to slower or less-efficient absorption of long-chain SFA [31]. Moreover, no difference in postprandial lipemia was observed following an oleate-rich meal (cis isomer) and an elaidate-rich meal (the trans isomer of oleic acid) [29]. Thus, most studies have shown that meals enriched with different proportion of SFA, MUFA or *n*-6 PUFA do not elicit marked differences in postprandial lipemia [7,32,33]. In healthy individuals, an *n*-3 PUFA-rich fat meal (fish oil) lowered postprandial lipemia compared to an SFA-rich fat meal (palm and coconut oils) [34]. However, two other studies did not show a difference in the incremental area under the curve (iAUC) plasma-TG after an SFA-rich meal compared to an *n*-3 PUFA-rich meal [28,35].

3.2.2. The Habitual Diet (Chronic)

Postprandial lipemia may be influenced by the habitual diet [7,19]. A short-term consumption (25 days) of isocaloric diets rich in SFA, *n*-6 PUFA or *n*-3 PUFA in healthy individuals resulted in greater TG and CM concentrations following SFA compared to *n*-3 PUFA, with intermediate concentrations with *n*-6 PUFA. CM from subjects on *n*-3 and *n*-6 PUFA diets were more susceptible to lipolysis in vitro [36]. After 15 and 29 days of dietary intervention (SFA-rich or *n*-6 PUFA-rich diet in healthy young men), postprandial response analysis suggested no change in the clearance of CM remnants but a prolonged accumulation of VLDL in individuals fed with the SFA-rich diet [37]. Two other short-term studies (six or four weeks) showed that *n*-3 PUFA supplementation (2.7 g/day or 4 g/day) in healthy subjects resulted in a significant reduction in postprandial TG compared with the control diet without supplementation [38,39]. In the second study, the supplementation with *n*-3 PUFA suppressed the increase in TG content in CM as well as in VLDL [39]. In contrast, consumption of a low-fat diet with *n*-3 PUFA (fish oil) supplementation for 16 weeks in healthy individuals led to a significant increase in postprandial TG concentration following a fat-rich test meal compared to chronic consumption of a low-fat diet alone [40] and another study did not show an effect of a six month *n*-3 PUFA supplementation in fasting or postprandial lipids compared to *n*-6 PUFA supplementation in moderately hyperlipidemic subjects [41]. In healthy individuals, an eight-week study of diet rich in SFA or MUFA (olive oil) showed a significant reduction in plasma total- and LDL-cholesterol concentrations but a higher postprandial plasma TG and TRL-apoB-48 concentrations with the MUFA diet [42]. In contrast, a 16-week moderate-MUFA diet or high-MUFA diet following an eight-week SFA-rich diet (reference diet) in healthy individuals resulted in the reduction of postprandial apoB-48 response without change in plasma TG concentration, suggesting that the CM formed carry larger amounts of dietary lipids per particle [43]. Similarly, a three-month SFA-rich diet compared to a MUFA-rich diet in healthy individuals did not show a difference in postprandial TG concentration but a reduction in both groups receiving a *n*-3 PUFA supplementation (3.6 g/day) versus placebo [44]. In this study, neither type of diet nor *n*-3 PUFA supplementation affected serum LDL size, but this parameter was measured in fasting state. Furthermore, in a cohort of 1048 subjects, hypertriglyceridemic (fasting serum TG > 150 mg/dL) participants had higher number of LDL particles, higher concentrations of small LDL particles and lower large LDL particles in baseline compared to normotriglyceridemic (fasting serum TG ≤ 150 mg/dL) participants. Following a high-fat meal challenge, both groups displayed similar patterns of change in LDL particle size concentrations with a small decrease in total LDL particle number, an increase in large LDL particle concentration, a decrease in small LDL particle concentration and no change in LDL particle size [45]. In another study, healthy individuals consumed three different diets for four weeks: a Western diet (38% fat of which 22% SFA), Mediterranean diet (38% fat, 24% MUFA) and a high carbohydrates diet with α-linolenic acid (ALA) (< 30% fat of which 8% PUFA). Consumption of the Mediterranean diet led to a decrease in the postprandial number of TRL compared with the other meals and also an increase in TRL particle size compared to the high carbohydrates with α-linolenic acid diet [46]. In the Medi-Rivage intervention study, a postprandial test was performed in individuals after either a three-month low fat or a Mediterranean-type diet (with SFA intake reduced by about half whereas MUFA increased). The consumption of the Mediterranean diet only lowered fasting TG concentration and both diets reduced TG and apoB-48 levels 5 h after the test meal. The overall 5 h postprandial apoB-48 response (AUC and iAUC) was lowered after both diets but this effect was more marked after the Mediterranean diet intervention [47].

A two-week trans FA-rich diet compared to a MUFA-rich diet (oleate) in healthy individuals did not show a difference in postprandial TG after a test meal [48].

Concerning the mechanisms, the reduction of postprandial lipemia following *n*-3 PUFA supplementation could be due to a decrease in CM synthesis/secretion and/or an increase in clearance. Several studies have shown an increase in LPL activity [44,49–51] and hepatic lipase activity [51] following supplementation with 3–5 g/day *n*-3 long-chain PUFA suggesting an effect on CM clearance. On the other hand, some studies are more in favor of an effect on CM production [52,53]. It is important not to forget that the reducing effect of *n*-3 PUFA on VLDL production [53–55] could also influence postprandial lipemia by the major link with CM metabolism. A lipoprotein kinetic study has examined the effect of the addition of *n*-3 FA ethyl esters (4 g/day: 46% eicosapentaenoic acid (EPA) and 38% docosahexaenoic acid (DHA)) to a weight-loss program for 12 weeks on postprandial apoB-48 kinetics in obese subjects after ingestion of an oral load. Compared with weight loss alone, weight loss plus *n*-3 supplementation significantly decreased fasting TG, apoB-48 concentrations, postprandial TG and apoB-48 total AUCs as well as postprandial TG iAUCs. This improvement of the postprandial profile was due to a decreased apoB-48 secretion in the basal state in the *n*-3 supplementation group without a significant effect during the postprandial period (3–6 hours) and no change in the clearance rate compared to the weight loss alone group [56]. A crossover study (but without a lipoprotein kinetic analysis) conducted by the same team showed a significant improvement of the postprandial lipemia after a fat load in a group of patients with familial hypercholesterolemia with *n*-3 PUFA supplementation (8 weeks; 4 g/day: 46% eicosapentaenoic acid and 38% docosahexaenoic acid) compared to no supplementation [57]. A review mainly focused on stable isotope tracer methodologies and compartmental modeling studies examined the mechanisms of action of dietary FA on lipoprotein metabolism [58]. Concerning n-3 PUFA, their effect on TG concentration reduction can be explained by several mechanisms: inhibition of diacylglycerol acyltransferase, FA synthase and acetyl CoA carboxylase enzymes; increase of FA β-oxydation via a peroxisome proliferator-activated receptor (PPAR) mediated pathway; inhibition of de novo lipogenesis by suppressing transcription of sterol regulatory element-binding protein-1c (SREBP-1c) gene; degradation of newly synthesized apolipoprotein B by stimulating the post-endoplasmic reticulum presecretory proteolysis pathway.

In vivo studies have confirmed that *n*-3 PUFA decrease the pool size (PS), the production rate (PR) of TRL-apoB-48 and VLDL-apoB-100, the fractional catabolic rate (FCR) of TRL-apoB-48 and increase the FCR of VLDL-apoB-100. Regarding *n*-6 PUFA, one lipoprotein kinetic study has shown after three weeks of *n*-6 PUFA supplementation a decrease in the PS of VLDL-apoB-100 due to an increase in FCR compared to a medium-chain FA supplementation. This effect could be due to an up-regulation of LPL activity and hepatic uptake of VLDL consequently to PPAR-activation by *n*-6 PUFA [59]. Lipoprotein kinetic studies on the impact of SFA on lipoprotein metabolism are lacking. One study did not show any effect of a four-week supplementation of medium-chain TG on TRL-apoB-48 and VLDL-apoB-100 metabolism in obese, insulin-resistant men [60]. For MUFA, one study conducted in twelve adults, has shown a decrease in PS and PR and an increase in FCR of VLDL-apoB-100 after a three-week MUFA-rich diet compared to a carbohydrate-rich diet [61]. Only one study has examined the impact of trans-FA on lipoprotein metabolism with no effect, in postmenopausal hypercholesterolemic women, on TRL-apoB-48 or TRL-apoB-100 metabolism [62].

3.2.3. Clinical Trials and Recommendations

A position paper from an international lipid expert panel concluded that *n*-3 EPA and DHA could be used efficiently as dietary supplements to reduce plasma TG (by 18–25%) whereas, their effects on LDL-C and HDL-C were clinically insignificant [63].

Despite the potential positive effect of *n*-3 PUFA on fasting TG and postprandial lipemia, two meta-analyses of randomized controlled trial, including a recent one (77,917 subjects; EPA supplementation doses between 226 and 1800 mg/day; mean follow-up: 4.4 years), did not show any effect of *n*-3 PUFA supplementation on mortality and cardiovascular events [64,65]. The most recent and larger meta-analysis of randomized controlled trial (112,059 subjects; 12 to 72 months duration; *n*-3 PUFA doses ranged from 0.5 g/day to > 5 g/day including EPA and DHA or ALA by supplementation or enriched food or dietary advice compared to placebo or usual diet) showed that increasing EPA and DHA has little or no effect on mortality or cardiovascular health and that low-quality evidence suggested ALA may slightly reduce cardiovascular disease and arrhythmia risk [66].

However, a recent randomized, double-blind, placebo-controlled trial (REDUCE-IT) involving 8179 patients with established cardiovascular disease or with diabetes and other risk factors, who had been receiving statin therapy and who had a fasting TG level of 135 to 499 mg/dL and a LDL-C level of 41 to 100 mg/dL with a median follow-up of 4.9 years, showed that a high-dose treatment of 4 g/d of EPA is accompanied by a significant decrease of 25% in the primary endpoint (cardiovascular death, non-fatal myocardial infarction, non-fatal stroke, coronary revascularization, unstable angina), of 20% in cardiovascular mortality, of 28% in fatal and non-fatal strokes, and a non-significant decrease of 13% in total mortality. More patients were hospitalized for cardiac arrhythmias in the EPA group [67].

Regarding dietary fat intake, the European guidelines on cardiovascular disease prevention in clinical practice advise: SFA to account for < 10% of total energy intake and should be reduced by an increase in PUFA; trans unsaturated FA as little as possible with preferably no intake from processed food and < 1% of total energy intake from natural origin; fish 1–2 servings per week, one of which to be oily fish; 30 g unsalted nuts per day [68]. The recent American guidelines on the primary prevention of cardiovascular disease recommend: the replacement of SFA with dietary MUFA and PUFA; the intake of trans FA should be avoided: a diet containing reduced amounts of cholesterol [69].

The European guidelines for the management of dyslipidemias recommend reducing TRL levels: to use supplements *n*-3 PUFA and replace SFA with MUFA or PUFA [70].

In line with the most recent meta-analysis, supplemental long-chain *n*-3 PUFA are probably not useful for preventing or treating cardiovascular disease, although they can help to reduce serum TG and raise HDL a little [66].

3.3. Dietary Cholesterol

The data available on the impact of dietary cholesterol on postprandial lipemia are limited. In healthy individuals, one single meal containing 0 or 140 mg cholesterol with a fixed amount of fat (45 g) elicited comparable postprandial lipemia whereas important doses of dietary cholesterol (280 or 710 mg) significantly increased postprandial lipemia compared to the lower amount [71]. A cholesterol-rich meal (1 g) compared to a cholesterol-free meal elicited increased postprandial CM-cholesterol and CM-TG in T2D patients and in matched non-diabetic control subjects. The increase in the postprandial VLDL-apoB-48 concentration was significantly higher in the diabetic patients (10-fold) compared to control individuals (three-fold) but the postprandial VLDL-apoB-100 concentration was not affected by dietary cholesterol, suggesting that intestinal CM production rather than their clearance explained these results [72]. However, chronic consumption (eight weeks) of diets varying in their amount of cholesterol (from 128 to 858 mg/day) by healthy individuals had no effect on postprandial lipemia [73].

To summarize, it appears that postprandial lipemia increases dose-dependently with the amount of dietary fat or cholesterol (at least above 15–20 g of fat and 140 mg of cholesterol) after a single meal. Acute test meals enriched with SFA, MUFA or PUFA do not generally elicit markedly different postprandial lipemia. Despite conflicting results, habitual diet studies show postprandial lipidemic responses to be in the order SFA > MUFA > *n*-6 PUFA > *n*-3 PUFA.

4. Effect of Dietary Carbohydrates

4.1. Effect of Chronic Consumption of Dietary Carbohydrates

High carbohydrate diets and especially highly digestible carbohydrate enriched diets have commonly been shown to alter lipid postprandial metabolism and to increase fasting plasma TG as a result of both intestinal CM and hepatic VLDL TRL and their remnants accumulation [7]. In 2 groups of normolipidemic and moderately hypertriglyceridemic subjects, Parks et al. compared the effects of two isoenergetic diets: a control (35% fat) diet for one week followed by a low-fat (15% fat) and high-carbohydrate diet for 5 weeks. Low fat/high-carbohydrate diet resulted in increased fasting TG concentration and decreased VLDL-TG clearance rate in both groups and increased fasting TRL-apoB-48 and TRL-apoB-100 concentrations in the hypertriglyceridemic group [74].

The monosaccharides have been extensively studied. As described by Livesey et al. in a meta-analysis, effect of fructose consumption on lipid profile was different depending on daily ingested amount: significant effect on postprandial TG was not evident unless > 50 g fructose/day was consumed, and no significant effect was seen for fasting TG with intakes of ≤ 100 g fructose/day [75]. In 66 overweight or obese men, the consumption of fructose sweetened beverages containing 75 g fructose per day for 12 weeks while continuing usual lifestyle and diet increased significantly fasting plasma TG and postprandial TG (AUC and iAUC) response after a mixed meal but did not impair glycemic control or incretin hormone responses during oral glucose or mixed meal challenge [76]. In contrast, another meta-analysis on the replacement of glucose or sucrose in foods or beverages by fructose (11 trials; length from 2 to 10 weeks; doses of fructose between 40 and 150 g/day) found no difference in fasting TG when fructose replaced glucose but a slight significant reduction when fructose replaced sucrose [77]. In a meta-analysis of 10-week to 26-week randomized trials of sugar-containing soft drinks, Bray concluded that plasma TG increase was due to fructose rather than glucose in sugar-containing soft drinks [78]. However, in a randomized control clinical study, Campos et al. showed in overweight subjects that substitution of high sugar-sweetened beverages providing large amounts of mono- or di-saccharides by artificially sweetened beverages during 12 weeks did not decrease postprandial TG despite of lower energy and fructose content of the meals [79]. Stanhope et al. showed, in overweight and obese subjects, that the consumption (eight weeks) of fructose-sweetened beverages significantly lowered glucose and insulin postmeal peaks and the AUC compared with the baseline diet (energy balanced diet containing 55% of energy as complex carbohydrates for 2 weeks) and with the consumption of glucose-sweetened beverages. Only fructose sweetened beverages diet consumption resulted in increased postprandial TG suggesting that the specific effect of fructose, but not of glucose and insulin excursions, contribute to the adverse effects of consuming sugar-sweetened beverages on lipids and insulin sensitivity [80]. Glucose supplementation had no effect on postprandial TG response as confirmed in 2 other studies of the same authors in young men and women [81] or in overweight or obese subjects [72] but a specific increase in fasting plasma TG concentration [82]. Despland et al. found a slightly decrease in postprandial blood glucose but no difference in postprandial TG plasma nor in hepatic insulin sensitivity in eight healthy male consuming a diet containing 25% energy as honey or pure fructose-glucose compared to an isocaloric starch diet [83]. Taken together, these studies suggest that isocaloric inclusion of fructose in mixed meal has inconsistent effects on postprandial TG despite its greater stimulation of de novo lipogenesis than other monosaccharides such as glucose [84]. However, in a systematic review and meta-analysis of controlled feeding trials, Wang et al. showed no significant postprandial TG increase when fructose was exchanged isocalorically for other carbohydrates in the diet but a significant postprandial TG raising-effect of fructose in studies in which fructose supplemented the background diet with excess energy from high-dose fructose compared with the background diet alone (without the excess energy) [85].

4.2. Effect of Acute Consumption of Dietary Carbohydrates

Both amount and nature of carbohydrates in a meal may alter postprandial lipid metabolism.

In healthy individuals, a high-fat/low-carbohydrate meal yielded a postprandial TG iAUC increase and an apoB-48 plasma iAUC reduction compared to a low-fat/high carbohydrate meal. This suggests difference in size and composition of CM depending on the meal composition [86]. In healthy individuals, the addition of 50 g or 100 g oral glucose to a fatty test meal diminished postprandial lipemia in a dose dependent manner compared to the meal containing fat alone. This effect was not due to increased clearance of TG from the circulation but appeared to reflect delayed gastric emptying and decreased hepatic secretion of TG. Starch ingestion had no discernible effect on postprandial lipemia [87]. Likewise, the addition of 75 g oral glucose to an oral fat meal delayed the gastric emptying and postponed the CM response compared to the fat meal in healthy individuals. The postprandial iAUC of serum TG and VLDL-TG were reduced but the CM-TG iAUC remained unchanged. The postprandial reduction of VLDL-TG iAUC may be due to the pronounced FFA depression during the glucose-induced rise in insulin [88]. In contrast, in healthy individuals, adding oral fructose as a monosaccharide [89] or as a disaccharide in sucrose [90] to an oral fat load led to an increase in postprandial lipemia.

In healthy individuals, physiological ranges of postprandial hyperglycemia and hyperinsulinemia as generated by starchy foods (white bread, pasta, beans) did not induce noticeable alterations in the overall postprandial TG response but delays and exacerbates postprandial accumulation of CM-apoB-48 in plasma [91]. Likewise, in healthy individuals, the intake of a standard fat dose meal (0.5 g/kg body weight) accompanied by either low-carbohydrate meal (17 g as lactose) or a high-carbohydrate meal (136 g of which 60 g was sucrose) did not show a difference in plasma TG or TRL-TG postprandial concentrations but a biphasic plasma TG response seen with the high-carbohydrate meal largely reflected the TRL-TG or CM fraction, which would tend to suggest a biphasic pattern of absorption. Higher insulin and GIP responses were seen with the high-carbohydrate meal [92]. Moreover, in obese, insulin-resistant subjects, the consumption of a high-glycemic index mixed meal, compared with a low-glycemic index one, increased the postprandial rise in plasma insulin and the accumulation of TRL-apoB-48 and TRL-apoB-100 thus increasing postprandial TG concentration as well as modifying the kinetics of peak occurrence. Thus, adding various digestible carbohydrates to a test meal can elicit a biphasic response of postprandial lipemia [93].

Besides the effects of dietary carbohydrates on hepatic VLDL metabolism [74,94], their effects on intestinal CM metabolism remain to be clarified, but have been reviewed in a recent paper pointing out the intestine as a contributor to carbohydrate-induced hyperlipidemia [95]. Oral fructose in a mixed meal can stimulate hepatic but also intestinal de novo lipogenesis, thereby increasing TG availability for CM and VLDL synthesis. This mechanism was increased for CM but not VLDL when glucose was added to the meal with a concomitant decrease in fructose oxidation and gluconeogenesis from fructose, suggesting the addition of glucose to the meal committed more fructose towards intestinal de novo lipogenesis [96]. In a lipoprotein kinetic study performed in healthy individuals, TRL-apoB-48 (CM) and TRL-apoB-100 (VLDL) metabolism was assessed after intraduodenal infusion (to avoid change in gastric emptying) of intralipid plus saline or glucose or fructose under pancreatic clamp conditions. Glucose markedly stimulated CM-PR with a moderate increase in CM-FCR resulting in net elevation of CM concentration but no effect on VLDL metabolism. Fructose significantly stimulated CM-PR and VLDL-PR but no effect on FCR [97]. The same team performed another lipoprotein kinetic study in healthy individuals to assess the effect of intravenous infusion of either 20% glucose or normal saline as control in a constant fed state. Compared with saline infusion; glucose infusion induced both hyperglycemia and hyperinsulinemia (despite pancreatic clamp conditions), FFA decrease and increased plasma TG, CM concentrations and CM-PR without affecting CM-FCR or VLDL metabolism [98]. Several studies have shown that a fructose-rich diet induced less insulin secretion than a glucose-rich diet, which could explain the lower postprandial LPL activity and the reduction in TG clearance after fructose compared to glucose [99].

At a cellular level, both luminal and basolateral glucose enhanced CM secretion with a greater effect of luminal glucose and a greater effect of luminal glucose than fructose [95]. In Caco-2/15 cells, basolateral exposure to glucose increased apical cholesterol uptake with increased expression of Niemann-Pick C1-like 1. This elevation of cholesterol uptake was associated with an increase in the transcription factors SREBP-2, carbohydrate-responsive element-binding protein (ChREBP) and liver X receptor (LXR)-β along with a fall in retinoid X receptor (RXR)-α [100]. Moreover, in Caco-2 cells, the incubation with glucose or fructose increased expression and protein abundance of microsomal triglyceride transfer protein (MTP) and fructose, but not glucose, activated SREBP-1 and ChREBP. These results show the link between carbohydrate and lipid pathways and suggest that these monosaccharides may play a role in enhancing TG synthesis and CM assembly [95]. Furthermore, several studies have shown that oral glucose compared to oral water can mobilize TG stored in cytosolic lipid droplets of enterocytes. This storage of droplets could contribute to CM appearance up to 16 h after the last meal. Moreover, the contractile activity of mesenteric lymphatics able to activate the secretion of extracellular CM, that reside between enterocytes, in lamina propria, lacteals and the mesenteric lymphatic system, was reduced with chronic high-fat and high-fructose feeding in rats [95].

4.3. Recommendations

Regarding dietary carbohydrate intake, the European guidelines on cardiovascular disease prevention in clinical practice advise: sugar-sweetened soft drinks must be discouraged [68]. The recent American guidelines on the primary prevention of cardiovascular disease recommend: to minimize the intake of refined carbohydrates and sweetened beverages [69].

The European guidelines for the management of dyslipidemias recommend reducing TRL levels: to reduce total amount of dietary carbohydrate and reduce intake of mono-disaccharides with a higher magnitude of the effect than the replacement of SFA with MUFA and PUFA. To illustrate this point, with a habitual fructose consumption between 15% and 20% of the total energy intake, plasma TG increase as much as 30–40% [70].

To summarize, it appears that postprandial lipemia increases more markedly with fructose than with glucose added to a single meal and in relation with the glycemic index of the carbohydrates. Acutely, a biphasic postprandial lipidemic response was described depending on the glycemic index of the carbohydrates. The chronic hypercaloric intake of fructose shows consistent results enhancing postprandial lipemia, but despite numerous studies, isocaloric chronic consumption of carbohydrates (fructose, glucose or starch) in mixed meals has led to discrepancies resulting in unclear divergent effects increasing or decreasing postprandial lipemia.

5. Effect of Dietary Proteins

In recent years, several studies have investigated the effect of protein quantity and quality on postprandial lipemia. The main mechanisms by which proteins have been hypothesized to affect postprandial lipid concentrations are through their slowing down of gastric emptying [101] and their potent effect on insulin release, notably via increased incretin secretion, i.e., GIP and GLP-1 [102,103]. Insulin is a well-known activator of LPL [104] but an increase in postprandial insulin also inhibits hormone-sensitive lipase and thereby suppresses the release of FFA from adipose tissue [105], which could limit the lipotoxicity associated with elevated FFA concentrations [106].

5.1. Effect of Acute Addition of Dietary Proteins

Early work by Cohen et al. showed that the addition of 23 g casein to a meal containing 40 g fat did not have any effect on postprandial lipemia in a group of 15 healthy adults [107]. In a group of 24 healthy adults, the addition of 50 g sodium caseinate to a fat meal (1 g fat/kg body weight as whipping cream) did not have any effect on the AUC of postprandial serum-, CM- or VLDL-TG concentrations, although a delay in serum-, CM- and VLDL-TG peak concentrations was observed, without any difference in gastric emptying rates. However, addition of casein led to a 20% reduction in FFA concentration over 8 h, together with a 30% increase in insulin release [108]. In contrast, in a group of 16 healthy adults, the addition of 50 g sodium caseinate or soy protein to a fat meal (1 g fat/kg body weight as whipping cream) led to a decrease in postprandial TG concentration (significant decrease at early time points but AUC was not calculated), with a 1 h delay in the peak time, together with a decrease in FFA associated with an increase in insulin secretion [109].

In a group of 11 patients with well-controlled T2D, the addition of 45 g casein to a control meal, consisting of energy-free soup with 80 g of fat, did not affect the postprandial TG or HDL response. However, when casein was added to the control meal plus 45 g carbohydrates (as white bread), it suppressed the increased postprandial TG concentration observed after the control meal plus carbohydrates alone, with an increase in insulin, glucagon and GIP release and a decrease in FFA concentration [110].

5.2. Effect of Dietary Protein Type

Mortensen et al. compared the acute effect of protein type on postprandial lipemia by providing 12 patients with T2D a test meal containing 100 g butter and 45 g carbohydrates in combination with 45 g casein, whey, cod, or gluten [111]. Compared to other sources of proteins, whey led to a decrease in the AUC of postprandial TG concentration after 360 min (−27% to −31%), in both plasma and CM-rich fraction, suggesting a concomitant lower production of CM, as illustrated by a lower retinyl palmitate concentration in the CM-rich fraction and a decrease in FFA secretion. No differences in insulin, glucagon and incretin concentrations or gastric emptying were observed. Using the same setting in 11 obese non-diabetic patients, the same group observed a significant lowering effect of whey proteins on postprandial plasma TG, notably in the CM-rich fraction, with an increase in insulin and glucagon secretion compared to cod and gluten but not compared to casein [112]. In a group of 20 overweight or obese postmenopausal women, the addition of 45 g whey to a breakfast meal significantly decreased postprandial TG concentrations as well as the exposure to smaller TG-enriched CM particles, as reflected by a decrease in the AUC of the TG:apoB-48 ratio, compared to the addition of 45 g glucose or casein to the same meal (−21% and −27%, respectively) [113]. In 11 obese non-diabetic subjects, Homer-Jenssen et al. did not observe any difference between the addition of 45 g of four whey fractions (alpha-lactalbumin, whey isolate, caseino-glycomacro-peptide and whey hydrolysate) to a high-fat meal on postprandial TG, insulin, glucagon or incretin concentrations although whey hydrolysate led to a smaller decrease in postprandial FFA production compared to the other proteins [114]. No difference in postprandial TG concentration was observed in a similar setting in 12 T2D subjects [115].

Mariotti et al. provided 10 healthy overweight men a high-fat meal plus 45 g casein, whey protein, or α-lactalbumin-enriched whey protein and observed a lower increase in postprandial plasma TG concentration following the meal that provided casein (AUC decreased by 22%), with no effect on FFA or insulin. As supported by their in vitro observations at pH values similar to those observed in the stomach during digestion, the authors hypothesized that this difference was due to the low solubility of casein at low pH leading to potential phase separation in the stomach, hence slowing down the digestion and absorption of fat [116]. These results are in disagreement with those of Mortensen et al. and Pal et al. Mariotti et al. put forward that they studied healthy individuals and that the components of the meals they used were pre-mixed, therefore allowing more interaction between the nutrients. When providing a pre-meal consisting of 17.6 g proteins (whey, casein or gluten) 15 to 30 min before a fat-rich meal to 16 subjects with metabolic syndrome, Bjornshave et al. did not observe any difference with regard to postprandial TG or FFA concentrations [117]. The authors hypothesized that the lower protein dose used compared to Mariotti et al. may explain the discrepancy between the results of the two studies. The same group also provided 12 matched subjects with and without T2D with 17.6 g whey proteins 15 min before a fat-rich meal or during the main meal. Although the whey protein pre-meal led to an increase in insulin, glucagon and GIP concentrations in both groups of subjects and a decrease in gastric emptying rates, the authors did not observe any effect on postprandial TG, apoB-48 or FFA concentrations [118].

5.3. Effect of Chronic Addition of Dietary Proteins

Mamo et al. investigated the long-term effect of a diet enriched in proteins from lean red meat, in place of carbohydrates, on postprandial lipemia [119]. Twenty moderately hypertriglyceridemic but otherwise healthy individuals consumed for six weeks two isocaloric diets (14%, 53% and 30% energy from proteins, carbohydrates and fats respectively vs. 25%, 30% and 35%) and then received a fat tolerance test meal. The protein-enriched diet led to a decrease in CM production, with a lower postprandial apoB-48 concentration, but no difference in fasting plasma or postprandial lipemia was observed. In a group of 52 patients with abdominal obesity, Bohl et al. did not observe any long-term effect of the addition of 60 g whey compared with casein to a high-fat test meal consumed daily for 12 weeks on postprandial TG or FFA concentrations but whey addition led to a decrease in CM production compared to casein addition, with a lower postprandial apoB-48 concentration [120].

To summarize, although the effects of dietary proteins on gastric emptying and FFA release seem well established, their effect on postprandial TG concentrations remains unclear. Nonetheless, at least three studies have pointed at a greater decrease in postprandial TG caused by the acute consumption of whey proteins compared to casein, the two most studied protein sources. Of note, one study has observed an opposite effect but it was the only one where food components were pre-mixed, and therefore not physically separated from the fat source, thereby underlining the potential effect of food structure on postprandial TG concentrations, as further discussed in chapter eight of the present review. Long-term studies that investigated dietary protein consumption suggest a possible decreasing effect on CM production, in particular by whey proteins.

6. Effect of Dietary Fibers

The effect of dietary fibers on postprandial lipemia has previously been reviewed [7,121] showing that some sources of fibers, particularly soluble fibers, at the level of 4–10 g/meal, can decrease postprandial TG and cholesterol concentrations following a mixed meal. Several interrelated mechanisms have been put forward, including slowed gastric emptying, alteration of TG hydrolysis, through increased viscosity decreasing the rate of hydrolysis or inhibition of pancreatic lipase activity, alteration of mixed micelle formation, and possibly that of intestinal secretion of CM [122] but also modification of insulin secretion [123].

6.1. Effect of Acute Consumption of Dietary Fibers

Cara et al. showed that enrichment of a high-fat meal with fibers from cereals, i.e., 10 g oat bran, wheat fibers or 4.2 g wheat germ, led to a decreased postprandial serum TG concentration in 6 healthy male adults compared to a low-fiber control meal, while only wheat fibers led to a decrease in postprandial CM-TG concentration. Rice bran did not have any effect on postprandial TG concentration. All fiber sources led to a decrease in postprandial CM-cholesterol concentration [124]. In another study involving six ileostomized subjects, oat bran added to a test meal (43.8 g fat) was shown to elicit a 37% reduction in postprandial CM-TG concentration and a 43% reduction in postprandial CM-cholesterol concentration compared to a low fiber test meal, although the limited sample size did not allow for these differences to reach statistical significance [125]. This was accompanied by an increase in ileum excretion of fat and cholesterol. Dubois et al. observed that the addition of 10 g pea or soybean fibers to a high-fat meal did not result in a decrease in postprandial TG concentration in six healthy male adults but both fibers led to a decrease in CM-cholesterol and -phospholipid concentrations [126]. However, Sandstrom et al. showed that consumption of pea fibers by eight healthy adults in two consecutive meals (containing 7.4 g and 9.3 g fibers respectively) resulted in a reduction in CM-TG and intermediate-density lipoprotein-TG concentrations compared to a low-fiber meal [127]. The meals provided in this study had less fat (19 and 43 g), and hence more carbohydrates, than those used by Dubois et al. (70 g), which could partly explain the discrepancies between the two studies.

Kristensen et al. showed that addition of flaxseed fibers to a test meal containing 50 g fat led to a decrease in postprandial plasma TG concentration in 18 healthy male adults (with 22 < body mass index < 30 kg/m^2) [128]. The difference (7 g dietary fibers) with the control meal was more marked with the meal that provided a high dose (17 g) of flaxseed fibers from mucilage than with the meals that provided 12 g dietary fibers from whole flaxseeds or low dose flaxseed mucilage. The addition of mucilage also led to a decrease in postprandial insulin secretion. Khossoussi et al. showed that addition of 12 g dietary fibers from psyllium husk to a standard meal (total fiber content = 15 g) led to a 21% decrease in postprandial serum TG concentration in 10 overweight and obese mean [129]. Moreover, apoB-48 concentration was lower 1h after consumption of the meal high in fibers compared to the meal low in fibers but the difference was not significant over a 6 h period. Resistant maltodextrin (5 or 10 g added to a test meal containing 50 g fat) has also been shown to elicit a decreased postprandial serum TG concentration in 13 healthy adults [130]. Kondo et al. provided 11 healthy male adults with moderate hypercholesterolemia 200 g yogurt with or without 6 g partially hydrolyzed guar gum together with a high-fat meal containing 43.5 g fat [131]. Ingestion of fibers led to a 15% decrease in postprandial serum TG concentration and a 23% decrease in postprandial remnant-like lipoprotein particle cholesterol concentration.

In contrast, some studies have shown no effect or even an increase in postprandial TG concentrations following acute consumption of fibers. For example, Bourdon et al. observed no effect of a barley pasta meal enriched with beta-glucan (15.7 g fibers versus 5 g fibers) on postprandial plasma or TRL-TG or -cholesterol concentrations or postprandial TRL-apoB-48 concentration in 11 healthy men [123]. Redard et al. observed a higher postprandial plasma TG concentration in females, but not in males, following consumption of a high-fiber (15.4 g mixture of oat bran and guar gum) versus a low-fiber (0.4 g) test meal [132]. Likewise, Ulmius et al. observed an increase in postprandial plasma TG concentration following consumption of meals enriched with fibers from oats, rye bran, sugar beet fibers or a mixture of these three fibers versus a low-fiber test meal in 13 healthy adults [133]. The meals providing fibers not only differed in their soluble fiber content but also in their insoluble fiber content and they were mixed with a blackcurrant beverage, hereby raising the question whether the way fibers are added to the food matrix can influence the postprandial response measured.

6.2. Effect of Chronic Consumption of Dietary Fibers

Maki et al. studied the effect of adding oat containing β-glucan or wheat cereal products to the usual diet of 27 healthy male adults for two weeks on their postprandial lipemia following consumption of a high-fat meal [134]. Both dietary treatments were matched for their energy and total fiber content. The postprandial serum peak TG concentration was lower and the postprandial serum TG concentration tended to be lower following consumption of the oat treatment but the postprandial FFA concentration was higher. Of note, Cara et al. did not observe any difference in postprandial serum-TG concentration following consumption of a meal containing 10 g β-glucan vs wheat fibers in healthy individuals [124]. Dubois et al. compared the postprandial effects of an oat bran test meal (12.8 g fibers) following consumption for 14 days of either an oat bran supplemented diet (23.8 g fibers/day) or a basal low-fiber diet (2.8 g fibers/day) in six normolipemic men [135]. Although it did not reach statistical significance, probably due to the low number of participants, the oat test meal following chronic consumption of oat elicited a slightly greater postprandial CM-TG concentration. However, the authors did observe significant differences in the postprandial concentrations of other lipid classes: postprandial plasma phospholipid, as well as plasma- and HDL-free cholesterol concentrations were increased while postprandial HDL-C ester concentration was decreased. The authors thus concluded that chronic consumption of oat exacerbates the acute effects of an oat meal on postprandial lipid concentrations. Wolever et al. provided 33 dyslipidemic participants for four months either a high soluble or insoluble fiber diet. The postprandial CM-TG concentration following consumption of a standardized fiber-free fatty liquid meal did not exhibit any difference when all subjects were considered but subjects with the APOε3 phenotype had a greater postprandial CM-TG concentration, due to an increase in the rate of fat absorption or CM synthesis or both, after soluble vs insoluble long-term fiber consumption [136]. Bozetto et al. provided 20 overweight/obese subjects with T2D with a high carbohydrate/fiber diet (26 g fibers/1000 kcal) for eight weeks and showed that participants exhibited a lower postprandial plasma TG concentration as well as a lower TG concentration in the CM+VLDL fraction following consumption of a test meal rich in saturated fat after the dietary intervention [137]. In another study using a relatively similar design, the same group also observed a decrease in postprandial CM-cholesterol concentration in T2D patients [138].

To summarize, most studies show that acute addition of soluble fibers to the meal leads to lower postprandial TG concentrations. Several interrelated mechanisms linked to lipid digestion have been implicated (see [121] for review). However, studies on this matter are relatively old and have thus not taken into consideration the possible effect of the interaction between the gut microbiota and fiber consumption on postprandial lipemia and more research is therefore needed to investigate if some strains of bacteria are associated with altered postprandial TG concentrations or lipoprotein profiles, as has been for example shown recently in the case of postprandial glucose concentrations [139].

7. Effect of Micronutrients and Phytochemicals

Some micronutrients, e.g., vitamins or trace elements, as well as polyphenols have been shown to modulate postprandial lipemia. However, mechanisms underlying these effects have not been fully elucidated and it is possible that not yet studied vitamins or trace metals, e.g., fat-soluble vitamins or selenium, or phytochemicals other than polyphenols, e.g., carotenoids, might modulate postprandial lipemia as well.

7.1. Niacin

The term niacin (also known as vitamin B3) refers to all molecules with the biological activity of nicotinamide. It functions in the body as a component of Nicotinamide Adenine Dinucleotide Phosphate (NADP) and Nicotinamide Adenine Dinucleotide Phosphate hydrogen (NADPH), which are involved in many metabolic processes, including glycolysis, FA metabolism, and tissue respiration. Its effect on fasting blood lipids is well established and pharmacological doses of niacin have been used for five decades to treat lipid disorders and try to prevent CVD. Indeed, niacin diminishes fasting blood TG concentration, likely by its inhibitory effect on VLDL production in T2D patients [140], as well as LDL-C concentration, and raises HDL-C concentration in hypercholesterolemic patients and in subjects with diabetes and peripheral arterial disease [141,142]. Nevertheless, its role on postprandial lipemia is still controversial because the only 3 studies dedicated to this topic showed different results. Indeed, in a first study, niacin treatment did not significantly changed CM kinetics in subjects with isolated low HDL-C [143] while in a second one, which was performed in normolipidemic men with hypoalphalipoproteinemia, it significantly diminished it [144]. In a third study performed in T2D patients, an intermediate effect was observed with a decrease in the postprandial secretion rate of apoB-48-containing particles without a significant change in iAUC of postprandial plasma TG and apoB-48 concentrations [145]. The reason for this discrepancy is not known but it might be due to the form of niacin used, i.e., crystalline or another formulation, e.g., extended-release or sustained-release [146], as well as to the characteristics of the subjects. Although it has been shown, in lean and obese subjects, that niacin reduces FFA mobilization from adipocytes, perhaps by suppressing lipolysis [147], and diminishes the liver formation of TG via noncompetitive inhibition of liver diacylglycerol acyltransferase-2 (DGAT2) [148], mechanisms that explain its role on postprandial lipemia have not been elucidated until the results of a recent study that give an idea on the potential mechanism. Indeed, in a study, which was performed in statin-treated T2D subjects, it was observed that extended-release niacin reduced postprandial secretion rate of apoB-48-containing particles [145].

However, a recent meta-analysis of randomized controlled trial (39,195 subjects; median duration of treatment 11.5 months; median dose of niacin 2 g/day in monotherapy or in combination with other component versus placebo/usual care or other component alone) showed no reduction with niacin in mortality, cardiovascular mortality, fatal or non-fatal myocardial infarction nor fatal or non-fatal strokes but niacin was associated with side effects [149].

7.2. Zinc

The pioneering studies on the effect of Zinc (Zn) on postprandial lipemia in rats were published in 1977 by Koo and Turk [150,151]. These researchers described in detail the consequences of Zn deficiency on lipid absorption in the rat. They notably found that the rate of TG absorption markedly decreased, with a huge accumulation of TG droplets in the mucosa and that these droplets were unstable and coalesced. They also observed that cell cytoplasm exhibited prominent cellular changes. They deduced that the enterocytes were not able, by an unknown mechanism, to secrete lipid droplets. They suggested that this was due to the failure of these cells to synthesize proteins required for the formation of CM, i.e., apolipoproteins. Later, it was found that marginal Zn depletion significantly diminished apo-C and -E concentrations in CM [152]. The nascent CM were also irregular and larger in shape and size. The same team further showed in rats that the CM from marginally Zn-deficient rats were less efficiently taken up by the liver [153]. This likely explained their delayed clearance from the blood. One year later, the same team showed, also in rats, that marginal Zn deficiency also significantly diminished CM-apo-B concentration [154]. About ten years later, Reaves et al. showed that the plasma ratio of apoB-48 to total apoB protein was significantly lower in Zn-deficient rats than in Zn-adequate rats [155]. They suggested that this is due to the editing of apoBmRNA that was impaired by Zn deficiency. Indeed, apoB mRNA editing is performed by a Zn-containing cytidine deaminase [156] and this enzyme determines whether apoB-48 or apoB-100 is synthesized. The same team later focused on the intestine and found that Zn deficiency modestly, but significantly, diminished intestinal apoB mRNA editing in hamsters [157]. Nevertheless, another team did not observe significant modification of apoB mRNA editing in rats upon zinc deficiency [158]. This suggests that this effect, if existing, is not very important. Nevertheless, the demonstrated inhibitory effect of Zn deficiency on the intestinal synthesis of CM in rats and hamsters has also been observed in Mongolian Gerbils [159], suggesting that this is a general phenomenon in rodents and, we assume, in mammals. Unfortunately, to our knowledge, there is no data in human. Zn deficiency has apparently also another effect on CM metabolism via a reduction of their lipolysis efficiency by LPL. Indeed, Zn-deficient rats exhibited a reduced LPL activity in postheparin serum and adipose tissue [160,161]. Koo and Lee suggested that this was not due to changes in the enzyme activity per se, but to Zn-deficiency-induced compositional alterations in CM, which modulate LPL activity. The key role of CM composition on LPL activity likely explains why the effect of Zn deficiency on LPL activity was observed in rats fed coconut oil, but not in rats fed fish oil [161].

7.3. Copper

Although there are few studies on the effect of copper (Cu) on CM metabolism, the two available studies performed in rats suggest an effect of Cu deficiency. In the first study, it was observed that Cu deficiency significantly diminished the activities of both endothelial LPL and hepatic lipase. This might explain the lower clearance rate of CM, which was observed in the beginning of the postprandial period in the Cu deficient rats [162]. The second study showed that TRL isolated from Cu-deficient rats were more fluid than those isolated from control rats [163]. This was apparently due to their low cholesterol/phospholipid ratio and their high TG content. The authors suggested that these modifications could affect the metabolism of these lipoproteins. Thus, dedicated clinical studies are required to assess the effect of Cu deficiency on postprandial lipemia.

7.4. Magnesium and Calcium

A pioneer study performed in inverted hamster intestine showed that a very low concentration of Mg in the intestinal lumen impairs the normal secretion of CM by the intestine [164]. Conversely, a clinical study performed in healthy individuals found that both the CM-TG response and the postprandial blood concentration of apoB-48 after a fat load were significantly lower after a meal that contained a Mg supplement (500 mg) than after a control meal [165]. Thus, as for Cu, further clinical studies are required to conclude on the effect of Mg status or Mg supplementation on postprandial lipemia. Concerning Calcium (Ca), a clinical study has suggested that dairy Ca, but not supplementary Ca carbonate, can attenuate postprandial lipemia in healthy moderately overweighted men [166]. It was suggested that this was due to impaired fat absorption because high Ca intake increases fecal fat excretion. However, further studies are needed to confirm this finding and to explain the different effect of these two chemical forms of Ca.

7.5. Polyphenols

The story on the effect of polyphenols on postprandial lipemia unusually started with a clinical study that found no significant effect of acute dealcoholized red wine, which is rich in polyphenols, on postprandial lipid metabolism in dyslipidemic postmenopausal women [167], suggesting that these polyphenols, at the tested dose, do not significantly affect lipid absorption and CM metabolism. However, a study performed in human Caco-2 cells led to an opposite conclusion by showing that red wine polyphenols significantly impaired the secretion of apoB-48 by these cells [168]. Conversely, in the same cell model, Vidal et al. did not find that wine polyphenols decreased the secretion of lipoproteins, contrarily to apple polyphenols that decreased it [169]. Tea polyphenols were also shown to decrease postprandial hypertriglyceridemia in rodents [170,171] and in men with mild or borderline hyperTG [172]. It was suggested that this was due to a decrease in TG absorption via an inhibition of pancreatic lipase activity [171]. However, as observed in mice, this could also be due to the fact that tea polyphenols decrease bile acid reabsorption, which results in lower intestinal bile acid levels, which might further decrease lipid absorption [173]. Interestingly, coffee polyphenols also inhibited pancreatic lipase activity, resulting in a lower postprandial increase in blood TG concentration. A study in mice suggested that this effect was apparently due mainly to one species of polyphenols among the 9 species that are recovered in coffee, i.e., di-cafeoylquinic acids [174]. Cinnamon extract, which is rich in polyphenols, was also able to diminish the secretion of apo-B48 and TRL in a fat load test performed in hamsters. Furthermore, it was observed that cinnamon extract reversed the expression of the impaired Insulin Receptor (IR), Insulin Receptor Substrate 1 (IRS1), IRS2 and AKT serine/threonin Kinase 1 (Akt1) mRNA levels and inhibited the overexpression of MTP and SREBP-1cin rodent enterocytes [175,176]. In another study, an anthocyanin-rich extract purified from a Haskap fruit significantly reduced the postprandial TG response measured in rats after a fat load [177]. Finally, a clinical study performed in overweight/obese subjects and components of the metabolic syndrome showed that subjects submitted to eight-week supplementation with a diet rich in polyphenols had lower postprandial TG response to a fat load than subjects who consumed a diet poor in polyphenols [178].

To summarize, it appears that some polyphenols, but not all, could significantly impair either the absorption of lipids or the intestinal secretion of CM. Nevertheless, additional studies, preferentially clinical ones, are required to identify which polyphenols and, at which dose, can significantly diminish postprandial lipemia.

8. Effect of the Food Structure (Matrix)

Although the study of the effects of single nutrients on postprandial lipemia is paramount in our understanding of the mechanisms involved, this approach bears limitations since human beings consume foods and not isolated nutrients. Indeed, most foods are complex, heterogeneous matrices and are defined not only by their qualitative and quantitative molecular composition but also by the organization of their molecules at multiple spatial length scales [179]. Moreover, the initial structure of a food is greatly modified by digestive processes, be it physical (e.g., mastication, antral grinding) or chemical (e.g., digestive enzymes, pH) ones. Hence, numerous interactions exist between the different components of each food and with other components from co-consumed foods. Jenkins and colleagues have long acknowledged this complexity in the case of postprandial glycaemia with the introduction of the glycemic index in 1981, which considered the postprandial effect of both nutrients, such as mono- or di-saccharides, as well as that of foods [180]. To date, such an approach has not been developed in the case of postprandial TG although guidelines have been proposed to assess postprandial TG concentrations in a standardized fashion [181]. Yet, several authors have pointed at the greater efficacy of food-based approaches in the prevention and treatment of some chronic diseases, including CVD [182–184], and thus advocate to switch the focus from nutrients to foods, for easier translation to the public but also to take into account the inherent complexity of food matrices. Numerous studies have shown that the distribution of FA in TG, the organization of lipids as oil droplet emulsions differing in their size and interfacial composition, the degree of crystallized fat or the permeability of the food matrix to digestive enzymes can influence lipid digestion and metabolism [185] but only a few studies have specifically investigated the effects of food structure on postprandial lipemia.

8.1. Effect of Dietary Lipid Physical State

Fats and oils in foods can be present either as a continuous phase or as emulsions, i.e., two immiscible phases dispersed as droplets, but they are also characterized by their crystallized/liquid TG ratio, which varies with temperature. Vors et al. provided nine normal weight and nine obese subjects with an identical breakfast containing 40 g milk fat either emulsified or non-emulsified [186]. Importantly, the two fats used had similar melting temperatures. The emulsified fat led to an earlier and greater CM-TG peak concentration, greater apoB-48 concentrations in all subjects, as well as larger CM size and iAUC of the CM-TG concentrations in obese subjects. Garaiova et al. also observed a 60% greater iAUC of the postprandial plasma TG concentrations following consumption by 24 healthy volunteers of a standardized meal comprising 30ml of an emulsified *n*-3-rich PUFA (EPA + DHA = 28% *w/w*) oil mixture compared to the same meal but with a non-emulsified oil mixture [187]. Nevertheless, only the postprandial AUC for plasma *n*-3 PUFA concentrations was affected by the emulsification, i.e., AUC for postprandial plasma SFA, MUFA and *n*-6 PUFA concentrations were not significantly different, strongly suggesting an increase in the absorption efficiency of *n*-3 PUFA with emulsification rather than a modification of postprandial CM metabolism.

Clemente et al. investigated postprandial TG concentrations in 8 T2D overweight patients after they received three test meals, identical in volume and macronutrient composition, but with fat originating from different sources, namely milk, butter and mozzarella cheese. No significant difference was observed in the increase in plasma TG concentration over the 6 h following the meal although the meal containing butter elicited a significantly delayed plasma TG peak time, not due to differences in gastric emptying rate. Unfortunately, it is not possible to conclude from the study design if this was due to the dispersion state of lipids (relatively small native milk fat globules in milk, aggregated milk fat globules dispersed in a protein matrix for mozzarella cheese, relatively larger fat droplets for butter) or to differences in viscosity (butter, mozzarella cheese and milk being respectively solid, semi-solid and liquid) [188]. Tholstrup et al. did not observe either any difference in postprandial plasma total TG, CM-TG and VLDL-TG concentrations when they provided 14 healthy young men with butter, cheese and milk [189].

8.2. Effect of the Droplet Size of the Oil Emulsion

Fats and oils in processed foods are mostly found as emulsions, and usually as oil-in-water emulsions. The initial oil droplet size has a major impact on lipid digestion, with smaller droplet size leading to faster digestion rate due to increased surface area. Armand et al. fed 8 healthy individuals with either a fine (surface-weighted mean diameter = 0.7 µm) or a coarse (surface-weighted mean diameter = 10 µm) emulsion and they observed a higher gastric and duodenal lipolysis, a slower gastric emptying, confirmed in [190], and a later postprandial serum- and CM-TG peak concentrations with the fine emulsion, but no significant difference was observed in the AUC of the postprandial serum- or CM-TG concentrations [191]. Tan et al. also studied the effect of emulsification and oil droplet size on postprandial TG concentrations [192]. Fifteen healthy Chinese males received a test meal containing olive oil as non-emulsified, finely emulsified (surface-weighted mean diameter = 0.7 µm) or coarsely emulsified (surface-weighted mean diameter = 10 µm). The meal with non-emulsified oil elicited the lowest iAUC of the postprandial plasma TG concentrations (although only with a trend against the coarse emulsion, $p = 0.07$), associated with the fastest gastric emptying, in agreement with the above-mentioned results from Vors et al. [186]. Moreover, a higher iAUC of the postprandial plasma TG concentration was observed following consumption of the test meal containing the fine emulsion compared to that containing the coarse emulsion (similar surface-weighted mean diameters as in Armand et al.) [192].

8.3. Effect of the Interfacial Film at the Oil-in-Water Emulsion Droplet Surface

The formation and stability of oil-in-water emulsions in the gastro-intestinal tract is influenced by the presence of emulsifiers, such as proteins, polysaccharides or phospholipids, which can in turn modulate oil droplet coalescence and hence fat digestion rate. Proteins differ in their emulsifying and stabilizing capacities, depending partly on their solubility, their hydrolysis rate by proteases and their displacement from the interfacial film by bile salts (see [193] for review). These characteristics could explain some of the differences observed in studies comparing the effect of protein sources on postprandial TG concentration but this has usually not been evaluated, with the exception of Mariotti et al. who observed in vitro a phase separation with the casein meal used in their clinical intervention study which they suggest could partly explain the associated lower increase in postprandial TG concentration [116]. Keogh et al. fed 10 men and 10 women (mean age = 59 years) two emulsions (iso-viscous, iso-caloric and same mean droplet size), containing 30 g of fat, differing in their emulsifier composition (namely sodium sterol lactylate or sodium caseinate/monoglycerides) [194]. Emulsions stabilized by sodium caseinate/monoglycerides elicited lower postprandial TG concentrations at 90 and 120 min compared to sodium sterol lactylate (no AUC calculated), with a concomitant faster gastric emptying and lower secretion of the gut hormones cholecystokinin, GLP-1 and peptide YY.

8.4. Effect of the Positional Distribution of FA in TG

Dietary TG can vary in FA chain length, degree of unsaturation but also in the distribution of FA on the glycerol backbone (stereospecificity), whether in naturally occurring TG or from technological processing by food industries, a technique termed interesterification. The isomers thus formed can lead to TG molecules with different physical properties, including melting temperature [195], digestion rates and biological effects, including postprandial TG concentrations. Dedicated reviews on the effects of interesterification on lipid metabolism have previously been published [196,197]. Berry reviewed 10 cross-over studies investigating the postprandial effects of stearic and palmitic acid-rich fats, the two major SFA in human diets, where test meals only differed from control meals by the stereospecificity of the TG sources and not by the FA composition. No conclusion could be drawn when only the positional distribution of FA was considered but she suggested that interesterified TG with higher melting points, i.e., crystalline at body temperature, led to a decrease in postprandial TG concentration, due to a slower assembling of micelles, leading to a slower rate of lipolysis in the gastro-intestinal tract [196]. This hypothesis has been confirmed in two subsequent studies by Berry's group where fat test meals containing interesterified palm olein led to lower postprandial plasma TG concentration compared to palm olein in healthy men and women [198] and in men aged 40–70 years with fasting plasma TG concentration > 1.2 mmol/L [199]. In these two studies, interesterified palm olein was characterized by a higher proportion of palmitic acid in the *sn*-2 position and a higher melting point (4.7% solid fat content at 37 °C whereas palm olein was fully melted at 37 °C).

8.5. Effect of Fat Localization within the Food Matrix

Berry et al. showed that a test meal containing 54 g fat provided as whole almond seed macroparticles elicited a 74% lower postprandial increase in plasma TG concentration compared to the same test meal containing almond oil and defatted almond flour (identical macronutrient composition) in 20 healthy adult men [200]. Oils bodies in almonds, as in many nuts, are found within thin-walled cells. These cell walls have been found to be highly resistant to digestion since almond microstructure has been shown to be only marginally affected by mastication, leading to low lipid bioaccessibility and hence lipolysis rates [201].

8.6. Effect of the Meal Consistency/Viscosity

The effect of the addition of fibers on postprandial TG concentrations has been specifically addressed in the dedicated chapter and is therefore left out of this section.

In a recent study, Dias et al. investigated the effect of three meals differing in their structure and form, namely solid, semi-solid and liquid, while having the same nutrient composition, on lipid digestion and postprandial TG concentration using an *in vitro* approach and a randomized, cross-over, dietary intervention trial in 26 healthy adults [202]. They showed that the liquid food elicited significantly higher postprandial TG concentration compared to the solid food, while the semi-solid food displayed an intermediate figure though not reaching statistical significance. This effect was partly attributed to the larger oil droplet size exhibited by the solid food compared to semi-solid and liquid food before and in the earlier stages of in vitro digestions as well as to the fact that solid food showed phase separation during gastric digestion together with a lower release of FA during intestinal digestion.

To summarize, food structure, whether native or manipulated, can significantly affect postprandial TG concentrations and can even override the effects of macronutrient composition. Lipid emulsification, particularly with a smaller droplet size, interesterification leading to TG with lower melting points or lower meal viscosity all elicit higher postprandial TG. Additional research is warranted to better characterize how manipulation of food structure can impact on postprandial TG concentration, e.g., effects of emulsifier type on the stability of oil-in-water emulsions. Nonetheless, consumption of foods with specific food structure, e.g., nuts, or technological modification of food structure certainly constitute relevant approaches in the prevention and management of elevated postprandial TG concentrations.

9. Conclusions

During the last decades, many clinical studies have highlighted the fact that healthy humans spend most of their time in a hyperlipidemic postprandial state due to the repetitive consumption of fat-containing meals and that this process is exacerbated in hyperlipidemic patients. Postprandial lipemia is characterized by the accumulation of both hepatic apoB-100 and intestinally-derived apoB-48 TRL in the circulation, which participate in atherosclerotic plaque progression. Accordingly, postprandial lipemia, in both its magnitude and duration, has been shown to constitute an independent risk factor for CVD, which confirms the central role of dietary modifications in the treatment and prevention of CVD.

Indeed, we have shown in this review that chylomicron production and postprandial lipemia are highly modulated by both habitual diet and single meal nutrient composition. Despite conflicting results between studies due to different methodological approaches and many potentially confounding factors, we have summarized in Table 1 the main acute and chronic effects of food components as well as food structure on chylomicron production and postprandial lipemia.

Table 1. Effects of nutrients and micronutrients on postprandial lipemia.

Dietary Components		Postprandial Lipemia	Level of Evidence
Fats	Amount	↑	+++
	Type (acute)	SFA = MUFA = PUFA	++
	Type (chronic)	SFA > MUFA > *n*-6 PUFA > *n*-3 PUFA	++
	Amount of cholesterol	↑	++
Carbohydrates	Acute	↑ (fructose > glucose)	+++
		↑ (glycemic index)	+++
	Chronic	↑ (fructose/dose dependent)	+++
Proteins	Whey proteins (acute)	↓	++
Fibers	Soluble (acute)	↓	++
Micronutrients	Ca supplement	↓	+
	Niacin supplement	↓	++
	Zn deficiency	↓	+
	Cu deficiency	↑	+
	Mg supplement	↓	+
	Polyphenol supplement	↓	+

Ca, calcium; Cu, copper; Mg, magnesium; MUFA, monounsaturated fatty acids; PUFA, polyunsaturated fatty acids; SFA, saturated fatty acids; Zn, zinc; +++: convincing; ++: probable; +: suggested.

Postprandial lipemia increases dose-dependently with the amount of dietary fat or cholesterol after a single meal, over a certain-amount threshold. However, due to interactions between long-chain fatty acid species and physico-chemical properties of fat structures, especially fat droplet characteristics, it is difficult to identify the acute effect of FA species on postprandial lipemia. Despite some conflicting results, studies of habitual diet show postprandial lipidemic responses in the order SFA > MUFA > *n*-6 PUFA > *n*-3 PUFA.

Although dietary fat has received the most attention, other nutrients and micronutrients can modulate postprandial lipemia. Dietary proteins could apparently display some effect but it depends on their nature and the few studies available do not allow us to conclude on their effect on postprandial lipemia. Carbohydrate sources added to a meal have been shown to modulate postprandial lipemia in relation with their glycemic index. Added to a fat-meal, glucose, and more markedly fructose, can noticeably increase postprandial lipemia. Chronic hypercaloric intake of fructose but not isocaloric consumption of carbohydrates (fructose, glucose or starch) resulted in constant increase postprandial lipemia. Some minerals (Ca, Zn, Cu) and some polyphenols, e.g., tea polyphenols, have also been shown to modulate postprandial lipemia but the number of studies is limited and the mechanisms suggested deserve more investigations. Dietary fibers, especially soluble fibers from various origins, can lower postprandial lipemia when added to a fatty meal in sufficient amount. Mechanisms involved are not fully understood and new studies should thus be performed to evaluate the possible interaction between the gut microbiota with the effect of fiber consumption on postprandial lipemia. Finally, nutrient composition alone cannot explain the effect of foods on postprandial lipemia and it is now clear that the food matrix is a key factor influencing fat digestion and hence postprandial lipemia.

The potential mechanisms of action have been reviewed and summarized in Table 2 but are not fully understood. Further studies and particularly lipoprotein kinetic studies in humans are needed.

Table 2. Potential mechanisms of nutrients and micronutrients action in postprandial lipemia.

Dietary Components		Potential Mechanisms
Fats	MUFA	↓ PR and ↑ FCR VLDL
	n-6 PUFA	↑ FCR VLDL
	n-3 PUFA	↓ PR CM; ↓ PR and ↑ FCR VLDL
Carbohydrates	Fructose	↑ PR CM; ↑ PR VLDL
	Glucose	↑ PR CM
Proteins	Whey protein	Slowed gastric emptying ↑ insulin release ↓ PR CM
Fibers		Slowed gastric emptying Alteration of TG hydrolysis Alteration of mixed micelle formation ↓ PR CM
Micronutrients	Ca supplement	↓ fat absorption
	Niacin supplement	↓ PR CM
	Zn deficiency	↓ TG absorption rate and PR CM
	Cu deficiency	↓ LPL and HL activities
	Mg supplement	↓ PR CM
	Polyphenol supplement	↓ TG absorption likely by ↓ pancreatic lipase activity

Ca, calcium; Cu, copper; Mg, magnesium; CM, chylomicrons; FCR, fractional catabolic rate; MUFA, monounsaturated fatty acids; PR, production rate; PUFA, polyunsaturated fatty acids; VLDL, very low, density lipoprotein; Zn, zinc.

Dietary approaches based on food nutrient composition and structural interactions represent relevant approaches to control postprandial lipemia.

A better understanding of the factors and mechanisms regulating chylomicron production and postprandial lipemia, and particularly diet, is essential to try to modulate their increases and thus reduce the risk of atherosclerotic cardiovascular diseases and potentially the risk of total mortality.

Author Contributions: Conceptualization, C.D., P.B., D.L., M.M., R.V.; Review Process, C.D., P.B., D.L., M.M., R.V.; Original Draft preparation, C.D., P.B., D.L., M.M., R.V.; Review and Editing of Final Manuscript, C.D., P.B., D.L., M.M., R.V.

Funding: This work received no funding.

Conflicts of Interest: The authors declare no conflict of interest.

References

1. Cardiovascular Diseases (CVDs). Available online: https://www.who.int/news-room/fact-sheets/detail/cardiovascular-diseases-(cvds) (accessed on 28 April 2019).
2. Ferrari, R.; Aguiar, C.; Alegria, E.; Bonadonna, R.C.; Cosentino, F.; Elisaf, M.; Farnier, M.; Ferrières, J.; Filardi, P.P.; Hancu, N.; et al. Current practice in identifying and treating cardiovascular risk, with a focus on residual risk associated with atherogenic dyslipidaemia. *Eur. Heart J. Suppl. J. Eur. Soc. Cardiol.* **2016**, *18*, C2–C12. [CrossRef] [PubMed]
3. Aguiar, C.; Alegria, E.; Bonadonna, R.C.; Catapano, A.L.; Cosentino, F.; Elisaf, M.; Farnier, M.; Ferrières, J.; Filardi, P.P.; Hancu, N.; et al. A review of the evidence on reducing macrovascular risk in patients with atherogenic dyslipidaemia: A report from an expert consensus meeting on the role of fenofibrate-statin combination therapy. *Atheroscler. Suppl.* **2015**, *19*, 1–12. [CrossRef]
4. Adiels, M.; Olofsson, S.-O.; Taskinen, M.-R.; Borén, J. Overproduction of very low-density lipoproteins is the hallmark of the dyslipidemia in the metabolic syndrome. *Arterioscler. Thromb. Vasc. Biol.* **2008**, *28*, 1225–1236. [CrossRef] [PubMed]
5. Duez, H.; Lamarche, B.; Uffelman, K.D.; Valero, R.; Cohn, J.S.; Lewis, G.F. Hyperinsulinemia is associated with increased production rate of intestinal apolipoprotein B-48-containing lipoproteins in humans. *Arterioscler. Thromb. Vasc. Biol.* **2006**, *26*, 1357–1363. [CrossRef] [PubMed]

6. Rashid, S.; Watanabe, T.; Sakaue, T.; Lewis, G.F. Mechanisms of HDL lowering in insulin resistant, hypertriglyceridemic states: The combined effect of HDL triglyceride enrichment and elevated hepatic lipase activity. *Clin. Biochem.* **2003**, *36*, 421–429. [CrossRef]

7. Lopez-Miranda, J.; Williams, C.; Lairon, D. Dietary, physiological, genetic and pathological influences on postprandial lipid metabolism. *Br. J. Nutr.* **2007**, *98*, 458–473. [CrossRef] [PubMed]

8. Nordestgaard, B.G. Triglyceride-rich lipoproteins and atherosclerotic cardiovascular disease: New insights from epidemiology, genetics, and biology. *Circ. Res.* **2016**, *118*, 547–563. [CrossRef] [PubMed]

9. Dash, S.; Xiao, C.; Morgantini, C.; Lewis, G.F. New insights into the regulation of chylomicron production. *Annu. Rev. Nutr.* **2015**, *35*, 265–294. [CrossRef]

10. Dias, C.B.; Moughan, P.J.; Wood, L.G.; Singh, H.; Garg, M.L. Postprandial lipemia: Factoring in lipemic response for ranking foods for their healthiness. *Lipids Health Dis.* **2017**, *16*, 178. [CrossRef]

11. Meikle, P.J.; Barlow, C.K.; Mellett, N.A.; Mundra, P.A.; Bonham, M.P.; Larsen, A.; Cameron-Smith, D.; Sinclair, A.; Nestel, P.J.; Wong, G. Postprandial plasma phospholipids in men are influenced by the source of dietary fat. *J. Nutr.* **2015**, *145*, 2012–2018. [CrossRef]

12. Bonham, M.P.; Linderborg, K.M.; Dordevic, A.; Larsen, A.E.; Nguo, K.; Weir, J.M.; Gran, P.; Luotonen, M.K.; Meikle, P.J.; Cameron-Smith, D.; et al. Lipidomic profiling of chylomicron triacylglycerols in response to high fat meals. *Lipids* **2013**, *48*, 39–50. [CrossRef] [PubMed]

13. Murphy, M.C.; Isherwood, S.G.; Sethi, S.; Gould, B.J.; Wright, J.W.; Knapper, J.A.; Williams, C.M. Postprandial lipid and hormone responses to meals of varying fat contents: Modulatory role of lipoprotein lipase? *Eur. J. Clin. Nutr.* **1995**, *49*, 578–588. [PubMed]

14. Dubois, C.; Beaumier, G.; Juhel, C.; Armand, M.; Portugal, H.; Pauli, A.M.; Borel, P.; Latgé, C.; Lairon, D. Effects of graded amounts (0–50 g) of dietary fat on postprandial lipemia and lipoproteins in normolipidemic adults. *Am. J. Clin. Nutr.* **1998**, *67*, 31–38. [CrossRef] [PubMed]

15. Vors, C.; Pineau, G.; Drai, J.; Meugnier, E.; Pesenti, S.; Laville, M.; Laugerette, F.; Malpuech-Brugère, C.; Vidal, H.; Michalski, M.-C. Postprandial endotoxemia linked with chylomicrons and lipopolysaccharides handling in obese versus lean men: A lipid dose-effect trial. *J. Clin. Endocrinol. Metab.* **2015**, *100*, 3427–3435. [CrossRef]

16. Maffeis, C.; Surano, M.G.; Cordioli, S.; Gasperotti, S.; Corradi, M.; Pinelli, L. A high-fat vs. a moderate-fat meal in obese boys: Nutrient balance, appetite, and gastrointestinal hormone changes. *Obesity* **2010**, *18*, 449–455. [CrossRef]

17. Cohen, J.C.; Noakes, T.D.; Benade, A.J. Serum triglyceride responses to fatty meals: Effects of meal fat content. *Am. J. Clin. Nutr.* **1988**, *47*, 825–827. [CrossRef] [PubMed]

18. Jackson, K.G.; Robertson, M.D.; Fielding, B.A.; Frayn, K.N.; Williams, C.M. Olive oil increases the number of triacylglycerol-rich chylomicron particles compared with other oils: An effect retained when a second standard meal is fed. *Am. J. Clin. Nutr.* **2002**, *76*, 942–949. [CrossRef] [PubMed]

19. Lairon, D. Macronutrient intake and modulation on chylomicron production and clearance. *Atheroscler. Suppl.* **2008**, *9*, 45–48. [CrossRef] [PubMed]

20. Panth, N.; Dias, C.B.; Wynne, K.; Singh, H.; Garg, M.L. Medium-chain fatty acids lower postprandial lipemia: A randomized crossover trial. *Clin. Nutr. Edinb. Scotl.* **2019**, in press. [CrossRef]

21. Jackson, K.G.; Wolstencroft, E.J.; Bateman, P.A.; Yaqoob, P.; Williams, C.M. Greater enrichment of triacylglycerol-rich lipoproteins with apolipoproteins E and C-III after meals rich in saturated fatty acids than after meals rich in unsaturated fatty acids. *Am. J. Clin. Nutr.* **2005**, *81*, 25–34. [CrossRef] [PubMed]

22. Dias, C.B.; Phang, M.; Wood, L.G.; Garg, M.L. Postprandial lipid responses do not differ following consumption of butter or vegetable oil when consumed with omega-3 polyunsaturated fatty acids. *Lipids* **2015**, *50*, 339–347. [CrossRef] [PubMed]

23. Tulk, H.M.F.; Robinson, L.E. Modifying the *n*-6/ *n*-3 polyunsaturated fatty acid ratio of a high-saturated fat challenge does not acutely attenuate postprandial changes in inflammatory markers in men with metabolic syndrome. *Metabolism* **2009**, *58*, 1709–1716. [CrossRef] [PubMed]

24. Masson, C.J.; Mensink, R.P. Exchanging saturated fatty acids for (*n*-6) polyunsaturated fatty acids in a mixed meal may decrease postprandial lipemia and markers of inflammation and endothelial activity in overweight men. *J. Nutr.* **2011**, *141*, 816–821. [CrossRef] [PubMed]

25. Thomsen, C.; Rasmussen, O.; Lousen, T.; Holst, J.J.; Fenselau, S.; Schrezenmeir, J.; Hermansen, K. Differential effects of saturated and monounsaturated fatty acids on postprandial lipemia and incretin responses in healthy subjects. *Am. J. Clin. Nutr.* **1999**, *69*, 1135–1143. [CrossRef] [PubMed]

26. Mekki, N.; Charbonnier, M.; Borel, P.; Leonardi, J.; Juhel, C.; Portugal, H.; Lairon, D. Butter differs from olive oil and sunflower oil in its effects on postprandial lipemia and triacylglycerol-rich lipoproteins after single mixed meals in healthy young men. *J. Nutr.* **2002**, *132*, 3642–3649. [CrossRef] [PubMed]

27. Sun, L.; Tan, K.W.J.; Lim, J.Z.; Magkos, F.; Henry, C.J. Dietary fat and carbohydrate quality have independent effects on postprandial glucose and lipid responses. *Eur. J. Nutr.* **2018**, *57*, 243–250. [CrossRef] [PubMed]

28. Svensson, J.; Rosenqvist, A.; Ohlsson, L. Postprandial lipid responses to an alpha-linolenic acid-rich oil, olive oil and butter in women: A randomized crossover trial. *Lipids Health Dis.* **2011**, *10*, 106. [CrossRef] [PubMed]

29. Sanders, T.A.; de Grassi, T.; Miller, G.J.; Morrissey, J.H. Influence of fatty acid chain length and *cis/trans* isomerization on postprandial lipemia and factor VII in healthy subjects (postprandial lipids and factor VII). *Atherosclerosis* **2000**, *149*, 413–420. [CrossRef]

30. Sanders, T.A.; Oakley, F.R.; Cooper, J.A.; Miller, G.J. Influence of a stearic acid-rich structured triacylglycerol on postprandial lipemia, factor VII concentrations, and fibrinolytic activity in healthy subjects. *Am. J. Clin. Nutr.* **2001**, *73*, 715–721. [CrossRef] [PubMed]

31. Tholstrup, T.; Sandström, B.; Bysted, A.; Hølmer, G. Effect of 6 dietary fatty acids on the postprandial lipid profile, plasma fatty acids, lipoprotein lipase, and cholesterol ester transfer activities in healthy young men. *Am. J. Clin. Nutr.* **2001**, *73*, 198–208. [CrossRef]

32. Roche, H.M.; Zampelas, A.; Jackson, K.G.; Williams, C.M.; Gibney, M.J. The effect of test meal monounsaturated fatty acid: Saturated fatty acid ratio on postprandial lipid metabolism. *Br. J. Nutr.* **1998**, *79*, 419–424. [CrossRef] [PubMed]

33. Burdge, G.C.; Powell, J.; Calder, P.C. Lack of effect of meal fatty acid composition on postprandial lipid, glucose and insulin responses in men and women aged 50–65 years consuming their habitual diets. *Br. J. Nutr.* **2006**, *96*, 489–500. [PubMed]

34. Zampelas, A.; Peel, A.S.; Gould, B.J.; Wright, J.; Williams, C.M. Polyunsaturated fatty acids of the *n*-6 and *n*-3 series: Effects on postprandial lipid and apolipoprotein levels in healthy men. *Eur. J. Clin. Nutr.* **1994**, *48*, 842–848. [PubMed]

35. Peairs, A.D.; Rankin, J.W.; Lee, Y.W. Effects of acute ingestion of different fats on oxidative stress and inflammation in overweight and obese adults. *Nutr. J.* **2011**, *10*, 122. [CrossRef] [PubMed]

36. Weintraub, M.S.; Zechner, R.; Brown, A.; Eisenberg, S.; Breslow, J.L. Dietary polyunsaturated fats of the W-6 and W-3 series reduce postprandial lipoprotein levels. Chronic and acute effects of fat saturation on postprandial lipoprotein metabolism. *J. Clin. Invest.* **1988**, *82*, 1884–1893. [CrossRef] [PubMed]

37. Bergeron, N.; Havel, R.J. Influence of diets rich in saturated and omega-6 polyunsaturated fatty acids on the postprandial responses of apolipoproteins B-48, B-100, E, and lipids in triglyceride-rich lipoproteins. *Arterioscler. Thromb. Vasc. Biol.* **1995**, *15*, 2111–2121. [CrossRef]

38. Williams, C.M.; Moore, F.; Morgan, L.; Wright, J. Effects of *n*-3 fatty acids on postprandial triacylglycerol and hormone concentrations in normal subjects. *Br. J. Nutr.* **1992**, *68*, 655–666. [CrossRef]

39. Miyoshi, T.; Noda, Y.; Ohno, Y.; Sugiyama, H.; Oe, H.; Nakamura, K.; Kohno, K.; Ito, H. Omega-3 fatty acids improve postprandial lipemia and associated endothelial dysfunction in healthy individuals—a randomized cross-over trial. *Biomed. Pharmacother. Biomedecine Pharmacother.* **2014**, *68*, 1071–1077. [CrossRef]

40. Roche, H.M.; Gibney, M.J. Postprandial triacylglycerolaemia: The effect of low-fat dietary treatment with and without fish oil supplementation. *Eur. J. Clin. Nutr.* **1996**, *50*, 617–624.

41. Finnegan, Y.E.; Minihane, A.M.; Leigh-Firbank, E.C.; Kew, S.; Meijer, G.W.; Muggli, R.; Calder, P.C.; Williams, C.M. Plant- and marine-derived *n*-3 polyunsaturated fatty acids have differential effects on fasting and postprandial blood lipid concentrations and on the susceptibility of LDL to oxidative modification in moderately hyperlipidemic subjects. *Am. J. Clin. Nutr.* **2003**, *77*, 783–795. [CrossRef]

42. Roche, H.M.; Zampelas, A.; Knapper, J.M.; Webb, D.; Brooks, C.; Jackson, K.G.; Wright, J.W.; Gould, B.J.; Kafatos, A.; Gibney, M.J.; et al. Effect of long-term olive oil dietary intervention on postprandial triacylglycerol and factor VII metabolism. *Am. J. Clin. Nutr.* **1998**, *68*, 552–560. [CrossRef] [PubMed]

43. Silva, K.D.R.R.; Kelly, C.N.M.; Jones, A.E.; Smith, R.D.; Wootton, S.A.; Miller, G.J.; Williams, C.M. Chylomicron particle size and number, factor VII activation and dietary monounsaturated fatty acids. *Atherosclerosis* **2003**, *166*, 73–84. [CrossRef]

44. Rivellese, A.A.; Maffettone, A.; Vessby, B.; Uusitupa, M.; Hermansen, K.; Berglund, L.; Louheranta, A.; Meyer, B.J.; Riccardi, G. Effects of dietary saturated, monounsaturated and *n*-3 fatty acids on fasting lipoproteins, LDL size and post-prandial lipid metabolism in healthy subjects. *Atherosclerosis* **2003**, *167*, 149–158. [CrossRef]

45. Wojczynski, M.K.; Glasser, S.P.; Oberman, A.; Kabagambe, E.K.; Hopkins, P.N.; Tsai, M.Y.; Straka, R.J.; Ordovas, J.M.; Arnett, D.K. High-fat meal effect on LDL, HDL, and VLDL particle size and number in the Genetics of Lipid-Lowering Drugs and Diet Network (GOLDN): An interventional study. *Lipids Health Dis.* **2011**, *10*, 181. [CrossRef] [PubMed]

46. Perez-Martinez, P.; Ordovas, J.M.; Garcia-Rios, A.; Delgado-Lista, J.; Delgado-Casado, N.; Cruz-Teno, C.; Camargo, A.; Yubero-Serrano, E.M.; Rodriguez, F.; Perez-Jimenez, F.; et al. Consumption of diets with different type of fat influences triacylglycerols-rich lipoproteins particle number and size during the postprandial state. *Nutr. Metab. Cardiovasc. Dis.* **2011**, *21*, 39–45. [CrossRef] [PubMed]

47. Defoort, C.; Vincent-Baudry, S.; Lairon, D. Effects of 3-month Mediterranean-type diet on postprandial TAG and apolipoprotein B48 in the Medi-RIVAGE cohort. *Public Health Nutr.* **2011**, *14*, 2302–2308. [CrossRef] [PubMed]

48. Sanders, T.A.B.; Oakley, F.R.; Crook, D.; Cooper, J.A.; Miller, G.J. High intakes of trans monounsaturated fatty acids taken for 2 weeks do not influence procoagulant and fibrinolytic risk markers for CHD in young healthy men. *Br. J. Nutr.* **2003**, *89*, 767–776. [CrossRef]

49. Khan, S.; Minihane, A.-M.; Talmud, P.J.; Wright, J.W.; Murphy, M.C.; Williams, C.M.; Griffin, B.A. Dietary long-chain *n*-3 PUFAs increase LPL gene expression in adipose tissue of subjects with an atherogenic lipoprotein phenotype. *J. Lipid Res.* **2002**, *43*, 979–985.

50. Park, Y.; Harris, W.S. Omega-3 fatty acid supplementation accelerates chylomicron triglyceride clearance. *J. Lipid Res.* **2003**, *44*, 455–463. [CrossRef]

51. Harris, W.S.; Lu, G.; Rambjør, G.S.; Wålen, A.I.; Ontko, J.A.; Cheng, Q.; Windsor, S.L. Influence of *n*-3 fatty acid supplementation on the endogenous activities of plasma lipases. *Am. J. Clin. Nutr.* **1997**, *66*, 254–260. [CrossRef]

52. Harris, W.S.; Muzio, F. Fish oil reduces postprandial triglyceride concentrations without accelerating lipid-emulsion removal rates. *Am. J. Clin. Nutr.* **1993**, *58*, 68–74. [CrossRef] [PubMed]

53. Westphal, S.; Orth, M.; Ambrosch, A.; Osmundsen, K.; Luley, C. Postprandial chylomicrons and VLDLs in severe hypertriacylglycerolemia are lowered more effectively than are chylomicron remnants after treatment with *n*-3 fatty acids. *Am. J. Clin. Nutr.* **2000**, *71*, 914–920. [CrossRef] [PubMed]

54. Harris, W.S.; Connor, W.E.; Illingworth, D.R.; Rothrock, D.W.; Foster, D.M. Effects of fish oil on VLDL triglyceride kinetics in humans. *J. Lipid Res.* **1990**, *31*, 1549–1558. [PubMed]

55. Nozaki, S.; Garg, A.; Vega, G.L.; Grundy, S.M. Postheparin lipolytic activity and plasma lipoprotein response to omega-3 polyunsaturated fatty acids in patients with primary hypertriglyceridemia. *Am. J. Clin. Nutr.* **1991**, *53*, 638–642. [CrossRef] [PubMed]

56. Wong, A.T.Y.; Chan, D.C.; Barrett, P.H.R.; Adams, L.A.; Watts, G.F. Effect of ω-3 fatty acid ethyl esters on apolipoprotein B-48 kinetics in obese subjects on a weight-loss diet: A new tracer kinetic study in the postprandial state. *J. Clin. Endocrinol. Metab.* **2014**, *99*, 1427–1435. [CrossRef] [PubMed]

57. Chan, D.C.; Pang, J.; Barrett, P.H.R.; Sullivan, D.R.; Burnett, J.R.; van Bockxmeer, F.M.; Watts, G.F. ω-3 Fatty Acid Ethyl Esters Diminish Postprandial Lipemia in Familial Hypercholesterolemia. *J. Clin. Endocrinol. Metab.* **2016**, *101*, 3732–3739. [CrossRef] [PubMed]

58. Ooi, E.M.M.; Watts, G.F.; Ng, T.W.K.; Barrett, P.H.R. Effect of dietary Fatty acids on human lipoprotein metabolism: A comprehensive update. *Nutrients* **2015**, *7*, 4416–4425. [CrossRef]

59. van Schalkwijk, D.B.; Pasman, W.J.; Hendriks, H.F.J.; Verheij, E.R.; Rubingh, C.M.; van Bochove, K.; Vaes, W.H.J.; Adiels, M.; Freidig, A.P.; de Graaf, A.A. Dietary medium chain fatty acid supplementation leads to reduced VLDL lipolysis and uptake rates in comparison to linoleic acid supplementation. *PLoS ONE* **2014**, *9*, e100376. [CrossRef]

60. Tremblay, A.J.; Lamarche, B.; Labonté, M.-È.; Lépine, M.-C.; Lemelin, V.; Couture, P. Dietary medium-chain triglyceride supplementation has no effect on apolipoprotein B-48 and apolipoprotein B-100 kinetics in insulin-resistant men. *Am. J. Clin. Nutr.* **2014**, *99*, 54–61. [CrossRef]

61. Zheng, C.; Khoo, C.; Furtado, J.; Ikewaki, K.; Sacks, F.M. Dietary monounsaturated fat activates metabolic pathways for triglyceride-rich lipoproteins that involve apolipoproteins E and C-III. *Am. J. Clin. Nutr.* **2008**, *88*, 272–281. [CrossRef]

62. Matthan, N.R.; Welty, F.K.; Barrett, P.H.R.; Harausz, C.; Dolnikowski, G.G.; Parks, J.S.; Eckel, R.H.; Schaefer, E.J.; Lichtenstein, A.H. Dietary hydrogenated fat increases high-density lipoprotein apoA-I catabolism and decreases low-density lipoprotein apoB-100 catabolism in hypercholesterolemic women. *Arterioscler. Thromb. Vasc. Biol.* **2004**, *24*, 1092–1097. [CrossRef] [PubMed]

63. Cicero, A.F.G.; Colletti, A.; Bajraktari, G.; Descamps, O.; Djuric, D.M.; Ezhov, M.; Fras, Z.; Katsiki, N.; Langlois, M.; Latkovskis, G.; et al. Lipid-lowering nutraceuticals in clinical practice: Position paper from an International Lipid Expert Panel. *Nutr. Rev.* **2017**, *75*, 731–767. [CrossRef] [PubMed]

64. Hooper, L.; Thompson, R.L.; Harrison, R.A.; Summerbell, C.D.; Ness, A.R.; Moore, H.J.; Worthington, H.V.; Durrington, P.N.; Higgins, J.P.T.; Capps, N.E.; et al. Risks and benefits of omega 3 fats for mortality, cardiovascular disease, and cancer: Systematic review. *BMJ* **2006**, *332*, 752–760. [CrossRef] [PubMed]

65. Aung, T.; Halsey, J.; Kromhout, D.; Gerstein, H.C.; Marchioli, R.; Tavazzi, L.; Geleijnse, J.M.; Rauch, B.; Ness, A.; Galan, P.; et al. Associations of Omega-3 Fatty Acid Supplement Use With Cardiovascular Disease Risks: Meta-analysis of 10 Trials Involving 77 917 Individuals. *JAMA Cardiol.* **2018**, *3*, 225–234. [CrossRef] [PubMed]

66. Abdelhamid, A.S.; Brown, T.J.; Brainard, J.S.; Biswas, P.; Thorpe, G.C.; Moore, H.J.; Deane, K.H.; AlAbdulghafoor, F.K.; Summerbell, C.D.; Worthington, H.V.; et al. Omega-3 fatty acids for the primary and secondary prevention of cardiovascular disease. *Cochrane Database Syst. Rev.* **2018**, *11*, CD003177. [PubMed]

67. Bhatt, D.L.; Steg, P.G.; Miller, M.; Brinton, E.A.; Jacobson, T.A.; Ketchum, S.B.; Doyle, R.T.; Juliano, R.A.; Jiao, L.; Granowitz, C.; et al. Cardiovascular Risk Reduction with Icosapent Ethyl for Hypertriglyceridemia. *N. Engl. J. Med.* **2019**, *380*, 11–22. [CrossRef] [PubMed]

68. Piepoli, M.F.; Hoes, A.W.; Agewall, S.; Albus, C.; Brotons, C.; Catapano, A.L.; Cooney, M.-T.; Corrà, U.; Cosyns, B.; Ian Graham, C.D.; et al. 2016 European Guidelines on cardiovascular disease prevention in clinical practice: The Sixth Joint Task Force of the European Society of Cardiology and Other Societies on Cardiovascular Disease Prevention in Clinical Practice (constituted by representatives of 10 societies and by invited experts) Developed with the special contribution of the European Association for Cardiovascular Prevention & Rehabilitation (EACPR). *Eur. Heart J.* **2016**, *37*, 2315–2381.

69. Arnett, D.K.; Blumenthal, R.S.; Albert, M.A.; Buroker, A.B.; Goldberger, Z.D.; Hahn, E.J.; Himmelfarb, C.D.; Khera, A.; Lloyd-Jones, D.; McEvoy, J.W.; et al. 2019 ACC/AHA Guideline on the Primary Prevention of Cardiovascular Disease. *Circulation* **2019**. [CrossRef]

70. Catapano, A.L.; Graham, I.; De Backer, G.; Wiklund, O.; Chapman, M.J.; Drexel, H.; Hoes, A.W.; Jennings, C.S.; Landmesser, U.; Pedersen, T.R.; et al. 2016 ESC/EAS Guidelines for the Management of Dyslipidaemias. *Eur. Heart J.* **2016**, *37*, 2999–3058. [CrossRef]

71. Dubois, C.; Armand, M.; Mekki, N.; Portugal, H.; Pauli, A.M.; Bernard, P.M.; Lafont, H.; Lairon, D. Effects of increasing amounts of dietary cholesterol on postprandial lipemia and lipoproteins in human subjects. *J. Lipid Res.* **1994**, *35*, 1993–2007.

72. Taggart, C.; Gibney, J.; Owens, D.; Collins, P.; Johnson, A.; Tomkin, G.H. The role of dietary cholesterol in the regulation of postprandial apolipoprotein B48 levels in diabetes. *Diabet. Med. J. Br. Diabet. Assoc.* **1997**, *14*, 1051–1058. [CrossRef]

73. Ginsberg, H.N.; Karmally, W.; Siddiqui, M.; Holleran, S.; Tall, A.R.; Rumsey, S.C.; Deckelbaum, R.J.; Blaner, W.S.; Ramakrishnan, R. A dose-response study of the effects of dietary cholesterol on fasting and postprandial lipid and lipoprotein metabolism in healthy young men. *Arterioscler. Thromb. J. Vasc. Biol.* **1994**, *14*, 576–586. [CrossRef]

74. Parks, E.J.; Krauss, R.M.; Christiansen, M.P.; Neese, R.A.; Hellerstein, M.K. Effects of a low-fat, high-carbohydrate diet on VLDL-triglyceride assembly, production, and clearance. *J. Clin. Invest.* **1999**, *104*, 1087–1096. [CrossRef] [PubMed]

75. Livesey, G.; Taylor, R. Fructose consumption and consequences for glycation, plasma triacylglycerol, and body weight: Meta-analyses and meta-regression models of intervention studies. *Am. J. Clin. Nutr.* **2008**, *88*, 1419–1437. [PubMed]

76. Matikainen, N.; Söderlund, S.; Björnson, E.; Bogl, L.H.; Pietiläinen, K.H.; Hakkarainen, A.; Lundbom, N.; Eliasson, B.; Räsänen, S.M.; Rivellese, A.; et al. Fructose intervention for 12 weeks does not impair glycemic control or incretin hormone responses during oral glucose or mixed meal tests in obese men. *Nutr. Metab. Cardiovasc. Dis. NMCD* **2017**, *27*, 534–542. [CrossRef] [PubMed]

77. Evans, R.A.; Frese, M.; Romero, J.; Cunningham, J.H.; Mills, K.E. Chronic fructose substitution for glucose or sucrose in food or beverages has little effect on fasting blood glucose, insulin, or triglycerides: A systematic review and meta-analysis. *Am. J. Clin. Nutr.* **2017**, *106*, 519–529. [CrossRef] [PubMed]

78. Bray, G.A. Fructose and risk of cardiometabolic disease. *Curr. Atheroscler. Rep.* **2012**, *14*, 570–578. [CrossRef] [PubMed]

79. Campos, V.; Despland, C.; Brandejsky, V.; Kreis, R.; Schneiter, P.; Boesch, C.; Tappy, L. Metabolic Effects of Replacing Sugar-Sweetened Beverages with Artificially-Sweetened Beverages in Overweight Subjects with or without Hepatic Steatosis: A Randomized Control Clinical Trial. *Nutrients* **2017**, *9*, 202. [CrossRef]

80. Stanhope, K.L.; Griffen, S.C.; Bremer, A.A.; Vink, R.G.; Schaefer, E.J.; Nakajima, K.; Schwarz, J.-M.; Beysen, C.; Berglund, L.; Keim, N.L.; et al. Metabolic responses to prolonged consumption of glucose- and fructose-sweetened beverages are not associated with postprandial or 24-h glucose and insulin excursions. *Am. J. Clin. Nutr.* **2011**, *94*, 112–119. [CrossRef]

81. Stanhope, K.L.; Bremer, A.A.; Medici, V.; Nakajima, K.; Ito, Y.; Nakano, T.; Chen, G.; Fong, T.H.; Lee, V.; Menorca, R.I.; et al. Consumption of fructose and high fructose corn syrup increase postprandial triglycerides, LDL-cholesterol, and apolipoprotein-B in young men and women. *J. Clin. Endocrinol. Metab.* **2011**, *96*, E1596–E1605. [CrossRef]

82. Stanhope, K.L.; Schwarz, J.M.; Keim, N.L.; Griffen, S.C.; Bremer, A.A.; Graham, J.L.; Hatcher, B.; Cox, C.L.; Dyachenko, A.; Zhang, W.; et al. Consuming fructose-sweetened, not glucose-sweetened, beverages increases visceral adiposity and lipids and decreases insulin sensitivity in overweight/obese humans. *J. Clin. Invest.* **2009**, *119*, 1322–1334. [CrossRef] [PubMed]

83. Despland, C.; Walther, B.; Kast, C.; Campos, V.; Rey, V.; Stefanoni, N.; Tappy, L. A randomized-controlled clinical trial of high fructose diets from either Robinia honey or free fructose and glucose in healthy normal weight males. *Clin. Nutr. ESPEN* **2017**, *19*, 16–22. [CrossRef]

84. Moore, J.B.; Gunn, P.J.; Fielding, B.A. The role of dietary sugars and de novo lipogenesis in non-alcoholic fatty liver disease. *Nutrients* **2014**, *6*, 5679–5703. [CrossRef] [PubMed]

85. David Wang, D.; Sievenpiper, J.L.; de Souza, R.J.; Cozma, A.I.; Chiavaroli, L.; Ha, V.; Mirrahimi, A.; Carleton, A.J.; Di Buono, M.; Jenkins, A.L.; et al. Effect of fructose on postprandial triglycerides: A systematic review and meta-analysis of controlled feeding trials. *Atherosclerosis* **2014**, *232*, 125–133. [CrossRef]

86. Smolders, L.; Mensink, R.P.; Plat, J. An acute intake of theobromine does not change postprandial lipid metabolism, whereas a high-fat meal lowers chylomicron particle number. *Nutr. Res.* **2017**, *40*, 85–94. [CrossRef] [PubMed]

87. Cohen, J.C.; Berger, G.M. Effects of glucose ingestion on postprandial lipemia and triglyceride clearance in humans. *J. Lipid Res.* **1990**, *31*, 597–602. [PubMed]

88. Westphal, S.; Leodolter, A.; Kahl, S.; Dierkes, J.; Malfertheiner, P.; Luley, C. Addition of glucose to a fatty meal delays chylomicrons and suppresses VLDL in healthy subjects. *Eur. J. Clin. Invest.* **2002**, *32*, 322–327. [CrossRef]

89. Jeppesen, J.; Chen, Y.I.; Zhou, M.Y.; Schaaf, P.; Coulston, A.; Reaven, G.M. Postprandial triglyceride and retinyl ester responses to oral fat: Effects of fructose. *Am. J. Clin. Nutr.* **1995**, *61*, 787–791. [CrossRef]

90. Grant, K.I.; Marais, M.P.; Dhansay, M.A. Sucrose in a lipid-rich meal amplifies the postprandial excursion of serum and lipoprotein triglyceride and cholesterol concentrations by decreasing triglyceride clearance. *Am. J. Clin. Nutr.* **1994**, *59*, 853–860. [CrossRef]

91. Harbis, A.; Defoort, C.; Narbonne, H.; Juhel, C.; Senft, M.; Latgé, C.; Delenne, B.; Portugal, H.; Atlan-Gepner, C.; Vialettes, B.; et al. Acute hyperinsulinism modulates plasma apolipoprotein B-48 triglyceride-rich lipoproteins in healthy subjects during the postprandial period. *Diabetes* **2001**, *50*, 462–469. [CrossRef]

92. Shishehbor, F.; Roche, H.M.; Gibney, M.J. The effect of acute carbohydrate load on the monophasic or biphasic nature of the postprandial lipaemic response to acute fat ingestion in human subjects. *Br. J. Nutr.* **1998**, *80*, 411–418. [CrossRef] [PubMed]

93. Harbis, A.; Perdreau, S.; Vincent-Baudry, S.; Charbonnier, M.; Bernard, M.-C.; Raccah, D.; Senft, M.; Lorec, A.-M.; Defoort, C.; Portugal, H.; et al. Glycemic and insulinemic meal responses modulate postprandial hepatic and intestinal lipoprotein accumulation in obese, insulin-resistant subjects. *Am. J. Clin. Nutr.* **2004**, *80*, 896–902. [CrossRef] [PubMed]

94. Roche, H.M. Dietary carbohydrates and triacylglycerol metabolism. *Proc. Nutr. Soc.* **1999**, *58*, 201–207. [CrossRef] [PubMed]

95. Stahel, P.; Xiao, C.; Lewis, G.F. Control of intestinal lipoprotein secretion by dietary carbohydrates. *Curr. Opin. Lipidol.* **2018**, *29*, 24–29. [CrossRef] [PubMed]

96. Theytaz, F.; de Giorgi, S.; Hodson, L.; Stefanoni, N.; Rey, V.; Schneiter, P.; Giusti, V.; Tappy, L. Metabolic fate of fructose ingested with and without glucose in a mixed meal. *Nutrients* **2014**, *6*, 2632–2649. [CrossRef]

97. Xiao, C.; Dash, S.; Morgantini, C.; Lewis, G.F. Novel role of enteral monosaccharides in intestinal lipoprotein production in healthy humans. *Arterioscler. Thromb. Vasc. Biol.* **2013**, *33*, 1056–1062. [CrossRef] [PubMed]

98. Xiao, C.; Dash, S.; Morgantini, C.; Lewis, G.F. Intravenous Glucose Acutely Stimulates Intestinal Lipoprotein Secretion in Healthy Humans. *Arterioscler. Thromb. Vasc. Biol.* **2016**, *36*, 1457–1463. [CrossRef]

99. Bidwell, A.J. Chronic Fructose Ingestion as a Major Health Concern: Is a Sedentary Lifestyle Making It Worse? A Review. *Nutrients* **2017**, *9*, 549. [CrossRef]

100. Grenier, E.; Mailhot, G.; Dion, D.; Ravid, Z.; Spahis, S.; Bendayan, M.; Levy, E. Role of the apical and basolateral domains of the enterocyte in the regulation of cholesterol transport by a high glucose concentration. *Biochem. Cell Biol. Biochim. Biol. Cell.* **2013**, *91*, 476–486. [CrossRef]

101. Karamanlis, A.; Chaikomin, R.; Doran, S.; Bellon, M.; Bartholomeusz, F.D.; Wishart, J.M.; Jones, K.L.; Horowitz, M.; Rayner, C.K. Effects of protein on glycemic and incretin responses and gastric emptying after oral glucose in healthy subjects. *Am. J. Clin. Nutr.* **2007**, *86*, 1364–1368. [CrossRef]

102. Geraedts, M.C.; Troost, F.J.; Fischer, M.A.; Edens, L.; Saris, W.H. Direct induction of CCK and GLP-1 release from murine endocrine cells by intact dietary proteins. *Mol. Nutr. Food Res.* **2011**, *55*, 476–484. [CrossRef]

103. Jakubowicz, D.; Froy, O. Biochemical and metabolic mechanisms by which dietary whey protein may combat obesity and Type 2 diabetes. *J. Nutr. Biochem.* **2013**, *24*, 1–5. [CrossRef] [PubMed]

104. Kersten, S. Physiological regulation of lipoprotein lipase. *Biochim. Biophys. Acta* **2014**, *1841*, 919–933. [CrossRef] [PubMed]

105. Jocken, J.W.; Langin, D.; Smit, E.; Saris, W.H.; Valle, C.; Hul, G.B.; Holm, C.; Arner, P.; Blaak, E.E. Adipose triglyceride lipase and hormone-sensitive lipase protein expression is decreased in the obese insulin-resistant state. *J. Clin. Endocrinol. Metab.* **2007**, *92*, 2292–2299. [CrossRef] [PubMed]

106. Bergman, R.N.; Ader, M. Free fatty acids and pathogenesis of type 2 diabetes mellitus. *Trends Endocrinol. Metab.* **2000**, *11*, 351–356. [CrossRef]

107. Cohen, J.C. Protein ingestion does not affect postprandial lipaemia or chylomicron-triglyceride clearance. *Eur. J. Clin. Nutr.* **1989**, *43*, 497–499. [PubMed]

108. Westphal, S.; Kastner, S.; Taneva, E.; Leodolter, A.; Dierkes, J.; Luley, C. Postprandial lipid and carbohydrate responses after the ingestion of a casein-enriched mixed meal. *Am. J. Clin. Nutr.* **2004**, *80*, 284–290. [CrossRef]

109. Westphal, S.; Taneva, E.; Kastner, S.; Martens-Lobenhoffer, J.; Bode-Boger, S.; Kropf, S.; Dierkes, J.; Luley, C. Endothelial dysfunction induced by postprandial lipemia is neutralized by addition of proteins to the fatty meal. *Atherosclerosis* **2006**, *185*, 313–319. [CrossRef]

110. Brader, L.; Holm, L.; Mortensen, L.; Thomsen, C.; Astrup, A.; Holst, J.J.; de Vrese, M.; Schrezenmeir, J.; Hermansen, K. Acute effects of casein on postprandial lipemia and incretin responses in type 2 diabetic subjects. *Nutr. Metab. Cardiovasc. Dis.* **2010**, *20*, 101–109. [CrossRef]

111. Mortensen, L.S.; Hartvigsen, M.L.; Brader, L.J.; Astrup, A.; Schrezenmeir, J.; Holst, J.J.; Thomsen, C.; Hermansen, K. Differential effects of protein quality on postprandial lipemia in response to a fat-rich meal in type 2 diabetes: Comparison of whey, casein, gluten, and cod protein. *Am. J. Clin. Nutr.* **2009**, *90*, 41–48. [CrossRef]

112. Holmer-Jensen, J.; Mortensen, L.S.; Astrup, A.; de Vrese, M.; Holst, J.J.; Thomsen, C.; Hermansen, K. Acute differential effects of dietary protein quality on postprandial lipemia in obese non-diabetic subjects. *Nutr. Res.* **2013**, *33*, 34–40. [CrossRef] [PubMed]

113. Pal, S.; Ellis, V.; Ho, S. Acute effects of whey protein isolate on cardiovascular risk factors in overweight, post-menopausal women. *Atherosclerosis* **2010**, *212*, 339–344. [CrossRef] [PubMed]

114. Holmer-Jensen, J.; Hartvigsen, M.L.; Mortensen, L.S.; Astrup, A.; de Vrese, M.; Holst, J.J.; Thomsen, C.; Hermansen, K. Acute differential effects of milk-derived dietary proteins on postprandial lipaemia in obese non-diabetic subjects. *Eur. J. Clin. Nutr.* **2012**, *66*, 32–38. [CrossRef] [PubMed]

115. Mortensen, L.S.; Holmer-Jensen, J.; Hartvigsen, M.L.; Jensen, V.K.; Astrup, A.; de Vrese, M.; Holst, J.J.; Thomsen, C.; Hermansen, K. Effects of different fractions of whey protein on postprandial lipid and hormone responses in type 2 diabetes. *Eur. J. Clin. Nutr.* **2012**, *66*, 799–805. [CrossRef] [PubMed]

116. Mariotti, F.; Valette, M.; Lopez, C.; Fouillet, H.; Famelart, M.H.; Mathe, V.; Airinei, G.; Benamouzig, R.; Gaudichon, C.; Tome, D.; et al. Casein Compared with Whey Proteins Affects the Organization of Dietary Fat during Digestion and Attenuates the Postprandial Triglyceride Response to a Mixed High-Fat Meal in Healthy, Overweight Men. *J. Nutr.* **2015**, *145*, 2657–2664. [CrossRef] [PubMed]

117. Bjornshave, A.; Johansen, T.N.; Amer, B.; Dalsgaard, T.K.; Holst, J.J.; Hermansen, K. Pre-meal and postprandial lipaemia in subjects with the metabolic syndrome: Effects of timing and protein quality (randomised crossover trial). *Br. J. Nutr.* **2019**, 1–10. [CrossRef] [PubMed]

118. Bjornshave, A.; Holst, J.J.; Hermansen, K. Pre-Meal Effect of Whey Proteins on Metabolic Parameters in Subjects with and without Type 2 Diabetes: A Randomized, Crossover Trial. *Nutrients* **2018**, *10*, 122. [CrossRef] [PubMed]

119. Mamo, J.C.; James, A.P.; Soares, M.J.; Griffiths, D.G.; Purcell, K.; Schwenke, J.L. A low-protein diet exacerbates postprandial chylomicron concentration in moderately dyslipidaemic subjects in comparison to a lean red meat protein-enriched diet. *Eur. J. Clin. Nutr.* **2005**, *59*, 1142–1148. [CrossRef] [PubMed]

120. Bohl, M.; Bjornshave, A.; Rasmussen, K.V.; Schioldan, A.G.; Amer, B.; Larsen, M.K.; Dalsgaard, T.K.; Holst, J.J.; Herrmann, A.; O'Neill, S.; et al. Dairy proteins, dairy lipids, and postprandial lipemia in persons with abdominal obesity (DairyHealth): A 12-wk, randomized, parallel-controlled, double-blinded, diet intervention study. *Am. J. Clin. Nutr.* **2015**, *101*, 870–878. [CrossRef] [PubMed]

121. Lairon, D.; Play, B.; Jourdheuil-Rahmani, D. Digestible and indigestible carbohydrates: Interactions with postprandial lipid metabolism. *J. Nutr. Biochem.* **2007**, *18*, 217–227. [CrossRef] [PubMed]

122. Grundy, M.M.; Edwards, C.H.; Mackie, A.R.; Gidley, M.J.; Butterworth, P.J.; Ellis, P.R. Re-evaluation of the mechanisms of dietary fibre and implications for macronutrient bioaccessibility, digestion and postprandial metabolism. *Br. J. Nutr.* **2016**, *116*, 816–833. [CrossRef] [PubMed]

123. Bourdon, I.; Yokoyama, W.; Davis, P.; Hudson, C.; Backus, R.; Richter, D.; Knuckles, B.; Schneeman, B.O. Postprandial lipid, glucose, insulin, and cholecystokinin responses in men fed barley pasta enriched with beta-glucan. *Am. J. Clin. Nutr.* **1999**, *69*, 55–63. [CrossRef]

124. Cara, L.; Dubois, C.; Borel, P.; Armand, M.; Senft, M.; Portugal, H.; Pauli, A.M.; Bernard, P.M.; Lairon, D. Effects of oat bran, rice bran, wheat fiber, and wheat germ on postprandial lipemia in healthy adults. *Am. J. Clin. Nutr.* **1992**, *55*, 81–88. [CrossRef]

125. Lia, A.; Andersson, H.; Mekki, N.; Juhel, C.; Senft, M.; Lairon, D. Postprandial lipemia in relation to sterol and fat excretion in ileostomy subjects given oat-bran and wheat test meals. *Am. J. Clin. Nutr.* **1997**, *66*, 357–365. [CrossRef] [PubMed]

126. Dubois, C.; Cara, L.; Armand, M.; Borel, P.; Senft, M.; Portugal, H.; Pauli, A.M.; Bernard, P.M.; Lafont, H.; Lairon, D. Effects of pea and soybean fibre on postprandial lipaemia and lipoproteins in healthy adults. *Eur. J. Clin. Nutr.* **1993**, *47*, 508–520. [PubMed]

127. Sandstrom, B.; Hansen, L.T.; Sorensen, A. Pea fiber lowers fasting and postprandial blood triglyceride concentrations in humans. *J. Nutr.* **1994**, *124*, 2386–2396. [CrossRef] [PubMed]

128. Kristensen, M.; Savorani, F.; Christensen, S.; Engelsen, S.B.; Bugel, S.; Toubro, S.; Tetens, I.; Astrup, A. Flaxseed dietary fibers suppress postprandial lipemia and appetite sensation in young men. *Nutr. Metab. Cardiovasc. Dis.* **2013**, *23*, 136–143. [CrossRef] [PubMed]

129. Khossousi, A.; Binns, C.W.; Dhaliwal, S.S.; Pal, S. The acute effects of psyllium on postprandial lipaemia and thermogenesis in overweight and obese men. *Br. J. Nutr.* **2008**, *99*, 1068–1075. [CrossRef] [PubMed]

130. Kishimoto, Y.; Oga, H.; Tagami, H.; Okuma, K.; Gordon, D.T. Suppressive effect of resistant maltodextrin on postprandial blood triacylglycerol elevation. *Eur. J. Nutr.* **2007**, *46*, 133–138. [CrossRef]

131. Kondo, S.; Xiao, J.Z.; Takahashi, N.; Miyaji, K.; Iwatsuki, K.; Kokubo, S. Suppressive effects of dietary fiber in yogurt on the postprandial serum lipid levels in healthy adult male volunteers. *Biosci. Biotechnol. Biochem.* **2004**, *68*, 1135–1138. [CrossRef]

132. Redard, C.L.; Davis, P.A.; Schneeman, B.O. Dietary fiber and gender: Effect on postprandial lipemia. *Am. J. Clin. Nutr.* **1990**, *52*, 837–845. [CrossRef] [PubMed]

133. Ulmius, M.; Johansson, A.; Onning, G. The influence of dietary fibre source and gender on the postprandial glucose and lipid response in healthy subjects. *Eur. J. Nutr.* **2009**, *48*, 395–402. [CrossRef] [PubMed]

134. Maki, K.C.; Davidson, M.H.; Witchger, M.S.; Dicklin, M.R.; Subbaiah, P.V. Effects of high-fiber oat and wheat cereals on postprandial glucose and lipid responses in healthy men. *Int. J. Vitam. Nutr. Res.* **2007**, *77*, 347–356. [CrossRef] [PubMed]

135. Dubois, C.; Armand, M.; Senft, M.; Portugal, H.; Pauli, A.M.; Bernard, P.M.; Lafont, H.; Lairon, D. Chronic oat bran intake alters postprandial lipemia and lipoproteins in healthy adults. *Am. J. Clin. Nutr.* **1995**, *61*, 325–333. [CrossRef] [PubMed]

136. Wolever, T.M.; Hegele, R.A.; Connelly, P.W.; Ransom, T.P.; Story, J.A.; Furumoto, E.J.; Jenkins, D.J. Long-term effect of soluble-fiber foods on postprandial fat metabolism in dyslipidemic subjects with apo E3 and apo E4 genotypes. *Am. J. Clin. Nutr.* **1997**, *66*, 584–590. [CrossRef] [PubMed]

137. Bozzetto, L.; Annuzzi, G.; Costabile, G.; Costagliola, L.; Giorgini, M.; Alderisio, A.; Strazzullo, A.; Patti, L.; Cipriano, P.; Mangione, A.; et al. A CHO/fibre diet reduces and a MUFA diet increases postprandial lipaemia in type 2 diabetes: No supplementary effects of low-volume physical training. *Acta Diabetol.* **2014**, *51*, 385–393. [CrossRef] [PubMed]

138. De Natale, C.; Annuzzi, G.; Bozzetto, L.; Mazzarella, R.; Costabile, G.; Ciano, O.; Riccardi, G.; Rivellese, A.A. Effects of a plant-based high-carbohydrate/high-fiber diet versus high-monounsaturated fat/low-carbohydrate diet on postprandial lipids in type 2 diabetic patients. *Diabetes Care* **2009**, *32*, 2168–2173. [CrossRef]

139. Zeevi, D.; Korem, T.; Zmora, N.; Israeli, D.; Rothschild, D.; Weinberger, A.; Ben-Yacov, O.; Lador, D.; Avnit-Sagi, T.; Lotan-Pompan, M.; et al. Personalized Nutrition by Prediction of Glycemic Responses. *Cell* **2015**, *163*, 1079–1094. [CrossRef]

140. Ooi, E.M.; Watts, G.F.; Chan, D.C.; Pang, J.; Tenneti, V.S.; Hamilton, S.J.; McCormick, S.P.; Marcovina, S.M.; Barrett, P.H. Effects of extended-release niacin on the postprandial metabolism of Lp(a) and ApoB-100-containing lipoproteins in statin-treated men with type 2 diabetes mellitus. *Arterioscler. Thromb. Vasc. Biol.* **2015**, *35*, 2686–2693. [CrossRef]

141. Guyton, J.R.; Goldberg, A.C.; Kreisberg, R.A.; Sprecher, D.L.; Superko, H.R.; O'Connor, C.M. Effectiveness of once-nightly dosing of extended-release niacin alone and in combination for hypercholesterolemia. *Am. J. Cardiol.* **1998**, *82*, 737–743. [CrossRef]

142. Elam, M.B.; Hunninghake, D.B.; Davis, K.B.; Garg, R.; Johnson, C.; Egan, D.; Kostis, J.B.; Sheps, D.S.; Brinton, E.A. Effect of niacin on lipid and lipoprotein levels and glycemic control in patients with diabetes and peripheral arterial disease: The ADMIT study: A randomized trial. Arterial Disease Multiple Intervention Trial. *JAMA* **2000**, *284*, 1263–1270. [CrossRef] [PubMed]

143. Benjo, A.M.; Maranhao, R.C.; Coimbra, S.R.; Andrade, A.C.; Favarato, D.; Molina, M.S.; Brandizzi, L.I.; da Luz, P.L. Accumulation of chylomicron remnants and impaired vascular reactivity occur in subjects with isolated low HDL cholesterol: Effects of niacin treatment. *Atherosclerosis* **2006**, *187*, 116–122. [CrossRef] [PubMed]

144. King, J.M.; Crouse, J.R.; Terry, J.G.; Morgan, T.M.; Spray, B.J.; Miller, N.E. Evaluation of effects of unmodified niacin on fasting and postprandial plasma lipids in normolipidemic men with hypoalphalipoproteinemia. *Am. J. Med.* **1994**, *97*, 323–331. [CrossRef]

145. Pang, J.; Chan, D.C.; Hamilton, S.J.; Tenneti, V.S.; Watts, G.F.; Barrett, P.H. Effect of niacin on triglyceride-rich lipoprotein apolipoprotein B-48 kinetics in statin-treated patients with type 2 diabetes. *Diabetes Obes. Metab.* **2016**, *18*, 384–391. [CrossRef] [PubMed]

146. Pieper, J.A. Understanding niacin formulations. *Am. J. Manag. Care* **2002**, *8*, S308–S314. [PubMed]

147. Nelson, R.H.; Vlazny, D.; Smailovic, A.; Miles, J.M. Intravenous niacin acutely improves the efficiency of dietary fat storage in lean and obese humans. *Diabetes* **2012**, *61*, 3172–3175. [CrossRef] [PubMed]

148. Ganji, S.H.; Tavintharan, S.; Zhu, D.; Xing, Y.; Kamanna, V.S.; Kashyap, M.L. Niacin noncompetitively inhibits DGAT2 but not DGAT1 activity in HepG2 cells. *J. Lipid Res.* **2004**, *45*, 1835–1845. [CrossRef] [PubMed]

149. Schandelmaier, S.; Briel, M.; Saccilotto, R.; Olu, K.K.; Arpagaus, A.; Hemkens, L.G.; Nordmann, A.J. Niacin for primary and secondary prevention of cardiovascular events. *Cochrane Database Syst. Rev.* **2017**, *6*, CD009744. [CrossRef]

150. Koo, S.I.; Turk, D.E. Effect of zinc deficiency on the ultrastructure of the pancreatic acinar cell and intestinal epithelium in the rat. *J. Nutr.* **1977**, *107*, 896–908. [CrossRef]

151. Koo, S.I.; Turk, D.E. Effect of zinc deficiency on intestinal transport triglyceride in the rat. *J. Nutr.* **1977**, *107*, 909–919. [CrossRef]

152. Koo, S.I.; Henderson, D.A.; Algilani, K.; Norvell, J.E. Effect of marginal zinc deficiency on the morphological characteristics of intestinal nascent chylomicrons and distribution of soluble apoproteins of lymph chylomicrons. *Am. J. Clin. Nutr.* **1985**, *42*, 671–680. [CrossRef] [PubMed]

153. Koo, S.I.; Algilani, K.; Norvell, J.E.; Henderson, D.A. Delayed plasma clearance and hepatic uptake of lymph chylomicron 14C-cholesterol in marginally zinc-deficient rats. *Am. J. Clin. Nutr.* **1986**, *43*, 429–437. [CrossRef] [PubMed]

154. Koo, S.I.; Lee, C.C.; Norvell, J.E. Effect of marginal zinc deficiency on the apolipoprotein-B content and size of mesenteric lymph chylomicrons in adult rats. *Lipids* **1987**, *22*, 1035–1040. [CrossRef] [PubMed]

155. Reaves, S.K.; Fanzo, J.C.; Wu, J.Y.; Wang, Y.R.; Wu, Y.W.; Zhu, L.; Lei, K.Y. Plasma apolipoprotein B-48, hepatic apolipoprotein B mRNA editing and apolipoprotein B mRNA editing catalytic subunit-1 mRNA levels are altered in zinc-deficient rats. *J. Nutr.* **1999**, *129*, 1855–1861. [CrossRef] [PubMed]

156. Scott, J.; Navaratnam, N.; Bhattacharya, S.; Morrison, J.R. The apolipoprotein B messenger RNA editing enzyme. *Curr. Opin. Lipidol.* **1994**, *5*, 87–93. [CrossRef] [PubMed]

157. Reaves, S.K.; Wu, J.Y.; Wu, Y.; Fanzo, J.C.; Wang, Y.R.; Lei, P.P.; Lei, K.Y. Regulation of intestinal apolipoprotein B mRNA editing levels by a zinc-deficient diet and cDNA cloning of editing protein in hamsters. *J. Nutr.* **2000**, *130*, 2166–2173. [CrossRef] [PubMed]

158. Nassir, F.; Blanchard, R.K.; Mazur, A.; Cousins, R.J.; Davidson, N.O. Apolipoprotein B mRNA editing is preserved in the intestine and liver of zinc-deficient rats. *J. Nutr.* **1996**, *126*, 860–864. [CrossRef]

159. Cerovic, A.; Miletic, I.; Sobajic, S.; Blagojevic, D.; Jones, D.R.; Poznanic, M.; Radusinovic, M. Effect of dietary zinc on the levels and distribution of Fatty acids and vitamin A in blood plasma chylomicrons. *Biol. Trace Elem. Res.* **2006**, *112*, 145–158. [CrossRef]

160. Koo, S.I.; Lee, C.C. Effect of marginal zinc deficiency on lipoprotein lipase activities in postheparin plasma, skeletal muscle and adipose tissues in the rat. *Lipids* **1989**, *24*, 132–136. [CrossRef]

161. Kettler, S.I.; Eder, K.; Kettler, A.; Kirchgessner, M. Zinc deficiency and the activities of lipoprotein lipase in plasma and tissues of rats force-fed diets with coconut oil or fish oil. *J. Nutr. Biochem.* **2000**, *11*, 132–138. [CrossRef]

162. Koo, S.I.; Lee, C.C.; Norvell, J.E. Effect of copper deficiency on the lymphatic absorption of cholesterol, plasma chylomicron clearance, and postheparin lipase activities. *Proc. Soc. Exp. Biol. Med.* **1988**, *188*, 410–419. [CrossRef] [PubMed]

163. Motta, C.; Gueux, E.; Mazur, A.; Rayssiguier, Y. Lipid fluidity of triacylglycerol-rich lipoproteins isolated from copper-deficient rats. *Br. J. Nutr.* **1996**, *75*, 767–773. [CrossRef] [PubMed]

164. Strauss, E.W. Effects of calcium and magnesium ions upon fat absorption by sacs of everted hamster intestine. *Gastroenterology* **1977**, *73*, 421–424. [CrossRef]

165. Kishimoto, Y.; Tani, M.; Uto-Kondo, H.; Saita, E.; Iizuka, M.; Sone, H.; Yokota, K.; Kondo, K. Effects of magnesium on postprandial serum lipid responses in healthy human subjects. *Br. J. Nutr.* **2010**, *103*, 469–472. [CrossRef] [PubMed]

166. Lorenzen, J.K.; Nielsen, S.; Holst, J.J.; Tetens, I.; Rehfeld, J.F.; Astrup, A. Effect of dairy calcium or supplementary calcium intake on postprandial fat metabolism, appetite, and subsequent energy intake. *Am. J. Clin. Nutr.* **2007**, *85*, 678–687. [CrossRef]

167. Naissides, M.; Mamo, J.C.; James, A.P.; Pal, S. The effect of acute red wine polyphenol consumption on postprandial lipaemia in postmenopausal women. *Atherosclerosis* **2004**, *177*, 401–408. [CrossRef] [PubMed]

168. Pal, S.; Ho, S.S.; Takechi, R. Red wine polyphenolics suppress the secretion of ApoB48 from human intestinal CaCo-2 cells. *J. Agric. Food Chem.* **2005**, *53*, 2767–2772. [CrossRef]

169. Vidal, R.; Hernandez-Vallejo, S.; Pauquai, T.; Texier, O.; Rousset, M.; Chambaz, J.; Demignot, S.; Lacorte, J.M. Apple procyanidins decrease cholesterol esterification and lipoprotein secretion in Caco-2/TC7 enterocytes. *J. Lipid Res.* **2005**, *46*, 258–268. [CrossRef]

170. Toyoda-Ono, Y.; Yoshimura, M.; Nakai, M.; Fukui, Y.; Asami, S.; Shibata, H.; Kiso, Y.; Ikeda, I. Suppression of postprandial hypertriglyceridemia in rats and mice by oolong tea polymerized polyphenols. *Biosci. Biotechnol. Biochem.* **2007**, *71*, 971–976. [CrossRef]

171. Kobayashi, M.; Ichitani, M.; Suzuki, Y.; Unno, T.; Sugawara, T.; Yamahira, T.; Kato, M.; Takihara, T.; Sagesaka, Y.; Kakuda, T.; et al. Black-tea polyphenols suppress postprandial hypertriacylglycerolemia by suppressing lymphatic transport of dietary fat in rats. *J. Agric. Food Chem.* **2009**, *57*, 7131–7136. [CrossRef]

172. Unno, T.; Tago, M.; Suzuki, Y.; Nozawa, A.; Sagesaka, Y.M.; Kakuda, T.; Egawa, K.; Kondo, K. Effect of tea catechins on postprandial plasma lipid responses in human subjects. *Br. J. Nutr.* **2005**, *93*, 543–547. [CrossRef]

173. Huang, J.; Feng, S.; Liu, A.; Dai, Z.; Wang, H.; Reuhl, K.; Lu, W.; Yang, C.S. Green Tea Polyphenol EGCG Alleviates Metabolic Abnormality and Fatty Liver by Decreasing Bile Acid and Lipid Absorption in Mice. *Mol. Nutr. Food Res.* **2018**. [CrossRef] [PubMed]

174. Murase, T.; Yokoi, Y.; Misawa, K.; Ominami, H.; Suzuki, Y.; Shibuya, Y.; Hase, T. Coffee polyphenols modulate whole-body substrate oxidation and suppress postprandial hyperglycaemia, hyperinsulinaemia and hyperlipidaemia. *Br. J. Nutr.* **2012**, *107*, 1757–1765. [CrossRef] [PubMed]

175. Qin, B.; Polansky, M.M.; Sato, Y.; Adeli, K.; Anderson, R.A. Cinnamon extract inhibits the postprandial overproduction of apolipoprotein B48-containing lipoproteins in fructose-fed animals. *J. Nutr. Biochem.* **2009**, *20*, 901–908. [CrossRef] [PubMed]

176. Qin, B.; Dawson, H.D.; Schoene, N.W.; Polansky, M.M.; Anderson, R.A. Cinnamon polyphenols regulate multiple metabolic pathways involved in insulin signaling and intestinal lipoprotein metabolism of small intestinal enterocytes. *Nutrition* **2012**, *28*, 1172–1179. [CrossRef]

177. Takahashi, A.; Okazaki, Y.; Nakamoto, A.; Watanabe, S.; Sakaguchi, H.; Tagashira, Y.; Kagii, A.; Nakagawara, S.; Higuchi, O.; Suzuki, T.; et al. Dietary anthocyanin-rich Haskap phytochemicals inhibit postprandial hyperlipidemia and hyperglycemia in rats. *J. Oleo. Sci.* **2014**, *63*, 201–209. [CrossRef]

178. Annuzzi, G.; Bozzetto, L.; Costabile, G.; Giacco, R.; Mangione, A.; Anniballi, G.; Vitale, M.; Vetrani, C.; Cipriano, P.; Della Corte, G.; et al. Diets naturally rich in polyphenols improve fasting and postprandial dyslipidemia and reduce oxidative stress: A randomized controlled trial. *Am. J. Clin. Nutr.* **2014**, *99*, 463–471. [CrossRef] [PubMed]

179. Ubbink, J.; Burbidge, A.; Mezzenga, R. Food structure and functionality: A soft matter perspective. *Soft. Matter* **2008**, *4*, 1569–1581. [CrossRef]

180. Jenkins, D.J.; Wolever, T.M.; Taylor, R.H.; Barker, H.; Fielden, H.; Baldwin, J.M.; Bowling, A.C.; Newman, H.C.; Jenkins, A.L.; Goff, D.V. Glycemic index of foods: A physiological basis for carbohydrate exchange. *Am. J. Clin. Nutr.* **1981**, *34*, 362–366. [CrossRef] [PubMed]

181. Kolovou, G.D.; Mikhailidis, D.P.; Kovar, J.; Lairon, D.; Nordestgaard, B.G.; Ooi, T.C.; Perez-Martinez, P.; Bilianou, H.; Anagnostopoulou, K.; Panotopoulos, G. Assessment and clinical relevance of non-fasting and postprandial triglycerides: An expert panel statement. *Curr. Vasc. Pharmacol.* **2011**, *9*, 258–270. [CrossRef]

182. Jacobs, D.R.; Tapsell, L.C. Food, not nutrients, is the fundamental unit in nutrition. *Nutr. Rev.* **2007**, *65*, 439–450. [CrossRef] [PubMed]

183. Mozaffarian, D. Dietary and Policy Priorities for Cardiovascular Disease, Diabetes, and Obesity: A Comprehensive Review. *Circulation* **2016**, *133*, 187–225. [CrossRef] [PubMed]

184. Mozaffarian, D.; Ludwig, D.S. Dietary guidelines in the 21st century–a time for food. *JAMA* **2010**, *304*, 681–682. [CrossRef] [PubMed]

185. Guo, Q.; Ye, A.; Bellissimo, N.; Singh, H.; Rousseau, D. Modulating fat digestion through food structure design. *Prog. Lipid Res.* **2017**, *68*, 109–118. [CrossRef] [PubMed]

186. Vors, C.; Pineau, G.; Gabert, L.; Drai, J.; Louche-Pelissier, C.; Defoort, C.; Lairon, D.; Desage, M.; Danthine, S.; Lambert-Porcheron, S.; et al. Modulating absorption and postprandial handling of dietary fatty acids by structuring fat in the meal: A randomized crossover clinical trial. *Am. J. Clin. Nutr.* **2013**, *97*, 23–36. [CrossRef] [PubMed]

187. Garaiova, I.; Guschina, I.A.; Plummer, S.F.; Tang, J.; Wang, D.; Plummer, N.T. A randomised cross-over trial in healthy adults indicating improved absorption of omega-3 fatty acids by pre-emulsification. *Nutr. J.* **2007**, *6*, 4. [CrossRef] [PubMed]

188. Clemente, G.; Mancini, M.; Nazzaro, F.; Lasorella, G.; Rivieccio, A.; Palumbo, A.M.; Rivellese, A.A.; Ferrara, L.; Giacco, R. Effects of different dairy products on postprandial lipemia. *Nutr. Metab. Cardiovasc. Dis.* **2003**, *13*, 377–383. [CrossRef]

189. Tholstrup, T.; Hoy, C.E.; Andersen, L.N.; Christensen, R.D.; Sandstrom, B. Does fat in milk, butter and cheese affect blood lipids and cholesterol differently? *J. Am. Coll. Nutr.* **2004**, *23*, 169–176. [CrossRef] [PubMed]

190. Steingoetter, A.; Radovic, T.; Buetikofer, S.; Curcic, J.; Menne, D.; Fried, M.; Schwizer, W.; Wooster, T.J. Imaging gastric structuring of lipid emulsions and its effect on gastrointestinal function: A randomized trial in healthy subjects. *Am. J. Clin. Nutr.* **2015**, *101*, 714–724. [CrossRef]

191. Armand, M.; Pasquier, B.; Andre, M.; Borel, P.; Senft, M.; Peyrot, J.; Salducci, J.; Portugal, H.; Jaussan, V.; Lairon, D. Digestion and absorption of 2 fat emulsions with different droplet sizes in the human digestive tract. *Am. J. Clin. Nutr.* **1999**, *70*, 1096–1106. [CrossRef]

192. Tan, K.W.; Sun, L.J.; Goh, K.K.; Henry, C.J. Lipid droplet size and emulsification on postprandial glycemia, insulinemia and lipidemia. *Food Funct.* **2016**, *7*, 4278–4284. [CrossRef] [PubMed]

193. Singh, H.; Ye, A. Structural and biochemical factors affecting the digestion of protein-stabilized emulsions. *Curr. Opin. Colloid Interface Sci.* **2013**, *18*, 360–370. [CrossRef]

194. Keogh, J.B.; Wooster, T.J.; Golding, M.; Day, L.; Otto, B.; Clifton, P.M. Slowly and rapidly digested fat emulsions are equally satiating but their triglycerides are differentially absorbed and metabolized in humans. *J. Nutr.* **2011**, *141*, 809–815. [CrossRef] [PubMed]

195. Small, D.M. The effects of glyceride structure on absorption and metabolism. *Annu. Rev. Nutr.* **1991**, *11*, 413–434. [CrossRef] [PubMed]

196. Berry, S.E. Triacylglycerol structure and interesterification of palmitic and stearic acid-rich fats: An overview and implications for cardiovascular disease. *Nutr. Res. Rev.* **2009**, *22*, 3–17. [CrossRef] [PubMed]

197. Alfieri, A.; Imperlini, E.; Nigro, E.; Vitucci, D.; Orru, S.; Daniele, A.; Buono, P.; Mancini, A. Effects of Plant Oil Interesterified Triacylglycerols on Lipemia and Human Health. *Int. J. Mol. Sci.* **2017**, *19*. [CrossRef]

198. Sanders, T.A.; Filippou, A.; Berry, S.E.; Baumgartner, S.; Mensink, R.P. Palmitic acid in the sn-2 position of triacylglycerols acutely influences postprandial lipid metabolism. *Am. J. Clin. Nutr.* **2011**, *94*, 1433–1441. [CrossRef]

199. Hall, W.L.; Brito, M.F.; Huang, J.; Wood, L.V.; Filippou, A.; Sanders, T.A.; Berry, S.E. An interesterified palm olein test meal decreases early-phase postprandial lipemia compared to palm olein: A randomized controlled trial. *Lipids* **2014**, *49*, 895–904. [CrossRef]

200. Berry, S.E.; Tydeman, E.A.; Lewis, H.B.; Phalora, R.; Rosborough, J.; Picout, D.R.; Ellis, P.R. Manipulation of lipid bioaccessibility of almond seeds influences postprandial lipemia in healthy human subjects. *Am. J. Clin. Nutr.* **2008**, *88*, 922–929. [CrossRef]

201. Grundy, M.M.; Grassby, T.; Mandalari, G.; Waldron, K.W.; Butterworth, P.J.; Berry, S.E.; Ellis, P.R. Effect of mastication on lipid bioaccessibility of almonds in a randomized human study and its implications for digestion kinetics, metabolizable energy, and postprandial lipemia. *Am. J. Clin. Nutr.* **2015**, *101*, 25–33. [CrossRef]

202. Dias, C.B.; Zhu, X.; Thompson, A.K.; Singh, H.; Garg, M.L. Effect of the food form and structure on lipid digestion and postprandial lipaemic response. *Food Funct.* **2019**, *10*, 112–124. [CrossRef] [PubMed]

Article

Postprandial Lipemic Responses to Various Sources of Saturated and Monounsaturated Fat in Adults

Christina M. Sciarrillo [1],*, Nicholas A. Koemel [1], Patrick M. Tomko [2], Katherine B. Bode [1] and Sam R. Emerson [1]

[1] Department of Nutritional Sciences, Oklahoma State University, Stillwater, OK 74078, USA; nick.koemel@okstate.edu (N.A.K.); katherine.bode@okstate.edu (K.B.B.); sam.emerson@okstate.edu (S.R.E.)
[2] School of Kinesiology, Applied Health and Recreation, Oklahoma State University, Stillwater, OK 74078, USA; ptomko@okstate.edu
* Correspondence: christina.sciarrillo@okstate.edu

Received: 15 April 2019; Accepted: 12 May 2019; Published: 16 May 2019

Abstract: Background: Postprandial lipemia (PPL) is a cardiovascular disease risk factor. However, the effects of different fat sources on PPL remain unclear. We aimed to determine the postprandial response in triglycerides (TG) to four dietary fat sources in adults. Methods: Participants completed four randomized meal trials. For each meal trial, participants (n = 10; 5M/5F) consumed a high-fat meal (HFM) (13 kcal/kg; 61% of total kcal from fat) with the fat source derived from butter, coconut oil, olive oil, or canola oil. Blood was drawn hourly for 6 h post-meal to quantify PPL. Results: Two-way ANOVA of TG revealed a time effect ($p < 0.0001$), but no time–meal interaction ($p = 0.56$), or meal effect ($p = 0.35$). Meal trials did not differ with regard to TG total ($p = 0.33$) or incremental ($p = 0.14$) area-under-the-curve. When stratified by sex and the TG response was averaged across meals, two-way ANOVA revealed a time effect ($p < 0.0001$), time–group interaction ($p = 0.0001$), and group effect ($p = 0.048$), with men exhibiting a greater response than women, although this difference could be attributed to the pronounced difference in BMI between men and women within the sample. Conclusion: In our sample of young adults, postprandial TG responses to a single HFM comprised of different fat sources did not differ.

Keywords: postprandial lipemia; coconut oil; butter; canola oil; olive oil; lipid; triglycerides; dietary fat; saturated fat; cardiovascular disease

1. Introduction

Cardiovascular disease (CVD) is a major public health concern and the leading cause of death in the United States [1]. Traditional risk factors for CVD include smoking, physical inactivity, poor dietary habits, overweight/obesity, dyslipidemia, diabetes, and hypertension [1]. In addition, emerging evidence has given rise to consideration of postprandial changes following single, high-fat meal (HFM) consumption as substantially impacting CVD risk [2]. In fact, postprandial triglycerides (TG) have been identified as a stronger predictor of CVD risk than fasting values [2]. This is partly because individuals are in a postprandial state for the majority of their day [3]. Adverse changes that occur in the postprandial period include increases in TG [3], oxidative stress [4], inflammation [5,6], oxidized low-density lipoprotein [7], and decreases in high-density lipoprotein (HDL-C) [8,9] and vascular dilation [9], all of which have been shown to contribute to the CVD pathology.

Postprandial lipemia (PPL) is the rise in blood TG response following a meal [3]. Several studies have shown that an altered or reduced ability to clear TG in the postprandial period (thus, a large postmeal TG response) is associated with CVD [3,10–12]. The connection between PPL and CVD has been demonstrated in a case-control study in men with coronary heart disease (CHD) compared to healthy controls and in the sons of men with CHD compared to sons of men without

CHD, where both disease-case groups exhibited significantly greater postprandial TG levels [10]. Similarly, data examined in women have revealed associations between greater postprandial TG and apolipoprotein B-48 (apoB-48) concentrations and CHD [2,12,13]. Moreover, in the Women's Health Study and in the Copenhagen City Heart Study, both large prospective cohort studies involving women, nonfasting TG concentrations were significantly associated with increased CVD risk, even after adjustment for various confounding variables [2,14]. It has been suggested that the mechanistic connection between individual PPL and CVD is the subendothelial penetration and retention of circulating TG-rich lipoproteins (TRL) [11,15].

Given that HFMs classically used to study PPL contain >50% fat [11,16], and the variability in different dietary fat sources ability to increase or decrease CVD risk, it is logical that the source of dietary fat can modulate the postprandial TG response. Several studies have found a reduced PPL response following meals rich in monounsaturated fatty acids (MUFA) and polyunsaturated fatty acids (PUFA) compared to meals rich in saturated fatty acids (SFA) in both healthy adults and those with characteristics of metabolic syndrome [17–19]. These findings are in line with classical dietary data showing that certain sources of SFA are generally associated with CVD [20,21]. Meanwhile, Schwingshackl and Hoffmann [22] found that MUFA and PUFA can induce a greater PPL response compared to SFA. Given these findings and classic dietary data, the effects of different dietary fats on PPL have been inconclusive to this point. Furthermore, since the effects of a given type of dietary fat on CVD risk can also depend on the source of the fat (animal- versus plant-based SFA [23]), it is reasonable to suspect different postprandial responses based on fat source, even when those foods are comprised of similar fatty acid contents and types. In support of this concept, recent data from Teng et al. and Panth et al. examining the effects of animal- vs. plant-based SFA have yielded inconsistent and contradictory results. Teng et al. observed a lower TG response after the consumption of animal-based SFA (lard) when compared to plant-based SFA (palm olein), while Panth et al. observed a greater TG response after the consumption of animal-based SFA (butter, lard) when compared to plant-based SFA (coconut oil) [24,25]. Considering the rising scientific data on ketogenic diets and low-carb eating patterns, and their recent popularity regarding the treatment of several phenotypes associated with CVD (diabetes, obesity), it is pertinent to understand further how various sources of dietary fat affect cardiometabolic health. Furthermore, considering the inconsistency with regard to current postprandial data, and that PPL is an independent risk factor for CVD [2], determining how various types and sources of dietary fat consumed may modify individual postprandial TG response would be valuable.

Therefore, the primary purpose of this investigation was to determine the effects of commonly consumed sources of dietary fat as part of a mixed meal on PPL in young adults. Specifically, this study compared the postprandial TG response to plant-based SFA (coconut oil), animal-based SFA (butter), MUFA-rich olive oil, and MUFA-rich canola oil.

2. Materials and Methods

2.1. Participants

Ten individuals (5 M/5 W) participated in the present study and were recruited via online survey, email, or flyer from the Oklahoma State University campus. Inclusion criteria were age 18–40 years, no evidence of dietary intolerances that precluded consumption of the test meals, no chronic disease, and not taking any lipid or blood pressure medications. The study protocol was approved by the Institutional Review Board at Oklahoma State University (HE-17-77) and carried out in accordance with the Declaration of Helsinki. All participants provided verbal and written consent prior to participating in the study.

2.2. Overall Study Design

Participants engaged in one initial assessment and four randomized meal trials. The initial assessment consisted of detailed paperwork (informed consent, medical history questionnaire, international physical activity questionnaire (IPAQ)) and anthropometric data measurements. Participants were also administered various lifestyle control instructions during the initial assessment. Meal trials began approximately one week after the initial assessment. Each meal trial was separated by a washout period of 1–3 weeks. The sequence in which a participant consumed the four test meals was randomized. Within each meal trial, participants arrived in the laboratory 10-h fasted, a baseline blood draw was taken, they consumed the test meal, and blood draws were taken every hour for six hours post-meal to determine the postprandial TG response.

2.3. Initial Assessment

The initial assessment entailed detailed copies of instructions for participants to follow, completion of written informed consent, a medical history questionnaire, the IPAQ, and anthropometric evaluations. Height was measured via stadiometer (Seca 213 portable stadiometer; Seca GmbH; Hamburg, Germany). Body mass was measured using a digital scale (Seca mBCA 514; Seca GmbH; Hamburg, Germany). Blood pressure was measured using an automatic blood pressure cuff (Omron 5 Series BP742N; Omron; Kyoto, Japan). Height, weight, and blood pressure were measured twice and the average of the two measures was recorded.

Lifestyle controls were assigned and all participants were instructed to follow and comply with the explained lifestyle instructions. Lifestyle controls consisted of a three-day food record, in which participants recorded their dietary intake for the three days prior to their first meal assessment; participants then were asked to replicate their first three-day food record for the remaining three meal assessments. Accelerometers (wGT3X-BT, Actigraph; Pensacola, FL, USA) were attached to each participant's nondominant wrist and recorded their physical activity for at least 48 h prior to each assessment. In addition, participants were asked to refrain from planned exercise for the 48 h prior to each assessment. Participants were given a 210-kcal snack, consisting of commercial peanut butter crackers (Snyder's-Lance, Inc.; Charlotte, NC, USA), to consume the evening before each assessment, after which the 10-h fast began. Participants were given a typed copy of all detailed instructions and lifestyle controls.

2.4. Meal Trials

After a 10-h overnight fast, participants arrived in the laboratory on the morning of each assessment. Each meal assessment began between 6:00–8:00 A.M., depending on the scheduling availability of the participant. An indwelling 24-gauge safelet catheter (Exel International; Redondo Beach, CA, USA) was inserted into a forearm vein and a slow infusion (~1 drip/s) of 0.9% NaCl solution was initiated. Once the catheter was set, a baseline blood draw was collected. First, a 3 mL syringe (BD; Franklin Lakes, NJ, USA) was used to clear the line of saline followed by a 5 mL syringe (BD; Franklin Lakes, NJ, USA) used to take the whole blood sample. Whole blood samples were collected for the assessment of metabolic outcomes: TG, glucose, LDL-C, HDL-C, and total cholesterol (TOTAL-C). Metabolic outcomes were determined by a Cholestech LDX analyzer (Alere Inc.; Waltham, MA, USA). For each individual blood draw, a few drops of whole blood were drawn into a capillary tube and plunged into a Cholestech LDX Lipid+Glu cassette (Alere Inc.; Waltham, MA, USA). The cassette was inserted into the Cholestech LDX analyzer and processed. The CV for TG assessment via the Cholestech LDX system is approximately 2–4%. Following the baseline blood draw, participants consumed the test meal within 20 min. Water was available for participant consumption ad libitum during the meal and throughout the postprandial period. Participants remained in the laboratory for 6 h following consumption of the test meal. The 6-h time period began after the last bite of the test meal. Additional blood draws were performed every hour for the 6 h after consumption of the test meal.

2.5. Test Meals

The test meal consisted of pasta sauce, whole-wheat spaghetti noodles, French bread, yellow onion, green bell pepper, sea salt, black pepper, and the specific fat source being tested. Each meal contained a test fat of either MUFA-based canola oil (CaO) (Great Value, Canola Oil), MUFA-based extra virgin olive oil (OO) (Great Value, Extra Virgin Olive Oil), SFA-based virgin unrefined coconut oil (CoO) (Organic Great Value, Unrefined Virgin Coconut Oil, expeller pressed), or SFA-based grass-fed butter (B) (Kerrygold, Grass-fed Pure Irish Butter, unsalted). The test meal contained 61% of total kcal from fat, 7% of total kcal from protein, and 32% of total kcal from carbohydrate (CHO). Each participant consumed a serving of the test meal that was relative to his or her body mass (13 kcal/kg body mass). The amount of meal consumed was designed to resemble a typical serving at a restaurant or social event (1–2 servings). For each assessment, the meal was prepared independently 1–2 days prior to the assessment. To prepare each meal, the test fat was added to a small saucepan and heated over medium heat for 2 min. Onion and bell pepper were diced finely and sautéed over medium heat in a large saucepan for 3 min. The pasta sauce was added to the saucepan and brought to a boil, after which the heat was reduced, the saucepan was covered, and the mixture cooked for 7 min until the internal temperature reached 165 °F. Once the pasta sauce mixture was finished cooking, it was removed from the heat, cooled for 20 min, labeled, and stored in a BPA-free food storage container at 0 °F until needed for each assessment. The night before each assessment, the pasta sauce was thawed at 36 °F overnight. On the morning of each assessment, the noodles were prepared separately by bringing four cups of water to a boil in a medium saucepan, after which the raw noodles were added, cooked uncovered for 9 min, and strained. The pasta sauce was reheated in a small saucepan until the internal temperature reached 165 °F. The pasta sauce and noodles were combined in a small serving bowl and the French bread was served on the side. All ingredients were weighed (g) using a digital food scale (Table 1).

Table 1. Test fats and meal composition. Data are representative of the test meal composition for a 60 kg participant.

	Weight (g)	Energy (kcal)	Protein (g)	Fat (g)	CHO (g)	Fiber (g)
Sauce	257	577	5	52	25	3.3
Bread	28	78	3	1	16	1.5
Pasta	33	124	6	1	24	1.5
Total	318	780	14	54	65	6.3

2.6. Statistical Analyses

An a priori sample size estimation, using standard deviations from previous studies [26,27], suggested that ten participants would need to be recruited to detect a clinically significant difference in the peak postprandial TG response of 0.5 mmol/L between meals with 80% power and alpha less than 0.05.

All data were assessed for normality via Shapiro–Wilk formal normality test and analysis of frequency distribution. The trapezoid method was used to calculate tAUC and incremental area under the curve (iAUC). Within each meal trial, tAUC, iAUC, peak value, and time to peak value were determined for each of the metabolic markers. These postprandial metabolic outcomes were compared across trials using a one-way analysis of variance (ANOVA) with Holm–Sidak adjustment for multiple comparisons. Time-course changes and sex-based differences in metabolic markers in the postprandial period were determined via two-way between and within (group × time) repeated measures ANOVA with a Tukey's adjustment for multiple comparisons.

Differences between participant characteristics were compared by sex via two-tailed paired t-test. Pearson's two-tailed correlation analysis was performed to assess the association between participant body mass index (BMI) and TG tAUC (averaged across meal trials).

A type 1 error rate of 0.05 was used in all analyses for the determination of statistically significant differences. Statistical analyses were conducted using GraphPad Prism statistical software (Version 7; GraphPad Software, Inc., La Jolla, CA, USA).

3. Results

3.1. Participant Characteristics and Premeal Physical Activity

Participant characteristics are presented in Table 2. Ten individuals participated in the present study (5 M/5 F; age: 23.8 ± 1.3 years; BMI: 25.5 ± 7.2 kg/m^2). Based on BMI, six participants (1 M/5 F) were healthy weight (18.5–24.9 kg/m^2), one participant (1 M) was overweight (25–29.9 kg/m^2), and three participants (3 M) were obese (>30 kg/m^2). One participant reported with fasting TG > 1.69 mmol/L on two occasions. Men were significantly older (mean difference: 1.2 years; $p = 0.03$) and had greater weight (mean difference: 36.2 kg; $p = 0.01$) and BMI (mean difference: 9.6 kg/m^2; $p = 0.02$) compared to women. Men had higher fasting LDL-C concentrations when compared to women (mean difference: 0.47 mmol/L; $p = 0.02$), but there were no differences in fasting TG ($p = 0.21$), glucose ($p = 0.96$), TOTAL-C ($p = 0.44$), or HDL-C ($p = 0.30$) between men and women. Additionally, fasting TG ($p = 0.39$), glucose ($p = 0.13$), TOTAL-C ($p = 0.07$), LDL-C ($p = 0.86$), and HDL-C ($p = 0.11$) were not different across meal trials. Physical activity, measured as moderate-vigorous physical activity (MVPA) and steps/day, was not different across meal trials ($p = 0.84$ and $p = 0.69$, respectively) and there was not a main effect by meal trial ($p = 0.69$; $p = 0.90$), sex ($p = 0.68$; $p = 0.51$), or meal–sex interaction ($p = 0.20$; $p = 0.67$) (Figure 1).

Table 2. Participant characteristics. Metabolic outcomes represent fasting data averaged across the four meal trials.

	Total	Men	Women	*p*-Value
Age	23.8 ± 1.3	24.4 ± 1.5 *	23.2 ± 0.8	0.03
Weight (kg)	76.52 ± 25.23	94.57 ± 23.09 *	58.42 ± 8.79	0.01
Height (cm)	171.5 ± 10.1	176.0 ± 9.8	167.0 ± 9.1	0.10
BMI (kg/m^2)	25.5 ± 7.2	30.3 ± 7.4 *	20.7 ± 1.7	0.02
Fasting TG (mmol/L)	0.78 ± 0.41	0.96 ± 0.53	0.61 ± 0.10	0.21
Fasting Glucose (mmol/L)	4.85 ± 0.32	4.86 ± 0.29	4.84 ± 0.38	0.96
Fasting TOTAL-C (mmol/L)	4.76 ± 0.91	5.05 ± 1.09	4.47±0.68	0.44
Fasting LDL-C (mmol/L)	2.41 ± 0.79	2.95 ± 0.42 *	1.87 ± 0.73	0.02
Fasting HDL-C (mmol/L)	1.62 ± 0.53	1.44 ± 0.68	1.79 ± 0.33	0.30
MVPA (minutes)	152.5 ± 17.3	166.9 ± 7.9	138 ± 24.4	0.51
Steps/day	7934.6 ± 747.7	7444.6 ± 1024.3	8424.6 ± 107.6	0.68

Data are presented as mean ± SD. *p*-value column indicates results of an unpaired *t*-test between men and women. * Indicates significant differences between men and women ($p < 0.05$). MVPA, moderate-vigorous physical activity; TG, triglycerides; HDL-C, high-density lipoprotein cholesterol; LDL-C, low-density lipoprotein; TOTAL-C, total cholesterol.

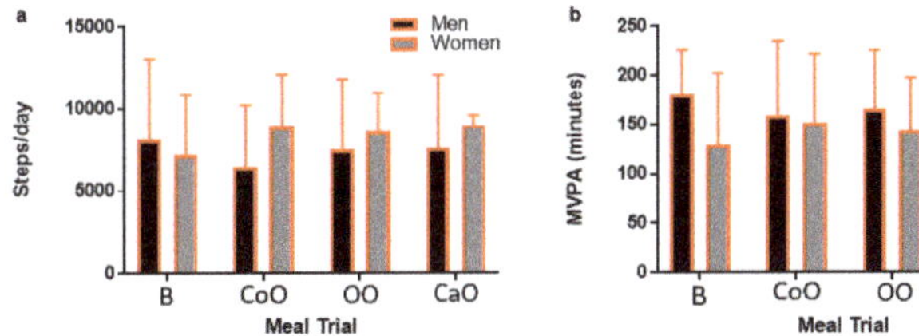

Figure 1. Pretrial physical activity. Physical activity in (**a**) steps/day and (**b**) MVPA for 48–72 h before each meal trial, stratified by sex. Data are presented as mean ± SD. Within each meal trial, there was no difference between men and women in physical activity in either MVPA or steps/day ($p > 0.05$). Similarly, within sex, there was no difference in physical activity across meal trials ($p > 0.05$). MVPA, moderate-vigorous physical activity; B, butter; CoO, coconut oil; OO, olive oil; CaO, canola oil.

3.2. Postprandial Metabolic Outcomes Were Similar Across Meal Trials

Metabolic outcomes are presented in Table 3 and Figure 2. Two-way ANOVA of TG revealed a significant time effect ($p < 0.0001$) but no time–meal interaction ($p = 0.56$) or overall meal effect

($p = 0.35$). One-way ANOVA revealed that TG peak ($p = 0.36$) and TG time to peak ($p = 0.23$) were not different across meal trials. Meal trials did not differ with regard to TG tAUC ($p = 0.33$) or TG iAUC ($p = 0.14$). Two-way ANOVA of glucose revealed no time effect ($p = 0.27$), meal effect ($p = 0.64$), or time–meal interaction ($p = 0.63$). Glucose peak ($p = 0.76$) and glucose time to peak ($p = 0.48$) were not different across meal trials. Meal trials did not differ with regard to glucose tAUC ($p = 0.60$) or iAUC ($p = 0.26$). Two-way ANOVA of metabolic load index (MLI; calculated as TG + glucose) revealed a time effect ($p < 0.0001$) but no meal effect ($p = 0.08$) or time–meal interaction ($p = 0.77$). MLI peak ($p = 0.24$) and time to peak ($p = 0.64$) were not different across meal trials. Meal trials did not differ with regard to MLI tAUC ($p = 0.12$) or MLI iAUC ($p = 0.08$). Two-way ANOVA of LDL-C revealed no time effect ($p = 0.27$), meal effect ($p = 0.83$), or time–meal interaction ($p = 0.72$). One-way ANOVA revealed that LDL-C peak ($p = 0.66$) and time to peak ($p = 0.59$) were not different across meal trials. Meal trials did not differ with regard to LDL-C tAUC ($p = 0.62$) or iAUC ($p = 0.72$). HDL-C results did not reveal a time effect ($p = 0.62$), meal effect ($p = 0.2$), or time–meal interaction ($p = 0.42$). One-way ANOVA revealed that HDL-C peak ($p = 0.19$) and time to peak ($p = 0.52$) were not different across meal trials. Meal trials did not differ with regard to HDL-C tAUC ($p = 0.23$) or iAUC ($p = 0.16$). Two-way ANOVA of TOTAL-C revealed no time effect ($p = 0.29$), meal effect ($p = 0.07$), or time–meal interaction ($p = 0.82$). One-way ANOVA revealed that TOTAL-C peak ($p = 0.12$) and TOTAL-C time to peak ($p = 0.09$) were not different across meal trials. Meal trials did not differ with regard to TOTAL-C tAUC ($p = 0.11$) or iAUC ($p = 0.37$).

Table 3. Postprandial metabolic outcomes for the four meal trials.

	Butter	Coconut Oil	Olive Oil	Canola Oil	p
Triglycerides					
Peak (mmol/L)	1.7 ± 1.1	1.4 ± 0.6	1.6 ± 0.9	1.6 ± 0.8	0.36
Time to peak (hours)	2.2 ± 0.8	3.0 ± 1.5	2.8 ± 1.1	2.7 ± 1.3	0.23
tAUC (mmol/L × 6 h)	7.6 ± 4.6	6.4 ± 2.8	7.4 ± 4.7	7.1 ± 2.8	0.33
iAUC (mmol/L × 6 h)	2.9 ± 2.5	2.0 ± 1.6	2.7 ± 2.4	1.9 ± 1.7	0.14
Glucose					
Peak (mmol/L)	5.4 ± 0.9	5.3 ± 0.7	5.4 ± 0.7	5.5 ± 0.8	0.76
Time to peak (hours)	3.2 ± 1.8	2.4 ± 1.9	2.5 ± 1.8	2.5 ± 2.3	0.48
tAUC (mmol/L × 6 h)	28.5 ± 3.7	27.5 ± 3.6	28.5 ± 2.2	28.0 ± 2.8	0.60
iAUC (mmol/L × 6 h)	0.3 ± 1.5	−0.9 ± 2.3	−1.1 ± 1.1	1.1 ± 2.6	0.26
Metabolic Load Index					
Peak (mg/dL)	241.0 ± 105.4	208.8 ± 60.5	230.4 ± 99.2	231.3 ± 231.3	0.24
Time to peak (hours)	2.7 ± 1.1	2.7 ± 1.4	2.9 ± 1.2	3.2 ± 1.9	0.64
tAUC (mg/dL 6 h)	1185.2 ± 447.7	1065.5 ± 302.4	1163.7 ± 448.9	1158.9 ± 367.8	0.12
iAUC (mg/dL 6 h)	260.7 ± 239.9	163.1 ± 151.6	219.1 ± 222.1	181.9 ± 228.6	0.08
TOTAL-C					
Peak (mmol/L)	4.6 ± 0.7	4.4 ± 1.2	5.1 ± 0.9	5.0 ± 0.9	0.12
Time to peak (hours)	3.2 ± 2.3	3.6 ± 2.4	4.3 ± 1.8	2.2 ± 2.3	0.09
tAUC (mmol/L × 6 h)	26.2 ± 4.2	24.7 ± 7.3	27.3 ± 5.0	28.7 ± 5.9	0.11
iAUC (mmol/L × 6 h)	0.4 ± 1.1	0.6 ± 0.9	1.1 ± 1.5	0.2 ± 1.6	0.37
LDL-C					
Peak (mmol/L)	2.7 ± 0.6	2.8 ± 0.8	2.8 ± 0.9	2.9 ± 0.8	0.66
Time to peak (hours)	2.4 ± 2.2	3.5 ± 2.8	3.3 ± 2.5	2.6 ± 2.1	0.59
tAUC (mmol/L × 6 h)	14.2 ± 3.4	14.3 ± 4.6	14.4 ± 4.6	15.5 ± 4.9	0.62
iAUC (mmol/L × 6 h)	−0.3 ± 2.8	0.9 ± 2.7	1.1 ± 4.7	1.6 ± 5.1	0.72
HDL-C					
Peak (mmol/L)	1.5 ± 0.4	1.7 ± 0.5	1.6 ± 0.5	1.7 ± 0.6	0.19
Time to peak (hours)	3.4 ± 2.7	3.7 ± 2.0	3.5 ± 2.4	2.2 ± 2.4	0.52
tAUC (mmol/L × 6 h)	8.3 ± 2.1	8.9 ± 2.8	8.8 ± 2.5	9.2 ± 3.0	0.23
iAUC (mmol/L × 6 h)	0.1 ± 0.5	0.5 ± 0.9	0.1 ± 0.6	−0.5 ± 1.2	0.16

Data are presented as mean ± SD. There were no differences between meals for all analyses ($p > 0.05$). TG, triglycerides; MLI, metabolic load index; HDL-C, high-density lipoprotein cholesterol; LDL-C, low-density lipoprotein; TOTAL-C, total cholesterol; tAUC, total area under curve; iAUC, incremental area under the curve.

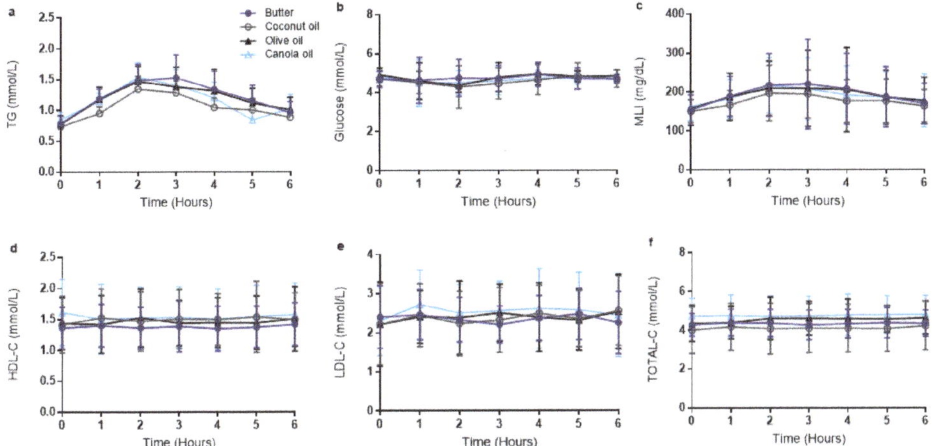

Figure 2. Postprandial metabolic responses. Metabolic responses in the four meal trials at baseline and hourly throughout the postprandial period for (**a**) TG, (**b**) glucose, (**c**) MLI, (**d**) HDL-C, (**e**) LDL-C, and (**f**) TOTAL-C. Data are presented as mean ± SD. Closed circles indicate B meal trial, open circles indicate CoO meal trial, closed triangles indicate OO meal trial, and open triangles indicate CaO meal trial. Error bars indicate SD. TG, triglycerides; MLI, metabolic load index; HDL-C, high-density lipoprotein cholesterol; LDL-C, low-density lipoprotein; TOTAL-C, total cholesterol.

3.3. Postprandial Lipemic Responses Were Different between Men and Women

When data were stratified by sex, a two-way ANOVA of TG revealed a significant time effect (men, $p < 0.0001$; women, $p = 0.0002$) but no time–meal interaction (men, $p = 0.20$; women, $p = 0.21$) or overall meal effect (men, $p = 0.53$; women, $p = 0.48$). When averaged across meal trials, men had a significantly higher TG peak ($p = 0.03$) when compared to women but there was no difference in TG time to peak between men and women ($p = 0.87$). Further, men had significantly higher TG peak ($p < 0.05$) within every meal trial (Mean sex difference: B, 1.49 mmol/L, $p = 0.0005$; CoO, 1.08 mmol/L, $p = 0.006$; OO, 1.05 mmol/L, $p = 0.007$; CaO, 1.23 mmol/L, $p = 0.002$) (Figure 3).

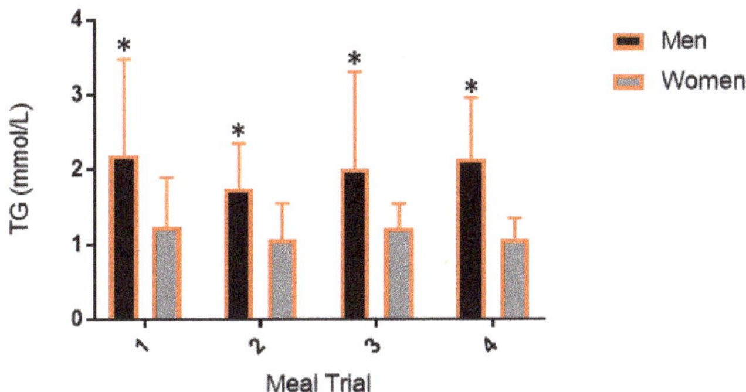

Figure 3. Peak response in triglycerides. Data are presented as mean ± SD. Peak TG responses for meal trials when stratified by sex. * Indicate differences between men and women for a specific meal trial ($p < 0.05$). TG, triglycerides.

When data were stratified by sex and the TG response was averaged for each participant, a two-way ANOVA revealed a significant time effect ($p < 0.0001$), time–group interaction ($p = 0.0001$), and group effect ($p = 0.048$) (Figure 4).

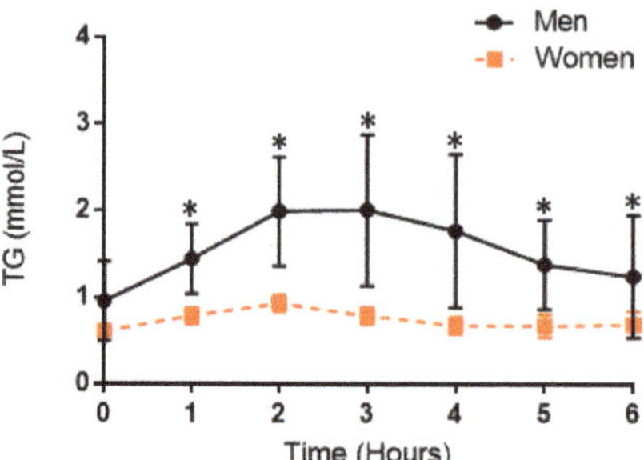

Figure 4. Consolidated postprandial responses in triglycerides in men and women. Average TG responses across meal trials at baseline and hourly throughout the postprandial period in men and women. Data are presented as mean ± SD. * Indicate differences between men and women at a specific time point ($p < 0.05$) based on post hoc pairwise comparison. TG, triglycerdes.

In post hoc pairwise testing, men had significantly higher TG than women at every time point in the postprandial period ($p < 0.05$). Postprandial TG responses in men and women within each meal trial are presented in Figure 5. When data were stratified by sex for men and women, there was a significant time effect ($p < 0.0001$, $p = 0.0002$), but no time–group interaction ($p = 0.19$, $p = 0.21$) or overall group effect ($p = 0.53$, $p = 0.47$), respectively.

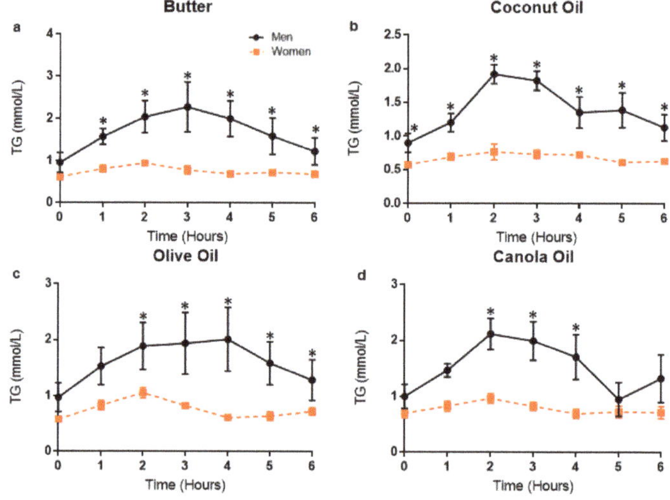

Figure 5. Postprandial responses in triglycerides in men and women based on meal trial. TG responses in men and women for each meal trial at baseline and hourly throughout the postprandial period. Data are presented as mean ± SD. (**a**) TG response for B meal trial; (**b**) TG response for CoO meal trial; (**c**) TG response for OO meal trial; (**d**) TG response for CaO meal trial. * Indicate differences between men and women at a specific time point ($p < 0.05$) based on *post hoc* pairwise comparison. TG, triglycerides; MLI, metabolic load index; HDL-C, high-density lipoprotein cholesterol; LDL-C, low-density lipoprotein; TOTAL-C, total cholesterol.

When data were stratified by sex for the B and OO meal trial, a two-way ANOVA of TG revealed a significant time effect (p's < 0.0001) and time–group interaction ($p = 0.001$ and $p = 0.002$, respectively), but no overall group effect ($p = 0.057$ and $p = 0.11$, respectively). When data were stratified by sex for the CoO and CaO meal trial, a two-way ANOVA of TG revealed a significant time effect ($p < 0.0001$,

$p = 0.0015$), time–group interaction ($p < 0.0001$, $p = 0.04$), and overall group effect ($p = 0.02$, $p = 0.047$). In post hoc pairwise testing, men had significantly higher TG at baseline, 1, 2, 3, 4, 5, and 6 h post-meal for the CoO meal trial and at 2, 3, and 4 h post-meal for the CaO meal trial. Although there was a nonsignificant group effect for the B and OO meal, in post hoc pairwise testing, men had significantly higher TG at 1, 2, 3, 4, 5, and 6 h post-meal for the B and OO meal trial. When Pearson's two-tailed correlation was performed, BMI was strongly associated with TG tAUC ($r = 0.79$, $R^2 = 0.63$, $p = 0.006$) (Figure 6).

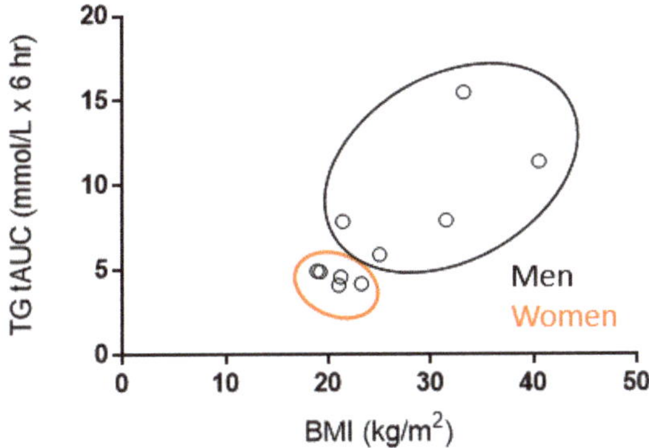

Figure 6. Correlation between TG tAUC and BMI. Data are means of TG tAUC for each participant (averaged across the four meal trials) and BMI (kg/m^2) value for each individual participant. There was a significant positive correlation between BMI and TG tAUC ($r = 0.79$, $R^2 = 0.63$, $p = 0.006$). BMI, body mass index; TG tAUC, triglycerides total area under the curve.

4. Discussion

4.1. Postprandial Responses in Triglycerides between Meals

The present study compared the effects of a high-fat mixed meal rich in butter, coconut oil, olive oil, or canola oil on the postprandial metabolic response in young adults. Peak postprandial TG concentrations were observed at 2–4 h post-meal (mean peak across meals: 1.59 mmol/L) and suggest that the HFM used in the present study induced a robust postprandial response. In our sample of young volunteers, consumption of a mixed HFM containing various sources of commonly consumed dietary fat did not result in different postprandial TG responses. Therefore, counter to our hypotheses, the results of this study do not support the notion that various sources of dietary fat result in markedly different PPL responses. As PPL has been identified as an independent and clinically relevant risk factor for CVD, these results advance understanding with regard to the effects of different dietary fats on cardiometabolic health.

In agreement with our findings, Lesser et al. examined the lipemic effects of a mixed breakfast meal with the fat derived from almonds (MUFA) or cream cheese (dairy-based SFA) in overweight/obese pregnant women and found no significant difference in the postprandial TG response between the two meal trials (MUFA versus dairy-based SFA) [28]. Notably, the test meal utilized by Lesser et al. [28] was a mixed meal, containing a heterogeneous mixture of macro- and micronutrients. Likewise, the HFM meal used in the present study contained moderate amounts of CHO (32% of total kcal) derived from fiber-rich whole grains, French bread, and vegetables. As part of a mixed meal, fiber has been shown to blunt the PPL response by interfering with lipid absorption and digestion via impairment of proper emulsification of lipids in the gastrointestinal tract [29]. In support of this concept, Lesser et al. utilized a mixed meal consisting of 46% of calories derived from CHO and found no differences between test meals [28]. The almond test meal contained 7 g more fiber when compared to the cream cheese test meal; therefore, the lack of detectable differences between meals may have been a result of the modifying effect of fiber on PPL. Kristensen et al. found that when participants consumed a mixed meal with added fiber from flax seed, the mean TG response was 18% lower when compared to the low-fiber control, reaffirming the notion that fiber interferes with the postprandial handling of lipids [29]. Consequently, in the present study, the presence of fiber and other nutrients besides fat in the test meal may have weakened our ability to detect differences between test meals, given the buffering effect that fiber has on the magnitude of PPL.

By contrast, some previous studies have found differences in PPL based on source of dietary fat. For example, researchers examined the effects of mixed meals containing low (basmati rice) or high (jasmine rice) glycemic index CHO and three different types of dietary fat sources (B, OO, grapeseed oil) on the postprandial metabolic response in healthy adults. The TG iAUC was significantly lower following the B (SFA) and grapeseed (PUFA) meals when compared to the OO (MUFA) meal, regardless of GI [30]. These results contrast the findings of our present study that found a similar postprandial TG response when comparing the B (SFA) meal with the OO (MUFA) meal.

Similarly, Mekki et al. [26] assessed the effects of various dietary fatty acids in a mixed HFM on PPL. The authors found that, when compared to the B meal, OO induced a greater PPL response, but a comparable postprandial response to the sunflower oil meal, concluding that B resulted in lower PPL than the OO and sunflower oil meals [26]. These results contrast to our results, but align with Sun et al. [30]. Mekki et al. [26] found that the size of circulating chylomicrons (CM) were consistently lower after the meal rich in B than those detected after the meals rich in vegetable oils (OO or sunflower oil). Although not an explanation as to why these authors found differences between various dietary fats and the present study did not, the lower TG response in the B trial could have been a result of greater or faster lipolysis of CM containing fatty acids from B or a reduced overall size of secreted CM due to the calcium present in B, contributing to the formation of calcium-soap complexes. For our present study, examining the size of circulating CM and concentrations of either intestinally derived apoB-48 present in CM and/or endogenous apoB-100 present in LDL-C and very low-density lipoprotein (VLDL-C) may have yielded detectable differences between meal trials. Additionally, Mekki et al. [26] did not standardize the test meals to body weight and used a homogeneous sample consisting of only men. These factors may also partially explain the disagreement between our study and Mekki et al. [26].

In contrast to Teng et al. [24], Sun et al. [30], Mekki et al. [26], and our present study, another study observed a lower PPL response following a meal consisting of 80 g of ingested OO when compared to 100 g of ingested B [27]. However, since the OO test meal had a lower amount of total fat (80 g) compared to the B meal (100 g), these test meals were not a uniform comparison of the independent effects of OO and B on PPL. In another study investigating acute PPL [31], participants consumed either 71 g of MCT oil, representative of the predominating fatty acid found in CoO, or CaO (MUFA), and the authors found that plasma TG concentrations increased 47% from baseline after the CaO ingestion, while they increased only 15% from baseline following MCT oil ingestion. Notably, only males were included in this study sample and the test meal was not standardized to body weight, nor was it a mixed meal.

Despite several studies comparing the effects of SFA with MUFA or PUFA, there are very few examining the acute effects of various sources of SFA (plant- and animal-based) on PPL. Teng et al. [24] compared the effects of animal-based SFA (lard) and plant-based SFA (palm olein) sources to oleic MUFA-rich dietary fat (virgin OO) as part of a mixed meal on postprandial TG. Researchers found that the lard (animal-based SFA) elicited a significantly lower TG response than the OO and palm olein (plant-based SFA). On the other hand, a recent study by Panth et al. [25] examined the effects of various sources of SFA on the PPL response in healthy adults. Researchers found that the PPL response was ~60% lower after the CoO meal (plant-based SFA) when compared to the B meal (animal-based SFA) and the lard meal (animal-based SFA). No difference was observed between the B and lard meal for PPL. These findings disagree with Teng et al. [24], who found that plant-based SFA (palm olein) elicited a greater PPL response when compared to animal-based SFA (lard). Teng et al. [24] found that animal-based SFA (lard) resulted in lower postprandial TG when compared to plant-based SFA (palm olein), whereas Panth et al. [25] found that plant-based SFA resulted in lower postprandial TG when compared to two sources of animal-based SFA (butter, lard).

These findings by Teng et al. [24] and Panth et al. [25] are contradictory and there were several key differences between the two study designs. First, Teng et al. [24] recruited an exclusively male sample and employed a three-day washout period between meal trials, while Panth et al. [25] recruited equal numbers of males and females and employed a one-week washout period between meal trials. Considering the brief washout period utilized by Teng et al. [24], the effects of the dietary fat in the preceding meal trial may have carried over to the subsequent meal trial, thus influencing the postprandial response and interfering with the evaluation of a singular source of dietary fat. Additionally, Teng et al. [24] utilized a meal higher in total kcal and percent of kcal from fat (~754 total kcal; 60% total kcal from fat, 33% total kcal from CHO, 7% total kcal from protein) when compared to Panth et al. [25] (~660 total kcal; 53% total kcal fat, 40% total kcal from CHO, 5–7% total kcal from protein). Teng et al. [24] also instructed participants to abstain from consuming high-fat foods the day before the meal trials and administered a low-fat meal for the dinner preceding the day of the meal trial, while Panth et al. [25] alternatively asked participants to consume the same meal the night before each meal trial. Lastly, the postprandial assessment period employed by Teng et al. [24] consisted of BL, 1, 2, 3, and 4 h post-meal, whereas Panth [25] measured TG at BL, 2, 3, 4, and 6 h post-meal. Teng et al. [24] may not have been able to capture the entire postprandial response, considering that postprandial TG tends to peak around 2–4 h post-meal consumption and return to postabsorptive values around 6 h post-meal [2].

4.2. Factors Influencing the Postprandial Lipemic Response

Mixed meals contain varying amounts of macronutrients and micronutrients, which modulate physiological processes of digestion, absorption, and metabolism of fatty acids [32–34]. The use of laboratory-derived fat mixtures and lipid emulsions in the place of mixed meals is a common feature in studies assessing PPL (e.g., Mekki et al. [26]), particularly in those evaluating the effects of specific types of fatty acids or sources of dietary fat on PPL. Several of the studies that have observed differences in PPL based on source or type of dietary fat have utilized laboratory-derived fat mixtures or lipid emulsions [31,35,36]. Considering that individuals do not consume these dietary fat sources in isolation or as a component of lipid emulsions in daily living, testing the effects of different fats within a mixed meal may be a more practical and appropriate approach. Our study, as well as others, tested the lipemic effects of different fat sources in the context of true-to-life mixed meal and did not observe differences across meal trials. If the various dietary fat sources used in this study were isolated in laboratory-derived fat mixtures, and thus the effects of macro- and micronutrients were removed, it is possible that differences in postprandial TG between various dietary fat sources may have been observed in the present study.

We observed a strong correlation between BMI and TG tAUC. Men had significantly higher BMI than women and no females were overweight or obese. In agreement with our findings, Kasai et al. found that men with a greater BMI ($\geq$23 kg/m^2) compared to men with a lower BMI (<23 kg/m^2) exhibit greater PPL in response to a HFM [37]. In contrast, Hansson et al. did not find that BMI or sex significantly altered the postprandial TG response to various types of dairy fat rich in SFA [38]. However, the study population (n = 31) consisted of 70% women and 30% men and therefore may not have been sufficient to detect an interaction between sex and postprandial TG in response to different fat sources. In addition, the median BMI was 23.6 kg/m^2 (range: 21.0–25.8). Consequently, the range of BMI may have been too narrow to establish a relationship between BMI and postprandial TG.

The majority of studies that found various sources of dietary fat influence PPL differently included a sample of only male participants [26,30,31,33]. We also observed greater PPL responses in men for all meal trials when compared to women. This finding adds to the notion that sex is an important modifying factor with regard to PPL. There are well-known sex-based differences in visceral adipose tissue accumulation, with women generally storing less adipose tissue in the visceral region than men [39,40]. Women tend to store fat in the gynoid regions (hips/breasts/thighs), while men tend to store fat primarily in the android regions (trunk/abdomen), and thus have a tendency to accumulate fat within visceral tissues [39]. One study has suggested that this difference in visceral adipose accumulation between men and women is the primary explanation for the amplified postprandial response observed in men compared to women [39]. Additionally, Blackburn et al. found that men with impaired glucose tolerance were characterized by greater visceral adiposity, waist circumference, and postprandial lipemia when compared to men with normal glucose tolerance, adding further evidence to the notion that visceral adiposity is an important modulator of the postprandial response [41]. Further, women with android obesity, both with normal and high fasting TG, exhibit a more pronounced and deleterious postprandial TG response when compared to women with gynoid obesity with normal fasting TG [40], further supporting the influence of sex on PPL via body composition differences. Considering that men had a greater BMI than women in our sample, these findings demonstrate one possible mechanism responsible for the marked sex difference in postprandial lipemia that we observed, as the anatomical location of fat storage clearly plays a significant role in determining postprandial lipemia. Thus, although our observed sex-based differences in postprandial lipemia are noteworthy, since there were sex differences in BMI (likely indicative of differences in body composition), it is not possible to form conclusions from our study about the independent role of sex on postprandial lipemic responses.

4.3. Strengths and Limitations

A strength of this study was the use of a "true-to-life" mixed HFM challenge, in contrast to many studies examining PPL that use lipid emulsions or laboratory-derived lipid formulations. The meal used in the present study was also scaled to body weight and resembled a meal that individuals might typically eat at a social gathering. This consideration is important because many postprandial studies utilize meals that are unrealistically high in calories, particularly calories from fat, and are not standardized to body weight. Therefore, this study allowed for the comparison of different dietary fats with regard to PPL in a realistic context. This study also consisted of a balanced sample with regard to sex (5 M/5 F). Several studies similar in design had a predominately or exclusively male sample population. Another strength of this study was the robust postprandial assessment protocol, whereby we quantified the postprandial response serially every hour for six hours post-meal.

A limitation of this study was only measuring blood lipids and glucose. Examining the size of circulating CM and concentrations of either intestinally derived apoB-48 and/or endogenous apoB-100, in addition to blood lipids and glucose, may have been valuable with regard to answering our hypotheses. Next, all of our participants were young and presented few CVD risk factors. Thus, features of atherosclerotic development, including exaggerated and prolonged PPL, may not have been prominent enough to detect differences between meal trials, especially when considering the "true-to-life" meal used. Additionally, although this study found consistent sex-based differences in postprandial TG, it was not designed to address these differences. In addition, considering that three male participants exhibited an obese BMI (BMI > 30 kg/m^2), it is not possible to conclude whether the greater postprandial TG response was due to sex or BMI. Similarly, the lack of body composition measurement beyond BMI was a limitation of the present study. Finally, while we conducted an a priori sample size estimation and our study featured the same sample size (n = 10) as similar previous studies [26,27], our null findings present the possibility that our study was not sufficiently powered to detect differences. A post hoc analysis revealed that, given our observed TG variations, the minimum difference in peak TG that our design could have detected was 0.54 mmol/L. Thus, while we view this to be reasonable, differences between meals less than 0.54 mmol/L could not have been statistically detected in our study.

5. Conclusions

In our study, the effect of various sources of dietary fat, namely plant- and animal-based SFA, on PPL did not differ. Sex-based differences regarding the PPL response to the meal trials were observed and there was a strong correlation between BMI and TG tAUC, supporting the notion that sex and BMI are important factors that modulate the acute PPL response. However, it is impossible to form conclusions about the role of sex-based differences versus differences regarding BMI within our sample, owing to the fact that BMI was different between men and women. Moreover, differences in the PPL responses to the various sources of dietary fat used in the present study were not observed, despite the inclusion of three obese individuals. It is worth noting that our null findings on the effects of different fat sources within a mixed meal on PPL were in a sample of young adults. Future studies should investigate the effects of these various dietary fat sources on PPL in populations at risk for CVD or with existing CVD. In more at-risk individuals with a larger postprandial response, differences in TG between different sources of fat may be more apparent. Overall, the magnitude of PPL in response to a realistic mixed meal is likely modulated by several interrelated dietary factors, such as the amount of fat, energy density, and the heterogeneous mixture of macro- and micronutrients, rather than the specific type or source of dietary fat alone. Future studies should continue to focus on delineating between various sources of animal-based SFA (dairy- vs. meat-based) with regard to CVD risk, both in an acute (postprandial) and chronic context.

Author Contributions: C.M.S. and S.R.E. contributed to the study design and conceptualization; C.M.S., N.A.K., P.M.T., and K.B.B. engaged in participant recruitment and data collection; C.M.S. and S.R.E. organized and analyzed data; C.M.S. and S.R.E. were primarily responsible for writing the manuscript; C.M.S., N.A.K., P.M.T., K.B.B. and S.R.E. reviewed, edited, and approved the final manuscript.

Funding: This project was funded internally at Oklahoma State University. This research received no external funding.

Conflicts of Interest: The authors declare no conflict of interest.

References

1. Benjamin, E.J.; Muntner, P.; Alonso, A.; Bittencourt, M.S.; Callaway, C.W.; Carson, A.P.; Chamberlain, A.M.; Chang, A.R.; Cheng, S.; Das, S.R.; et al. Heart Disease and Stroke Statistics—2019 Update. A Report from the American Heart Association. *Circulation* **2019**, *139*, e56–e528. [CrossRef] [PubMed]

2. Bansal, S.; Buring, J.E.; Rifai, N.; Mora, S.; Sacks, F.M.; Ridker, P.M. Fasting Compared with Nonfasting Triglycerides and Risk of Cardiovascular Events in Women. *J. Am. Med. Assoc.* **2007**, *298*, 309–315. [CrossRef]

3. Hyson, D.; Rutledge, J.C.; Berglund, L. Postprandial Lipemia and Cardiovascular Disease. *Curr. Atheroscler. Rep.* **2003**, *5*, 437–444. [CrossRef]

4. Herieka, M.; Erridge, C. High-fat meal induced postprandial inflammation. *Mol. Nutr. Food Res.* **2014**, *58*, 136–146. [CrossRef]

5. Gower, R.M.; Wu, H.; Foster, G.A.; Devaraj, S.; Jialal, I.; Ballantyne, C.M.; Knowlton, A.A.; Simon, S.I. CD11c/CD18 expression is upregulated on blood monocytes during hypertriglyceridemia and enhances adhesion to vas- cular cell adhesion molecule-1. *Arterioscler. Thromb. Vasc. Biol.* **2011**, *31*, 160–166. [CrossRef]

6. Kim, K.A.; Gu, W.; Lee, I.A.; Joh, E.H.; Kim, D.H. High fat diet-induced gut microbiota exacerbates inflammation and obesity in mice via the TLR4 signaling pathway. *PLoS ONE* **2012**, *7*, e47713. [CrossRef]

7. Graner, M.; Kahri, J.; Nakano, T.; Sarna, S.J.; Nieminen, M.S.; Syvanne, M.; Taskinen, M.R. Impact of postprandial lipaemia on low-density lipoprotein (LDL) size and oxidized LDL in patients with coronary artery disease. *Eur. J. Clin. Investig.* **2006**, *36*, 764–770. [CrossRef]

8. Cortés, B.; Núñez, I.; Cofán, M.; Gilabert, R.; Pérez-Heras, A.; Casals, E.; Deulofeu, R.; Ros, E. Acute effects of high-fat meals enriched with walnuts or olive oil on postprandial endothelial function. *J. Am. Coll. Cardiol.* **2006**, *48*, 1666–1671. [CrossRef]

9. Tushuizen, M.E.; Nieuwland, R.; Scheffer, P.G.; Sturk, A.; Heine, R.J.; Diamant, M. Two consecutive high-fat meals affect endothelial-dependent vasodilation, oxidative stress and cellular microparticles in healthy men. *J. Thromb. Haemost.* **2006**, *4*, 1003–1010. [CrossRef]

10. Lopez-Miranda, J.; Williams, C.; Lairon, D. Dietary, physiological, genetic and pathological influences on postprandial lipid metabolism. *Br. J. Nutr.* **2007**, *98*, 458–473. [CrossRef]

11. Boren, J.; Matikainen, N.; Adiels, M.; Taskinen, M.R. Postprandial hypertriglyceridemia as a coronary risk factor. *Clin. Chim. Acta* **2014**, *431*, 131–142. [CrossRef]

12. Jackson, K.G.; Poppitt, S.D.; Minihane, A.M. Postprandial lipemia and cardiovascular disease risk: Interrelationships between dietary, physiological and genetic determinants. *Atherosclerosis* **2012**, *220*, 22–33. [CrossRef]

13. Meyer, E.; Westerveld, H.T.; de Ruyter-Meijstek, F.C.; van Greevenbroek, M.M.; Rienks, R.; van Rijn, H.J.; Erkelens, D.W.; de Bruin, T.W. Abnormal Postprandial Apolipoprotein B48 and TG responses in normolipidemic women with greater than 70% stenotic coronary artery disease. *Atherosclerosis* **1996**, *124*, 221–235. [CrossRef]

14. Langsted, A.; Freiberg, J.J.; Tybjaerg-Hansen, A.; Schnohr, P.; Jensen, G.B.; Nordestgaard, B.G. Nonfasting cholesterol and triglycerides and association with risk of myocardial infarction and total mortality: The Copenhagen City Heart Study with 31 years of follow-up. *J. Intern. Med.* **2011**, *270*, 65–75. [CrossRef]

15. Linton, M.F.; Fazio, S. Macrophages, inflammation, and atherosclerosis. *Int. J. Obes. Relat. Metab. Disord.* **2003**, *27* (Suppl. 3), S35. [CrossRef]

16. Chan, D.C.; Pang, J.; Romic, G.; Watts, G.F. Postprandial hypertriglyceridemia and cardiovascular disease: Current and future therapies. *Curr. Atheroscler. Rep.* **2013**, *15*, 309. [CrossRef]

17. Masson, C.J.; Mensink, R.P. Exchanging saturated fatty acids for (n-6) polyunsaturated fatty acids in a mixed meal may decrease postprandial lipemia and markers of inflammation and endothelial activity in overweight men. *J. Nutr.* **2011**, *141*, 816–821. [CrossRef]

18. Kruse, M.; von Loeffelholz, C.; Hoffmann, D.; Pohlmann, A.; Seltmann, A.C.; Osterhoff, M.; Hornemann, S.; Pivovarova, O.; Rohn, S.; Jahreis, G.; et al. Dietary rapeseed/canola-oil supplementation reduces serum lipids and liver enzymes and alters postprandial inflammatory responses in adipose tissue compared to olive-oil supplementation in obese men. *Mol. Nutr. Food Res.* **2015**, *59*, 507–519. [CrossRef]

19. Jones, P.J.; Senanayake, V.K.; Pu, S.; Jenkins, D.J.; Connelly, P.W.; Lamarche, B.; Couture, P.; Charest, A.; Baril-Gravel, L.; West, S.G.; et al. DHA-enriched high-oleic acid canola oil improves lipid profile and lowers predicted cardiovascular disease risk in the canola oil multicenter randomized controlled trial. *Am. J. Clin. Nutr.* **2014**, *100*, 88–97. [CrossRef]

20. Boniface, D.R.; Tefft, M.E. Dietary Fats and 16-year CHD mortality in a cohort of men and women in Great Britain. *Eur. J. Clin. Nutr.* **2002**, *56*, 786–792. [CrossRef]

21. Mann, J.I.; Appleby, P.N.; Key, T.J.; Thorogood, M. Dietary Determinants of IHD in Health Conscious Individuals. *Heart* **1997**, *78*, 450–455. [CrossRef] [PubMed]

22. Schwingshackl, L.; Hoffmann, G. Dietary fatty acids in the secondary prevention of coronary heart disease: A systematic review, meta-analysis and metaregression. *BMJ Open* **2014**, *4*, e004487. [CrossRef]

23. de Oliveira Otto, M.C.; Mozaffarian, D.; Kromhout, D.; Bertoni, A.G.; Sibley, C.T.; Jacobs, D.R., Jr.; Nettleton, J.A. Dietary intake of saturated fat by food source and incident cardiovascular disease: The Multi-Ethnic Study of Atherosclerosis. *Am. J. Clin. Nutr.* **2012**, *96*, 397–404. [CrossRef]

24. Teng, K.T.; Nagapan, G.; Cheng, H.M.; Nesaretnam, K. Palm olein and olive oil cause a higher increase in postprandial lipemia compared with lard but had no effect on plasma glucose, insulin and adipocytokines. *Lipids* **2011**, *46*, 381–388. [CrossRef] [PubMed]

25. Panth, N.; Dias, C.B.; Wynne, K.; Singh, H.; Garg, M.L. Medium-chain fatty acids lower postprandial lipemia: A randomized crossover trial. *Clin. Nutr.* **2019**, in press. [CrossRef]

26. Mekki, N.; Charbonnier, M.; Borel, P.; Leonardi, J.; Juhel, C.; Portugal, H.; Lairon, D. Butter Differs from Olive Oil and Sunflower Oil in Its Effects on Postprandial Lipemia and Triacylglycerol-Rich Lipoproteins after Single Mixed Meals in Healthy Young Men. *J. Nutr.* **2002**, *132*, 3642–3649. [CrossRef]

27. Thomsen, C.; Rasmussen, O.; Lousen, T.; Holst, J.J.; Fenselau, S.; Schrezenmeir, J.; Hermansen, K. Differential effects of saturated and monounsaturated fatty acids on postprandial lipemia and incretin responses in healthy subjects. *Am. J. Clin. Nutr.* **1999**, *69*, 1135–1143. [CrossRef] [PubMed]

28. Lesser, M.; Mauldin, K.; Sawrey-Kubicek, L.; Gildengorin, V.; King, J. The Type of Dietary Fat in an Isocaloric Breakfast Meal Does Not Modify Postprandial Metabolism in Overweight/Obese Pregnant Women. *Nutrients* **2019**, *11*, 490. [CrossRef]

29. Kristensen, M.; Savorani, F.; Christensen, S.; Engelsen, S.B.; Bügel, S.; Toubro, S.; Tetens, I.; Astrup, A. Flaxseed dietary fibers suppress postprandial lipemia and appetite sensation in young men. *Nutr. Metab. Cardiovasc. Dis.* **2013**, *23*, 136–143. [CrossRef]

30. Sun, L.; Tan, K.W.; Lim, J.Z.; Magkos, F.; Henry, C.J. Dietary fat and carbohydrate quality have independent effects on postprandial glucose and lipid responses. *Eur. J. Nutr.* **2016**, *57*, 243–250. [CrossRef]

31. Calabrese, C.; Myer, S.; Munson, S.; Turet, P.; Birdsall, T.C. A cross-over study of the effect of a single oral feeding of medium chain triglyceride oil vs. canola oil on post-ingestion plasma triglyceride in healthy men. *Altern. Med. Rev.* **1999**, *4*, 23–28. [PubMed]

32. Lorenzen, J.K.; Nielsen, S.; Holst, J.J.; Tetens, I.; Rehfeld, J.F.; Astrup, A. Effect of dairy calcium or supplementary calcium intake on postprandial fat metabolism, appetite, and subsequent energy intake. *Am. J. Clin. Nutr.* **2007**, *85*, 678–687. [CrossRef] [PubMed]

33. Hazim, J.; Hlais, S.; Ghattas, H.; Shatila, D.; Bassil, M.; Obeid, O. Phosphorus supplement alters postprandial lipemia of healthy male subjects: A pilot cross-over trial. *Lipids Health Dis.* **2014**, *13*, 109. [CrossRef] [PubMed]

34. Kay, C.D.; Holub, B.J. The effect of wild blueberry (*Vaccinium angustifolium*) consumption on postprandial serum antioxidant status in human subjects. *Br. J. Nutr.* **2002**, *88*, 389. [CrossRef]

35. Dubois, C.; Armand, M.; Azais-Braesco, V.; Portugal, H.; Pauli, A.M.; Bernard, P.M.; Latge, C.; Lafont, H.; Borel, P.; Lairon, D. Effects of moderate amounts of emulsified dietary fat on postprandial lipemia and lipoproteins in normolipidemic adults. *Am. J. Clin. Nutr.* **1994**, *60*, 374–382. [CrossRef]

36. Dubois, C.; Beaumier, G.; Juhel, C.; Armand, M.; Portugal, H.; Pauli, A.M.; Borel, P.; Latgé, C.; Lairon, D. Effects of graded amounts of dietary fat on postprandial lipemia and lipoproteins and in normolipidemic adults. *Am. J. Clin. Nutr.* **1998**, *67*, 31–38. [CrossRef]

37. Kasai, M.; Maki, H.; Nosaka, N.; Aoyama, T.; Ooyama, K.; Uto, H.; Okazaki, M.; Igarashi, O.; Kondo, K. Effect of Medium-chain Triglycerides on the Postprandial Triglyceride Concentration in Healthy Men. *Biosci. Biotechnol. Biochem.* **2003**, *67*, 46–53. [CrossRef]

38. Hansson, P.; Holven, K.B.; Øyri, L.K.; Brekke, H.K.; Biong, A.S.; Gjevestad, G.O.; Raza, G.S.; Herzig, K.H.; Thoresen, M.; Ulven, S.M. Meals with Similar Fat Content from Different Dairy Products Induce Different Postprandial Triglyceride Responses in Healthy Adults: A Randomized Controlled Cross-Over Trial. *J. Nutr.* **2019**, *149*, 422–431. [CrossRef] [PubMed]

39. Couillard, C.; Bergeron, N.; Prud'homme, D.; Bergeron, J.; Tremblay, A.; Bouchard, C.; Mauriege, P.; Després, J.P. Gender Difference in Postprandial Lipemia. *Arterioscler. Thromb. Vasc. Biol.* **1999**, *19*, 2448–2455. [CrossRef]

40. Mekki, N. Influence of Obesity and Body Fat Distribution on Postprandial Lipemia and Triglyceride-Rich Lipoproteins in Adult Women. *J. Clin. Endocrinol. Metab.* **1999**, *84*, 184–191. [CrossRef]

41. Blackburn, P.; Lamarche, B.; Couillard, C.; Pascot, A.; Tremblay, A.; Bergeron, J.; Lemieux, I.; Després, J.P. Contribution of Visceral Adiposity to the Exaggerated Postprandial Lipemia of Men with Impaired Glucose Tolerance. *Diabetes Care* **2003**, *26*, 3303–3309. [CrossRef] [PubMed]

Article

Postprandial Endotoxin Transporters LBP and sCD14 Differ in Obese vs. Overweight and Normal Weight Men during Fat-Rich Meal Digestion

Fabienne Laugerette [1,*], Cécile Vors [1], Maud Alligier [1,2], Gaëlle Pineau [1], Jocelyne Drai [1,3], Carole Knibbe [1], Béatrice Morio [1,2], Stéphanie Lambert-Porcheron [2,4], Martine Laville [1,2,4], Hubert Vidal [1,2,4] and Marie-Caroline Michalski [1,2]

[1] Univ Lyon, CarMeN Laboratory, INRAE, UMR1397, INSERM, UMR1060, Université Claude Bernard Lyon 1, 69310 Pierre Bénite, France; cecile.vors@univ-lyon1.fr (C.V.); maud.alligier@chu-lyon.fr (M.A.); gaelle.pineau@gmail.com (G.P.); jocelyne.drai@gmail.com (J.D.); carole.knibbe@insa-lyon.fr (C.K.); beatrice.morio@clermont.inra.fr (B.M.); martine.laville@univ-lyon1.fr (M.L.); hubert.vidal@univ-lyon1.fr (H.V.); marie-caroline.michalski@insa-lyon.fr (M.-C.M.)

[2] Centre de Recherche en Nutrition Humaine Rhône-Alpes, Univ-Lyon, CarMeN Laboratory, Université Claude Bernard Lyon1, Hospices Civils de Lyon, CENS, FCRIN/FORCE Network, 69310 Pierre-Bénite, France; stephanie.lambert-porcheron@chu-lyon.fr

[3] Laboratoire de Biochimie, Centre Hospitalier Lyon Sud, 69600 Oullins, France

[4] Hospices Civils de Lyon, 69000 Lyon, France

[*] Correspondence: fabienne.laugerette@univ-lyon1.fr; Tel.: +33-4-26-23-61-74

Received: 29 April 2020; Accepted: 15 June 2020; Published: 18 June 2020

Abstract: Circulating levels of lipopolysaccharide-binding protein (LBP) and soluble cluster of differentiation 14 (sCD14) are recognized as clinical markers of endotoxemia. In obese men, postprandial endotoxemia is modulated by the amount of fat ingested, being higher compared to normal-weight (NW) subjects. Relative variations of LBP/sCD14 ratio in response to overfeeding are also considered important in the inflammation set-up, as measured through IL-6 concentration. We tested the hypothesis that postprandial LBP and sCD14 circulating concentrations differed in obese vs. overweight and NW men after a fat-rich meal. We thus analyzed the postprandial kinetics of LBP and sCD14 in the context of two clinical trials involving postprandial tests in normal-, over-weight and obese men. In the first clinical trial eight NW and 8 obese men ingested breakfasts containing 10 vs. 40 g of fat. In the second clinical trial, 18 healthy men were overfed during 8 weeks. sCD14, LBP and Il-6 were measured in all subjects during 5 h after test meal. Obese men presented a higher fasting and postprandial LBP concentration in plasma than NW men regardless of fat load, while postprandial sCD14 was similar in both groups. Irrespective of the overfeeding treatment, we observed postprandial increase of sCD14 and decrease of LBP before and after OF. In obese individuals receiving a 10 g fat load, whereas IL-6 increased 5h after meal, LBP and sCD14 did not increase. No direct association between the postprandial kinetics of endotoxemia markers sCD14 and LBP and of inflammation in obese men was observed in this study.

Keywords: LBP; sCD14; postprandial kinetics; high-fat diet

---

1. Introduction

Metabolic diseases, including obesity and type 2 diabetes, are associated with a chronic inflammatory state and increased plasma levels of lipopolysaccharides (LPS), also called endotoxins [1]. Endotoxins are components of the outer membrane of Gram-negative bacteria, which are dominant in the healthy human gut microbiome [1]. During lipid digestion, some endotoxins are translocated in the bloodstream, and thereby contribute to the onset and maintenance of low-grade inflammation during the postprandial phase following a fat-rich meal [2–5]. Indeed, Clemente-Postigo et al. (2012) reported an increase in the plasma level of endotoxins after a fat overload in morbidly obese humans [6]. This includes the binding of LPS to LPS-binding protein (LBP) and its transfer to the receptor CD14, present in both soluble (sCD14) and membrane-bound (mCD14) forms [7]. The activation of the LPS-LBP-CD14 complex leads to the secretion of pro-inflammatory markers and contributes to the inflammatory state [8]. LBP and sCD14 are now recognized as clinical markers of endotoxin exposure [9] and have been associated with obesity and metabolic disorders [10]. LBP is synthesized and released into the bloodstream in the presence of LPS, and is considered as a surrogate biomarker for the activation of LPS-induced innate immune response regarding its relative long half-life (24–48 h) [11]. Moreno-Navarrete et al. have also shown that the increase of plasma LBP in vivo may contribute to a vicious cycle that prevents white adipose tissue (WAT) expansion, and exacerbates the inflammatory response in WAT [12]. It was further shown that serum LBP and sCD14 are markers of Crohn's disease [13]. The relative variations of LBP/sCD14 ratio are also considered important in the inflammation set-up [14]. Furthermore, higher plasma LBP concentrations have been observed in obesity [7,13,15] and plasma LBP concentration is associated to abdominal obesity [16]. We have recently shown that increase of plasma IL-6 is linked to a rise of LBP/sCD14 ratio in both humans [14] and mice [17]. More recently, Sakura et al. have suggested that plasma LBP concentration is associated with arterial stiffness, independently of traditional cardiovascular risk factors and especially in men with type 2 diabetes [18,19]. In children and adults, obesity and obstructive sleep apnea are also associated with increased LBP concentrations, and the presence of both conditions may enhance LBP concentration [20,21]. Concerning sCD14, it has been reported in mice that sCD14 presents protective effects in inflammatory bowel disease [22]. Therefore, the literature shows that LBP and sCD14 are important factors implicated in low-grade inflammation during metabolic diseases. We have previously demonstrated in obese men that postprandial endotoxemia is modified by ingested fat amount with a higher postprandial endotoxemia compared with normal-weight subjects after a higher fat load (40 g vs. 10 g) [23]. More recently, a study of healthy premenopausal women has suggested that the consumption of a pre-meal yogurt improves the postprandial metabolism and decreases metabolic endotoxemia, LBP and sCD14 [24]. However, the postprandial variations of LBP and sCD14 after a mixed meal with different amount of lipids remain poorly described in humans of different weight status. We thus analyzed the postprandial kinetics of both markers of endotoxin exposure in the context of two clinical trials involving postprandial tests in normal weight, overweight and obese men. We hypothesized that postprandial LBP and sCD14 after a fat-rich meal differed in obese vs. overweight and normal weight men, hence, contributing to different IL-6 responses after a fat load or OF.

2. Materials and Methods

2.1. Clinical Trials and Subjects

The present work relies on two previous clinical trials performed in the Human Nutrition Research Centers (CRNH) of Rhône-Alpes and Auvergne. The first clinical trial, called the Lipinflox study, was approved by the Ethics Committee of Lyon-Sud-Est-II and AFSSAPS and was registered at ClinicalTrials.gov (NCT01249378). The second one, called the Overfeeding Study, was approved by the Ethics Committee of Lyon (RG/FL-2005-067) and registered at clinicalTrials.gov (NCT00905892). The eligibility criteria of included subjects and postprandial metabolic explorations for both trials have been described elsewhere [23,25,26]. Briefly, in the crossover Lipinflox Study, 8 normal weight (20 < BMI < 25 kg/m^2) and 8 obese (30 < BMI < 35 kg/m^2 and waist circumference > 94 cm) men were submitted to a fat load of 10 g or 40 g at breakfast. The test breakfast contained 10 or 40 g of anhydrous milk fat with bread and a glass of skim milk (282 kcal and 551 kcal, respectively). Blood samples were collected from an antecubital arm vein through a catheter at baseline and at regular intervals during 5 h, following ingestion of fat load. Plasma was separated by centrifugation (1500 g, 10 min, 4 °C) and stored at −80 °C until further analysis. In the overfeeding clinical trial, 18 healthy young men (lean to overweight, BMI 25.8 ± 0.8 kg/m^2, age 30.6 ± 2.1 years) were submitted to an overfeeding during 56 days. During this period, the subjects added to their daily diet +760 kcal/day as previously described [25]. Fasting plasma sampling was performed at Day 0 and Day 56. In this article, we focused on a subcohort of eight subjects that consumed a test meal (882 kcal) and performed a postprandial test including endotoxemia analyses, as described previously [25]. The mixed meal contained 33 g of fat (291 kcal) and was composed of 200 mL of Fortimel (enteral emulsion), 23 g of margarine, 9.4 g of butter, 1 g of olive oil, 85 g of bread, 20 g of jam and 200 g of banana. The body composition (fat and lean mass) was performed in 18 subjects. All subjects gave written consent after being informed of the nature, purpose, and possible risks of the clinical trial. Both clinical trials were approved by the ethics committee of Lyon Sud-Est, according to the French"Huriet-Serusclat" law and the Second Declaration of Helsinki.

2.2. Plasma sCD14 and LBP

To measure the plasma concentrations of circulating LBP and sCD14, plasma samples were assayed using sandwich ELISA kits (CliniSciences and R&D Systems; Nanterre, France), following the manufacturer's instructions.

2.3. Plasma Analyses

Endotoxins were determined using the limule amoebocyte lysate assay in kinetic chromogenic conditions (Biogenic; Pérols, France) [15]. Serum high-sensitive C-reactive protein was assessed by immunonephelometry on an image analyzer (Beckman-Coulter; Villepinte, France). For the overfeeding trial, IL-6 was measured in serum using ELISA kits (Quantikine; Abingdon, UK). For Lipinflox, plasma hsIL-6 levels were measured using a sandwich Ultrasensitive ELISA kit (Invitrogen; Illkirch, France). The liver enzymes ASAT and ALAT were routinely measured during the screening visit (Pentra C400; Horiba, Kyoto, Japan) in order to exclude patients if the values were not biologically normal. In fact, as patients were submitted to a fat load, or to an overfeeding, it was important that they had no liver disease.

2.4. Anthropometry and Body Composition

As previously described by Alligier et al., body composition was determined before and after overfeeding by dual-energy X-ray absorptiometry (Hologic, Inc.; Bedford, MA, USA), and abdominal adipose tissue distribution by magnetic resonance imaging (Magnetom Symphonie 1.5 Tesla; Siemens AG, Munich, Germany) [25].

2.5. Statistical Analysis

Data are presented as means ± SEM and were analyzed with Graph Pad Prism$^{\circledR}$ (version 7.0, San Diego, CA, USA) and with R (version 3.6.3, Saint Louis, USA). Clinical characteristics of the subjects of the two clinical trials (before vs. after overfeeding and lean vs. obese subjects) were performed using Student's *t*-test with Graph Pad Prism. To evaluate possible relationships among the various outcomes, Spearman correlations were performed using Graph Pad Prism Software. For the Overfeeding Study, postprandial kinetics of plasma LBP, sCD14, LBP/sCD14 ratio and IL-6 were each analyzed by two-way ANOVA for repeated measurements in both factors (time and overfeeding), followed by Bonferroni's post hoc test, using Graph Pad Prism. For the Lipinflox study, postprandial kinetics of plasma LBP, sCD14 and IL-6 were each analyzed with a linear mixed-effects model, analogous to a 3-way ANOVA for repeated measurements, where the factors with fixed effects were time, obese status, fat load and their 2- and 3-way interactions, while the factor with random effects was the subject identifier. This analysis was carried out with R, using the lme function of the nlme package. The main effects and the interactions effects were tested using marginal (Type III) sums of squares like in GraphPad Prism. For the Lipinflox study, postprandial kinetics of plasma LBP, sCD14 and IL-6 were each analyzed with a linear mixed-effects model, where the factors with fixed effects were time, obese status, fat load and their 2- and 3-way interactions, while the factor with random effects was the subject identifier. This approach is similar to a three-way repeated-measures ANOVA but is more robust to missing data (we lacked IL6 data for one of the ten lean subjects). A posthoc analysis was carried out to assess the effect of obesity on the estimated marginal means at all time points. Specifically, the significance of the difference of the two marginal means (lean versus obese, collapsing the fat load) was assessed at each time point with a t-test based on the standard errors of the estimated marginal means. The five raw p-values obtained for each response in this posthoc analysis were adjusted using the False Discovery Rate (FDR) correction, in order to correct for multiple testing. This posthoc analysis was carried out with the emmeans package. A p-value (or a FDR-corrected p-value) lower than 0.05 was considered significant.

3. Results

3.1. Characteristics of the Subjects

The characteristics of the eight subjects, lean to overweight, from the overfeeding clinical trial (OF) and from the 16 subjects (normal weight and obese) from Lipinflox clinical trial are shown Table 1. As expected weight, BMI and waist circumference were significantly higher after OF, compared to before, and higher in the obese than in normal weight (NW) subjects. Concerning body composition, lean mass and fat mass were significantly enhanced after OF. CRP, IL-6 and LPS did not vary significantly in the OF clinical trial. CRP and IL-6 were higher in obese compared to NW subjects.

Table 1. Clinical characteristics of study subjects. Data are means ± SEM. Groups are compared using unpaired Student t-test, * $p < 0.05$, ** $p < 0.01$; $n = 8$ OF clinical trial, $n = 8$ normal weight, $n = 8$ obese.

	Overfeeding Clinical Trial			Lipinflox Clinical Trial		
	Before OF	**After OF**	*p* **Value**	**Normal Weight**	**Obese**	*p* **Value**
Age (year)	26 ± 2			29 ± 1	31 ± 2	0.426
Body weight (kg)	78.9 ± 4.9	81.7 ± 4.9	0.005 **	72.5 ± 2.1	101.1 ± 2.1	0.001 **
BMI (kg/m^2)	24.9 ± 1.5	25.7 ± 1.4	0.004 **	22.4 ± 0.5	31.8 ± 0.3	0.001 **
Waist circumference (cm)	84.9 ± 3.3	86.9 ± 3.3	0.038 *	83.6 ± 1.7	105 ± 0.8	0.001 **
hsCRP (µg/mL)	0.6 ± 0.1	1.1 ± 0.5	0.38	1.96 ± 0.01	2.98 ± 0.47	0.036 *
IL-6 (pg/mL)	1.23 ± 0.15	1.43 ± 0.35	0.53	0.18 ± 0.04	0.45 ± 0.16	0.006 *
LPS (EU/mL)	0.11 ± 0.03	0.16 ± 0.10	0.59	0.19 ± 0.05	0.18 ± 0.04	0.729
Lean mass (kg)	57,3 ± 2.5	58.2 ± 2.60	0.032 *	-	-	
Fat mass (kg)	15.8 ± 2.9	17.1 ±3.1	0.048 *	-	-	
Visceral fat (kg)	66.0 ± 18.1	73.5 ± 16.5	0.469	-	-	

3.2. Postprandial Kinetics of LBP, sCD14 and IL-6 in Normal Weight and Obese Subjects (Lipinflox Study)

We evaluated the postprandial variations of plasma LBP, sCD14 and IL-6 in NW and obese subjects during 5 h after mixed meals differing only by fat amount: 10 vs. 40 g of milkfat spread on bread.

Figure 1A,B shows that, regardless of lipid amount in the meal, the concentration of LBP in plasma was higher in obese subjects than in lean subjects (obese status, $p = 0.027$), with no impact of postprandial time. No significant difference was observed between lean and obese subjects along the postprandial kinetics of sCD14, regardless of fat load in the meal (Figure 1C,D).

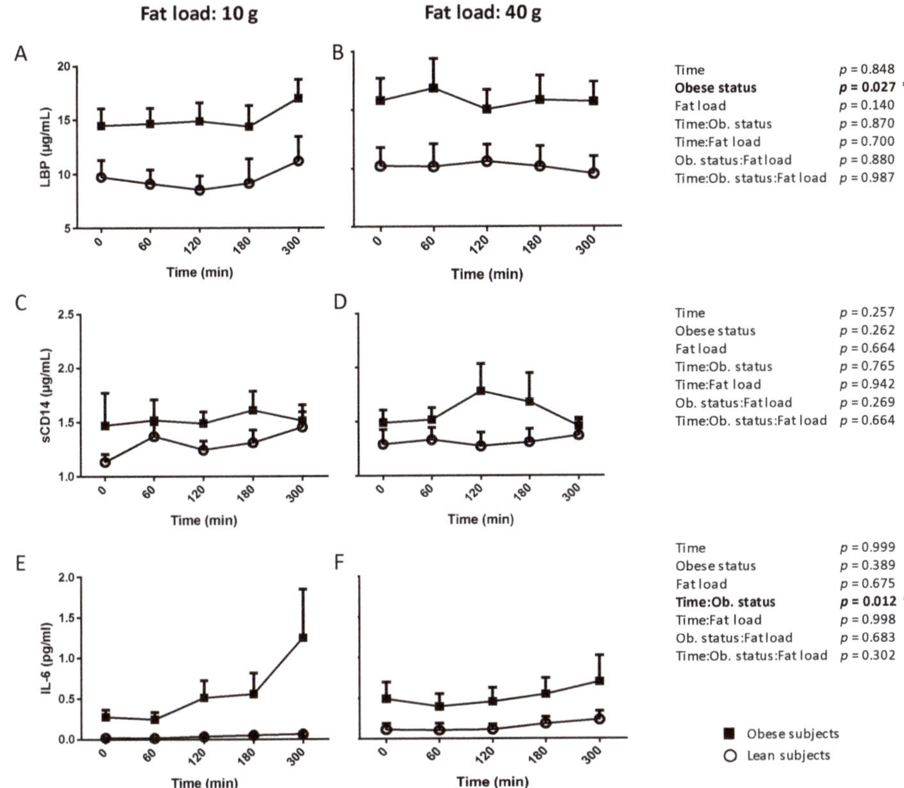

Figure 1. Postprandial kinetics in plasma of lipopolysaccharides (LPS)-binding protein (LBP) (**A,B**); sCD14 (**C,D**); and Il-6 (**E,F**) in normal-weight and obese subjects after 10 g vs. 40 g fat load. Data are means ± SEM. * $p < 0.05$.

Regarding IL-6, we observed an interaction effect between time and obese status ($p = 0.012$) during the postprandial kinetics (Figure 1E,F). The post-hoc analysis revealed that the significant two-way interaction between obesity and time occurred mostly at $t = 300$ min. Indeed, IL-6 plasma concentration at 300 min was on average 5.6 times higher in obese vs. lean men ($p = 0.02$) (Figure 1E).

There was no impact of BMI and postprandial time on the LBP/sCD14 ratio in all subjects for each fat amount (Appendix A).

3.3. Postprandial Kinetics of LBP, sCD14 and IL-6 Before and After Overfeeding

Previously, we have shown in 18 subjects of the OF clinical trial that the LBP/sCD14 ratio was enhanced in the fasting state after OF [14]. Here, we completed our exploration by analyzing the postprandial evolution of LBP, sCD14 and LBP/sCD14 in the subgroup of eight subjects who performed the postprandial explorations. Altogether, there was no significant effect of OF on the postprandial kinetics of LBP and sCD14 (Figure 2A,B). Concerning LBP before OF, we observed a significant time effect with an increase of LBP concentration in plasma 120 min after the mixed meal (Figure 2A, $p < 0.05$). The concentration of LBP was decreased at 240 min vs. 120 min (Figure 2A, $p < 0.05$). After OF, no significant increase was observed at 120 min after the mixed meal but a significant decrease was observed at 240 min compared to 120 min ($p < 0.05$). Notably, even if after OF LBP concentration was higher at fasting compared to before OF ($p = 0.07$), at 240 min LBP concentrations were similar before and after OF (18.3 ± 1.9 μg/mL and 18.2 ± 0.8 μg/mL, respectively). A significant increase of sCD14 at 240 min compared to 0 and 60 min ($p < 0.01$ and 0.01 respectively) was observed both before and after OF (Figure 2B). The postprandial kinetics of the LBP/sCD14 ratio was not impacted by OF (Figure 2C). Both before and after OF, the LBP/sCD14 ratio decreased at 240 min ($p < 0.001$ vs. other postprandial times). We have measured IL-6 plasma concentration at 0, 120 and 240 min before and after OF (Figure 2D). There was no significant OF effect but of note, after OF plasma IL-6 concentration increased at 240 min compared to 0 min ($p < 0.05$). No interaction (between time and OF) was observed for LBP, sCD14, LBP/sCD14 and IL-6 (Figure 2A–D).

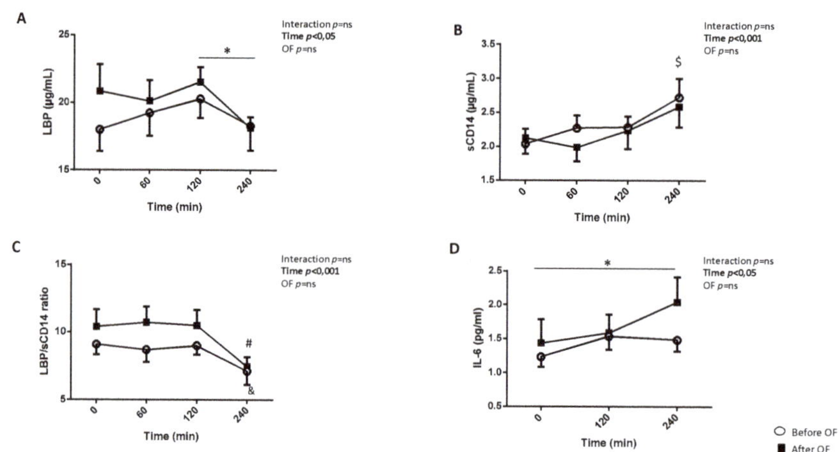

Figure 2. Postprandial kinetics of LBP (**A**), sCD14 (**B**), LBP/sCD14 ratio (**C**) and IL-6 (**D**) before and after overfeeding. Data are means ± SEM. The effects of time (postprandial kinetics) and overfeeding (OF) were determined by ANOVA for repeated measurements followed by post hoc test (Bonferroni). * $p < 0.05$ (after OF), & $p < 0.05$ vs. T0 and T120 (before OF); # $p < 0.001$ vs. other time (after OF); $ $p < 0.05$; 0.01; 0.001 vs. other time before and after OF.

3.4. Associations between Fasting and Postprandial LBP and Selected Parameters of Dietary Trials and Intervention

We examined the association of plasma LBP concentrations with some parameters related to obesity and metabolic disorders in subjects from both the overfeeding and Lipinflox studies. Subjects of the Lipinflox study who have consumed a test meal with 40 g of lipids presented no significant correlation between circulating LBP and the selected parameters, age, weight, BMI, waist circumference (WC), aspartame amino transferase (AST), alanine amino transferase (ALT), sCD14 (data not shown).

The correlations between LBP and selected parameters of the Lipinflox postprandial clinical trial after the test meal containing 10 g of fat are reported Table 2. At all times of the postprandial kinetics, the LBP concentration was significantly and positively associated with waist circumference (WC), but not with weight and BMI in all subjects from Lipinflox clinical trial (lean and obese, Table 2). After stratification according to BMI group, this positive correlation was maintained for lean subjects, but not for obese subjects (Table 2). LBP concentration was also significantly and positively associated with ALT but not with AST. These correlations were also maintained for lean subjects at 0, 60 and 120 min, a trend was observed at 180 and 300 min (Table 2). Conversely, no association of LBP with WC or with ALT was observed for obese subjects (Table 2).

Concerning the overfeeding clinical trial, we also used some mass parameters, such as abdominal visceral fat and subcutaneous WAT depots, measured by magnetic resonance imaging [27]. As shown in Table 3, LBP concentration in plasma was significantly and positively associated with trunk lean mass, lean mass and waist circumference before and after OF. No correlation with subject age was observed (data not shown). Notably, the LBP/sCD14 ratio was significantly and positively associated with lean mass, fat trunk and BMI, before and after overfeeding (Table 3).

Table 2. Correlation analyses (Spearman) with postprandial plasma LBP concentrations as dependent variable after a 10 g fat load (Lipinflox clinical trial in men). * $p < 0.05$; ** $p < 0.01$.

Postprandial Time	All ($n = 16$)									
	0 min		60 min		120 min		180 min		300 min	
	r	p	r	p	r	p	r	p	r	p
Age (years)	0.66	ns	−0.06	ns	0.065	ns	−0.10	0.07 ns	0.23	ns
Weight (kg)	0.26	ns	0.42	ns	0.43	0.058	0.37	ns	0.36	ns
BMI (kg/m^2)	0.24	ns	0.35	ns	0.44	0.052	0.37	ns	0.36	ns
Waist circumference (cm)	0.58	0.007 **	0.58	0.006 **	0.63	0.003 **	0.41	0.07	0.51	0.02 *
AST (U/L)	0.20	ns	0.1	ns	0.24	ns	0.27	ns	0.14	ns
ALT (U/L)	0.49	0.02 *	0.57	0.008 **	0.63	0.003 **	0.53	0.016 *	0.43	0.05 *
sCD14 T0 (μg/mL)	0.41	0.06	0.45	0.04 *	0.28	ns	−0.15	ns	0.058	ns
	Lean ($n = 8$)									
	T0 min		T60 min		T120 min		T180 min		T300 min	
	r	p	r	p	r	p	r	p	r	p
Age (years)	0.7	0.02 *	0.43	ns	0.43	ns	0.35	ns	0.47	ns
Weight (kg)	−0.31	ns	−0.16	ns	−0.02	ns	0.09	ns	0.07	ns
BMI (kg/m^2)	−0.23	ns	−0.18	ns	−0.08	ns	0.02	ns	0.05	ns
Waist circumference (cm)	0.66	0.04 *	0.60	0.06	0.71	0.02 *	0.72	0.02 *	0.58	0.08
AST (U/L)	0.55	ns	0.18	ns	0.22	ns	0.27	ns	0.23	ns
ALT (U/L)	0.69	0.03 *	0.62	0.05 *	0.68	0.03 *	0.51	ns	0.38	ns
sCD14 T0 (μg/mL)	0.05	ns	0.05	ns	−0.01	ns	0.17	ns	−0.02	ns
	Obese ($n = 8$)									
	T0 min		T60 min		T120 min		T180 min		T300 min	
	r	p	r	p	r	p	r	p	r	p
Age (years)	−0.48	ns	−0.56	ns	−0.17	ns	−0.09	ns	−0.04	ns
Weight (kg)	−0.13	ns	0.18	ns	−0.27	ns	0.25	ns	0.06	ns
BMI (kg/m^2)	−0.31	ns	−0.46	ns	−0.13	ns	0.07	ns	0.17	ns
Waist circumference (cm)	0.25	ns	0.05	ns	0.13	ns	−0.13	ns	0.35	ns
AST (U/L)	−0.33	ns	−0.27	ns	−0.07	ns	0.23	ns	−0.13	ns
ALT (U/L)	−0.11	ns	0.04	ns	0.18	ns	0.53	ns	0.16	ns
sCD14 T0 (μg/mL)	0.22	ns	0.4	ns	0.10	ns	−0.13	ns	−0.29	ns

Table 3. Correlation analyses (Spearman) of fasting LBP and LBP/sCD14 with mass parameters in the overfeeding clinical trial ($n = 18$ men). * $p < 0.05$; ** $p < 0.01$.

	Before OF		
Variable 1	Variable 2	r	p
LBP (μg/mL)	Trunk lean mas (kg)	0.65	0.002 **
	Lean mass (kg)	0.57	0.009 **
	Waist circumference (cm)	0.53	0.02 *
	BMI (kg/m^2)	0.47	0.04 *
	Fat trunk (kg)	0.41	ns
	Visceral fat (kg)	0.52	0.03 *
LBP/sCD14	Lean mass (kg)	0.62	0.004 **
	Fat trunk (kg)	0.48	0.04 *
	BMI (kg/m^2)	0.62	0.004 **
	After OF		
Variable 1	Variable 2	r	p
LBP (μg/mL)	Trunk lean mas (kg)	0.62	0.003 **
	Lean mass (kg)	0.56	0.01 *
	Waist circumference (cm)	0.51	0.02 *
	BMI (kg/m^2)	0.32	ns
	Fat trunk (kg)	0.47	0.04 *
	Visceral fat (kg)	0.07	ns
LBP/sCD14	Lean mass (kg)	0.49	0.03 *
	Fat trunk (kg)	0.49	0.03 *
	BMI (kg/m^2)	0.49	0.03 *

4. Discussion

The novel feature of the present study is the postprandial explorations of LBP and sCD14 concentrations in plasma in two different clinical trials in obese vs. overweight and normal-weight men. In the first clinical trial (Lipinflox), we have shown that obese men presented a higher LBP concentration in plasma than lean men, regardless of fat load (10 g vs. 40 g), with no time effect during the postprandial period. However, no significant differences in sCD14 concentrations were observed between lean and obese men. In the second clinical trial (overfeeding), we demonstrated, in lean to overweight men, a time effect on LBP, sCD14, IL-6 concentrations and LBP/sCD14 ratio during the postprandial period. Only a few human studies have focused on the effect of an OF on endotoxemia [24] and, to our knowledge, no study has been performed to date in order to evaluate the impact of dietary fat intake on the postprandial kinetics of LBP and sCD14 in NW, overweight and obese men.

It is now well established that the intestinal barrier function is modified in obesity, leading to low-grade inflammation and endotoxemia [9]. Endotoxins have been described as contributors to the inflammation observed in hypercaloric diets [5,28,29]. LBP and sCD14, two important actors in the endotoxin metabolic pathway, are now considered indirect markers of gut permeability in metabolic diseases, but also in other diseases, such as HIV [9,22]. Indeed, both reductions of these markers and gut permeability were shown after 12 weeks of fish oil supplementation, compared to placebo in HIV+ patients [30]. We show in the Lipinflox study that LBP concentration was higher in obese men at all times of the postprandial kinetics, independently of fat load (10 vs. 40 g of fat).

The relationship between inflammation and "metabolically healthy" and "non-metabolic healthy" status remains relatively unknown, even though some studies have shown relationships between inflammatory markers and obesity or metabolic syndrome. Aguilar-Salinas CA et al. demonstrated that obese individuals depict similar adiponectin levels to normal-weight subjects and this may be associated with the "metabolically healthy" obese phenotype [31]. In turn, patients with severe obesity were reported to have higher LBP concentration, which can be reduced after bariatric surgery [32]. In a recent study, it was suggested that serum LBP is related to abdominal obesity more than to metabolic health [16]. Serum LBP is also known to be associated with the carotid intima media thickness [19], suggesting that the pro-inflammatory action of LBP might be a contributor to the progression of cardiovascular events [11,33]. In hemodialysis patients, for whom gut barrier and microbiota are impaired, a positive association between LBP and chronic inflammation and metabolic syndrome was reported [34]. The present study shows a strong relationship between plasma LBP and waist circumference. It is interesting to note that plasma LBP was correlated significantly with WC only in lean subjects. Another point is the correlation of LBP with ALT only in lean subjects. Patients with non-alcoholic fatty liver disease are known to present higher LBP concentrations when they develop steatohepatitis [15]. An association between LBP and ALT was noted in hepatitis C virus, and these authors suggested that, as with ALT, LBP might serve as another hepatic inflammatory biomarker [35]. Another study has shown that endotoxemia reflects the hepatic functional reserve capacity of end-stage liver disease [36]. Interestingly, in obese subjects of the present study, no correlation of circulating LBP with WC or ALT was observed. Naghizadek et al., 2018 have shown that WC is important for the association among TLR4, serum LBP and IFNβ and metabolic state only for the highest WC range [16]. This finding is not in agreement with the present study, and could be explained by the fact that obese subjects were not morbid.

In the Lipinflox study, the fat load (10 vs. 40 g) did not modify the kinetics of the LBP concentration in the plasma of lean and obese subjects. However, the LBP concentration in lean subjects was lower than that in obese subjects for both fat loads. Moreover, the dietary intervention in the OF study did not affect postprandial LBP concentrations. Umoh F. et al. also suggested that obesity might not result in enhanced exposure to intestinal bacteria, as the effect of BMI was no longer significant in multiple linear regression models [37]. It is well known that obesity is associated with modifications in the secretion of cytokines from adipose tissue and liver. However, only few studies have investigated the impact of overfeeding on the production of cytokines/adipokines. In the present OF study, the concentrations of LBP, sCD14 and Il-6 were not significantly affected by a two-month overfeeding (+760 kcal/day by adding 70 g of lipids to the usual daily diet). More precisely, the fat OF consisted in daily addition of 20 g of butter, 100 g of cheese (Emmental) and 40 g of almonds. Another study of acute overfeeding (+1250 kcal/day with a nutrient composition of 45% fat, 15% protein and 40% carbohydrate) has shown that 3 days are sufficient to increase body weight and HOMA-IR, without affecting MCP-1 and CRP plasma concentrations [38]. Tam C.S. et al. also demonstrated that moderate weight gain after 28 days of overfeeding (+1250 kcal via high-fat snacks composed of 45% fat, 15% protein, and 40% carbohydrate) in healthy humans resulted in a significant weight gain and increased circulating levels of CRP and MCP-1, without changes in subcutaneous WAT mass [39]. Those results demonstrate that different compositions of overfeeding diets may differentially impact cytokines/adipokine concentrations in plasma and tissues.

The present study has some limitations. Firstly, the sample size was small, owing the cumbersome aspects of postprandial explorations, and only one plasma inflammatory marker was measured. However, the postprandial results were obtained from two different trials, and the timing of plasma collection was not the same. Another limitation of this study is the measure of IL-6 in plasma taken serially from a catheter in an arm vein. Indeed some authors have shown that the presence of the catheter irritate the vein and cause an increase in IL-6 levels in plasma samples taken over several hours compared to fresh samples taken from the opposite arm [40,41]. Finally, ensuring adherence to dietary instructions given during the OF clinical trial is difficult in a feeding trial.

To conclude, we have shown in obese men that a fat load of 10 g of lipids can drive an increase of IL-6 concentration in plasma at 300 min post-meal, maybe due to a lesser clearance of LPS by lipoproteins. Indeed Vors et al. has shown that obese subject chylomicrons were more enriched with LPS compared to NW, which could contribute to LPS clearance and to a lesser IL-6 concentration in plasma [23]. This result should be confirmed in further studies with the measure of other inflammatory markers. The LBP concentration in plasma was higher in obese subjects than in NW subjects both at fasting and along the postprandial period (Lipinflox clinical trial), and was significantly and positively associated with trunk lean mass, lean mass and waist circumference before and after OF. Further studies are now necessary to better understand the relative role of LBP and sCD14 during the postprandial phase after diets containing different amounts and types of fats in different food matrixes.

Author Contributions: F.L., C.V., M.-C.M. did the conception and design of the research; F.L., C.V., M.A., G.P., J.D., B.M., S.L.-P., M.L., H.V. and M.-C.M. performed the experiments. F.L., C.V., C.K., M.-C.M. interpreted the results of the experiments; F.L., M.-C.M. prepared the figures and drafted the manuscript. F.L., C.V., M.-C.M. edited and revised the manuscript. All authors have read and agree to the published version of the manuscript.

Funding: This work was funded by the French National Research Agency (ANR), projects FLORINFLAM ANR-07-PNRA-0007 and METAPROFILE ANR-06-PNRA-0007. F.L. received a postdoctoral grant from INRA and a research grant from ALFEDIAM-SFD. M.-C.M. received financial support from INRA. C.V. received a PhD grant from INRA and CNIEL (French Dairy Interbranch Organization). The Lipinflox study was funded by CNIEL. The funding agencies had no role in the data analysis, interpretation and publication.

Acknowledgments: We gratefully thank all the volunteers for their involvement and the clinical teams of the CRNH-Rhône-Alpes and CRNH-Auvergne for their help in subjects' recruitment and metabolic explorations: J. Peyrat, C. Maitrepierre, N. Torche, M. Sothier, N. Feugier.

Conflicts of Interest: M.-C.M. received other research fundings on other topics from Sodiaal-Candia R&D, the Centre National Interprofessionnel de l'Economie Laitière (CNIEL) and Nutricia Research and has consultancy activities for food & dairy companies. These activities had no link with the present study. Other authors have no conflict of interest to disclose.

Abbreviations

ALT, alanine amino transferase; AST, aspartame amino transferase; LBP, lipopolysaccharides-binding protein; LPS, lipopolysaccharides; NW, normal weight; OF, overfeeding; sCD14, soluble cluster of differentiation; WAT, white adipose tissue; WC, waist circumference.

Appendix A. Postprandial Kinetics of LBP/sCD14 after 10 g vs. 40 g Fat Load. Data Are Means ± SEM

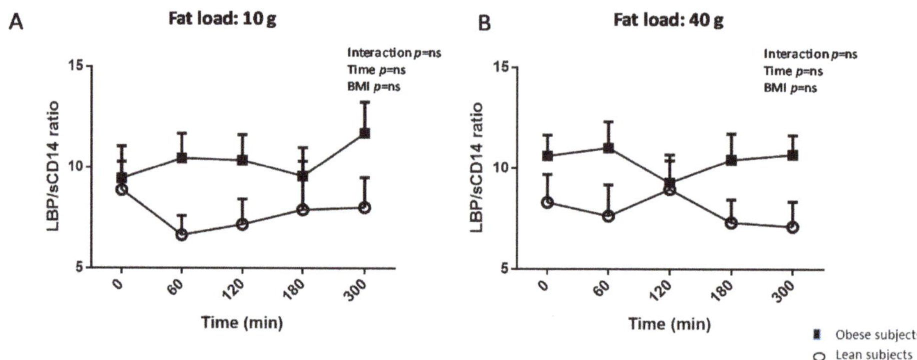

References

1. Hotamisligil, G.S. Inflammation and metabolic disorders. *Nature* **2006**, *444*, 860–867. [CrossRef] [PubMed]
2. Erridge, C.; Attina, T.; Spickett, C.M.; Webb, D.J. A high-fat meal induces low-grade endotoxemia: Evidence of a novel mechanism of postprandial inflammation. *Am. J. Clin. Nutr.* **2007**, *86*, 1286–1292. [CrossRef] [PubMed]
3. Ghoshal, S.; Witta, J.; Zhong, J.; de Villiers, W.; Eckhardt, E. Chylomicrons promote intestinal absorption of lipopolysaccharides. *J. Lipid Res.* **2009**, *50*, 90–97. [CrossRef] [PubMed]
4. Laugerette, F.; Vors, C.; Geloen, A.; Chauvin, M.A.; Soulage, C.; Lambert-Porcheron, S.; Peretti, N.; Alligier, M.; Burcelin, R.; Laville, M.; et al. Emulsified lipids increase endotoxemia: Possible role in early postprandial low-grade inflammation. *J. Nutr. Biochem.* **2011**, *22*, 53–59. [CrossRef] [PubMed]
5. Herieka, M.; Erridge, C. High-fat meal induced postprandial inflammation. *Mol. Nutr. Food Res.* **2014**, *58*, 136–146. [CrossRef]
6. Clemente-Postigo, M.; Queipo-Ortuno, M.I.; Murri, M.; Boto-Ordonez, M.; Perez-Martinez, P.; Andres-Lacueva, C.; Cardona, F.; Tinahones, F.J. Endotoxin increase after fat overload is related to postprandial hypertriglyceridemia in morbidly obese patients. *J. Lipid Res.* **2012**, *53*, 973–978. [CrossRef]
7. Hailman, E.; Lichenstein, H.S.; Wurfel, M.M.; Miller, D.S.; Johnson, D.A.; Kelley, M.; Busse, L.A.; Zukowski, M.M.; Wright, S.D. Lipopolysaccharide (LPS)-binding protein accelerates the binding of LPS to CD14. *J. Exp. Med.* **1994**, *179*, 269–277. [CrossRef]
8. Stoll, L.L.; Denning, G.M.; Weintraub, N.L. Potential role of endotoxin as a proinflammatory mediator of atherosclerosis. *Arterioscler. Thromb. Vasc. Biol.* **2004**, *24*, 2227–2236. [CrossRef]
9. Sun, L.; Yu, Z.; Ye, X.; Zou, S.; Li, H.; Yu, D.; Wu, H.; Chen, Y.; Dore, J.; Clement, K.; et al. A marker of endotoxemia is associated with obesity and related metabolic disorders in apparently healthy Chinese. *Diabetes Care* **2010**, *33*, 1925–1932. [CrossRef]
10. Gonzalez-Quintela, A.; Alonso, M.; Campos, J.; Vizcaino, L.; Loidi, L.; Gude, F. Determinants of serum concentrations of lipopolysaccharide-binding protein (LBP) in the adult population: The role of obesity. *PLoS ONE* **2013**, *8*, e54600. [CrossRef]
11. Lepper, P.M.; Schumann, C.; Triantafilou, K.; Rasche, F.M.; Schuster, T.; Frank, H.; Schneider, E.M.; Triantafilou, M.; von Eynatten, M. Association of lipopolysaccharide-binding protein and coronary artery disease in men. *J. Am. Coll. Cardiol.* **2007**, *50*, 25–31. [CrossRef] [PubMed]
12. Moreno-Navarrete, J.M.; Escote, X.; Ortega, F.; Camps, M.; Ricart, W.; Zorzano, A.; Vendrell, J.; Vidal-Puig, A.; Fernandez-Real, J.M. Lipopolysaccharide binding protein is an adipokine involved in the resilience of the mouse adipocyte to inflammation. *Diabetologia* **2015**, *58*, 2424–2434. [CrossRef] [PubMed]
13. Lakatos, P.L.; Kiss, L.S.; Palatka, K.; Altorjay, I.; Antal-Szalmas, P.; Palyu, E.; Udvardy, M.; Molnar, T.; Farkas, K.; Veres, G.; et al. Serum lipopolysaccharide-binding protein and soluble CD14 are markers of disease activity in patients with Crohn's disease. *Inflam. Bowel Dis.* **2011**, *17*, 767–777. [CrossRef] [PubMed]

14. Laugerette, F.; Alligier, M.; Bastard, J.P.; Drai, J.; Chanseaume, E.; Lambert-Porcheron, S.; Laville, M.; Morio, B.; Vidal, H.; Michalski, M.C. Overfeeding increases postprandial endotoxemia in men: Inflammatory outcome may depend on LPS transporters LBP and sCD14. *Mol. Nutr. Food Res.* **2014**, *58*, 1513–1518. [CrossRef] [PubMed]

15. Ruiz, A.G.; Casafont, F.; Crespo, J.; Cayon, A.; Mayorga, M.; Estebanez, A.; Fernadez-Escalante, J.C.; Pons-Romero, F. Lipopolysaccharide-binding protein plasma levels and liver TNF-alpha gene expression in obese patients: Evidence for the potential role of endotoxin in the pathogenesis of non-alcoholic steatohepatitis. *Obes. Surg.* **2007**, *17*, 1374–1380. [CrossRef]

16. Naghizadeh, M.; Baradaran, B.; Saghafi-Asl, M.; Amiri, P.; Shanehbandi, D.; Karamzad, N.; Mohamed-Khosroshahi, L. Toll-like receptor signaling and serum levels of interferon beta and lipopolysaccharide binding protein are related to abdominal obesity: A case-control study between metabolically healthy and metabolically unhealthy obese individuals. *Nutr. Res.* **2018**, *55*, 11–20. [CrossRef]

17. Laugerette, F.; Furet, J.P.; Debard, C.; Daira, P.; Loizon, E.; Geloen, A.; Soulage, C.O.; Simonet, C.; Lefils-Lacourtablaise, J.; Bernoud-Hubac, N.; et al. Oil composition of high-fat diet affects metabolic inflammation differently in connection with endotoxin receptors in mice. *Am. J. Physiol. Endocrinol. Metab.* **2012**, *302*, E374–E386. [CrossRef]

18. Sakura, T.; Morioka, T.; Shioi, A.; Kakutani, Y.; Miki, Y.; Yamazaki, Y.; Motoyama, K.; Mori, K.; Fukumoto, S.; Shoji, T.; et al. Lipopolysaccharide-binding protein is associated with arterial stiffness in patients with type 2 diabetes: A cross-sectional study. *Cardiovasc. Diabetol.* **2017**, *16*, 62. [CrossRef]

19. Serrano, M.; Moreno-Navarrete, J.M.; Puig, J.; Moreno, M.; Guerra, E.; Ortega, F.; Xifra, G.; Ricart, W.; Fernandez-Real, J.M. Serum lipopolysaccharide-binding protein as a marker of atherosclerosis. *Atherosclerosis* **2013**, *230*, 223–227. [CrossRef]

20. Kheirandish-Gozal, L.; Peris, E.; Wang, Y.; Tamae Kakazu, M.; Khalyfa, A.; Carreras, A.; Gozal, D. Lipopolysaccharide-binding protein plasma levels in children: Effects of obstructive sleep apnea and obesity. *J. Clin. Endocrinol. Metab.* **2014**, *99*, 656–663. [CrossRef]

21. Kong, Y.; Li, Z.; Tang, T.; Wu, H.; Liu, J.; Gu, L.; Zhao, T.; Huang, Q. The level of lipopolysaccharide-binding protein is elevated in adult patients with obstructive sleep apnea. *BMC Pulm. Med.* **2018**, *18*, 90. [CrossRef] [PubMed]

22. de Buhr, M.F.; Hedrich, H.J.; Westendorf, A.M.; Obermeier, F.; Hofmann, C.; Zschemisch, N.H.; Buer, J.; Bumann, D.; Goyert, S.M.; Bleich, A. Analysis of Cd14 as a genetic modifier of experimental inflammatory bowel disease (IBD) in mice. *Inflam. Bowel Dis.* **2009**, *15*, 1824–1836. [CrossRef] [PubMed]

23. Vors, C.; Pineau, G.; Drai, J.; Meugnier, E.; Pesenti, S.; Laville, M.; Laugerette, F.; Malpuech-Brugere, C.; Vidal, H.; Michalski, M.C. Postprandial Endotoxemia Linked With Chylomicrons and Lipopolysaccharides Handling in Obese Versus Lean Men: A Lipid Dose-Effect Trial. *J. Clin. Endocrinol. Metab.* **2015**, *100*, 3427–3435. [CrossRef] [PubMed]

24. Pei, R.; DiMarco, D.M.; Putt, K.K.; Martin, D.A.; Chitchumroonchokchai, C.; Bruno, R.S.; Bolling, B.W. Premeal Low-Fat Yogurt Consumption Reduces Postprandial Inflammation and Markers of Endotoxin Exposure in Healthy Premenopausal Women in a Randomized Controlled Trial. *J. Nutr.* **2018**, *148*, 910–916. [CrossRef]

25. Alligier, M.; Meugnier, E.; Debard, C.; Lambert-Porcheron, S.; Chanseaume, E.; Sothier, M.; Loizon, E.; Hssain, A.A.; Brozek, J.; Scoazec, J.Y.; et al. Subcutaneous adipose tissue remodeling during the initial phase of weight gain induced by overfeeding in humans. *J. Clin. Endocrinol. Metab.* **2012**, *97*, E183–E192. [CrossRef] [PubMed]

26. Vors, C.; Drai, J.; Gabert, L.; Pineau, G.; Laville, M.; Vidal, H.; Guichard, E.; Michalski, M.C.; Feron, G. Salivary composition in obese vs normal-weight subjects: Towards a role in postprandial lipid metabolism? *Int. J. Obes. (Lond.)* **2015**, *39*, 1425–1428. [CrossRef] [PubMed]

27. Alligier, M.; Gabert, L.; Meugnier, E.; Lambert-Porcheron, S.; Chanseaume, E.; Pilleul, F.; Debard, C.; Sauvinet, V.; Morio, B.; Vidal-Puig, A.; et al. Visceral fat accumulation during lipid overfeeding is related to subcutaneous adipose tissue characteristics in healthy men. *J. Clin. Endocrinol. Metab.* **2013**, *98*, 802–810. [CrossRef] [PubMed]

28. Amar, J.; Burcelin, R.; Ruidavets, J.B.; Cani, P.D.; Fauvel, J.; Alessi, M.C.; Chamontin, B.; Ferrieres, J. Energy intake is associated with endotoxemia in apparently healthy men. *Am. J. Clin. Nutr.* **2008**, *87*, 1219–1223. [CrossRef]

29. Zhang, J.; Wang, G.; Liu, J.; Gao, L.R.; Liu, M.; Wang, C.J.; Chuai, M.; Bao, Y.; Li, G.; Li, R.M.; et al. Gut microbiota-derived endotoxin enhanced the incidence of cardia bifida during cardiogenesis. *J. Cell. Physiol.* **2018**, *233*, 9271–9283. [CrossRef]

30. Fernandez-Real, J.M.; Perez del Pulgar, S.; Luche, E.; Moreno-Navarrete, J.M.; Waget, A.; Serino, M.; Sorianello, E.; Sanchez-Pla, A.; Pontaque, F.C.; Vendrell, J.; et al. CD14 modulates inflammation-driven insulin resistance. *Diabetes* **2011**, *60*, 2179–2186. [CrossRef]

31. Aguilar-Salinas, C.A.; Garcia, E.G.; Robles, L.; Riano, D.; Ruiz-Gomez, D.G.; Garcia-Ulloa, A.C.; Melgarejo, M.A.; Zamora, M.; Guillen-Pineda, L.E.; Mehta, R.; et al. High adiponectin concentrations are associated with the metabolically healthy obese phenotype. *J. Clin. Endocrinol. Metab.* **2008**, *93*, 4075–4079. [CrossRef]

32. Yang, P.J.; Lee, W.J.; Tseng, P.H.; Lee, P.H.; Lin, M.T.; Yang, W.S. Bariatric surgery decreased the serum level of an endotoxin-associated marker: Lipopolysaccharide-binding protein. *Surg. Obes. Relat. Dis. Off. J. Am. Soc. Bariatr. Surg.* **2014**, *10*, 1182–1187. [CrossRef] [PubMed]

33. Lepper, P.M.; Kleber, M.E.; Grammer, T.B.; Hoffmann, K.; Dietz, S.; Winkelmann, B.R.; Boehm, B.O.; Marz, W. Lipopolysaccharide-binding protein (LBP) is associated with total and cardiovascular mortality in individuals with or without stable coronary artery disease—Results from the Ludwigshafen Risk and Cardiovascular Health Study (LURIC). *Atherosclerosis* **2011**, *219*, 291–297. [CrossRef] [PubMed]

34. Lim, P.S.; Chang, Y.K.; Wu, T.K. Serum Lipopolysaccharide-Binding Protein is Associated with Chronic Inflammation and Metabolic Syndrome in Hemodialysis Patients. *Blood Purif.* **2019**, *47*, 28–36. [CrossRef] [PubMed]

35. Nien, H.C.; Hsu, S.J.; Su, T.H.; Yang, P.J.; Sheu, J.C.; Wang, J.T.; Chow, L.P.; Chen, C.L.; Kao, J.H.; Yang, W.S. High Serum Lipopolysaccharide-Binding Protein Level in Chronic Hepatitis C Viral Infection Is Reduced by Anti-Viral Treatments. *PLoS ONE* **2017**, *12*, e0170028. [CrossRef] [PubMed]

36. Okada, N.; Sanada, Y.; Urahashi, T.; Ihara, Y.; Yamada, N.; Hirata, Y.; Katano, T.; Otomo, S.; Ushijima, K.; Mizuta, K. Endotoxin Metabolism Reflects Hepatic Functional Reserve in End-Stage Liver Disease. *Transplant. Proc.* **2018**, *50*, 1360–1364. [CrossRef]

37. Umoh, F.I.; Kato, I.; Ren, J.; Wachowiak, P.L.; Ruffin, M.T.; Turgeon, D.K.; Sen, A.; Brenner, D.E.; Djuric, Z. Markers of systemic exposures to products of intestinal bacteria in a dietary intervention study. *Eur. J. Nutr.* **2016**, *55*, 793–798. [CrossRef]

38. Chen, M.; Liu, B.; Thompson, C.H.; Wittert, G.A.; Heilbronn, L.K. Acute Overfeeding Does Not Alter Liver or Adipose Tissue-Derived Cytokines in Healthy Humans. *Ann. Nutr. Metab.* **2016**, *69*, 165–170. [CrossRef]

39. Tam, C.S.; Viardot, A.; Clement, K.; Tordjman, J.; Tonks, K.; Greenfield, J.R.; Campbell, L.V.; Samocha-Bonet, D.; Heilbronn, L.K. Short-term overfeeding may induce peripheral insulin resistance without altering subcutaneous adipose tissue macrophages in humans. *Diabetes* **2010**, *59*, 2164–2170. [CrossRef]

40. Dixon, N.C.; Hurst, T.L.; Talbot, D.C.; Tyrrell, R.M.; Thompson, D. Active middle-aged men have lower fasting inflammatory markers but the postprandial inflammatory response is minimal and unaffected by physical activity status. *J. Appl. Physiol.* **2009**, *107*, 63–68. [CrossRef]

41. Thompson, D.; Dixon, N. Measurement of postprandial interleukin-6 via a catheter: What does it tell us? *Eur. J. Appl. Physiol.* **2009**, *107*, 621–622. [CrossRef] [PubMed]

Article

Asymmetric and Symmetric Protein Arginine Dimethylation: Concept and Postprandial Effects of High-Fat Protein Meals in Healthy Overweight Men

Alexander Bollenbach [1], Jean-François Huneau [2], François Mariotti [2,†] and Dimitrios Tsikas [1,*,†]

[1] Institute of Toxicology, Core Unit Proteomics, Hannover Medical School, 30623 Hannover, Germany
[2] UMR PNCA, AgroParisTech, INRA, Université Paris-Saclay, 75005 Paris, France
* Correspondence: tsikas.dimitros@mh-hannover.de; Tel.: +49-511-532-3984
† The two authors contributed equally to the manuscript.

Received: 20 May 2019; Accepted: 24 June 2019; Published: 27 June 2019

Abstract: Asymmetric and symmetric dimethylarginine (ADMA and SDMA, respectively) are risk factors for the cardiovascular and renal systems. There is a paucity of data in humans regarding variations of protein L-arginine (Arg) methylation leading to ADMA and SDMA. In this study, we introduced and used Arg dimethylation indices based on the creatinine-corrected urinary excretion of SDMA and ADMA, and its major metabolite dimethylamine (DMA). The main objective of the present study was to assess whether, and to which extent, a high-fat protein meal (HFM), a classical allostatic load eliciting various adverse effects, may contribute to Arg dimethylation in proteins in humans. Reliable gas chromatography–mass spectrometry methods were used to measure the concentration of ADMA, DMA, SDMA, and creatinine in spot urine samples collected before (0 h), and after (2, 4, 6 h) three HFM sessions in 10 healthy overweight individuals. At baseline, urinary ADMA, DMA, and SDMA excretion correlated positively with circulating TNF-α and IL-6. Arg dimethylation indices did not change postprandially. Our study shows that three HFMs do not contribute to Arg dimethylation in proteins. The proposed indices should be useful to determine extent and status of the whole-body Arg dimethylation in proteins in humans under various conditions.

Keywords: ADMA; arginine; SDMA; DMA; PRMT

1. Introduction

L-Arginine (Arg) is a nutritionally semi-essential proteinogenic amino acid. It is involved in many pathways and numerous physiological processes [1,2]. Arg is the substrate of all known nitric oxide synthase (NOS; EC 1.14.13.39) isoforms, which are present in virtually all cell types and oxidize the guanidine (N^G) imine group of free Arg to nitric oxide (NO) via an N^G-hydroxy-L-arginine intermediate. NO is one of the most potent endogenous vasodilators and inhibitors of platelet aggregation, and has many other biological functions [3]. The NOS-catalyzed conversion of Arg to NO and L-citrulline (Cit) is inhibited by three endogenous Arg derivates: L-N^G-monomethylarginine (MMA), L-N^G,N^G-dimethylarginine (asymmetric dimethylarginine, ADMA), and L-N^G,$N^{'G}$-dimethylarginine (symmetric dimethylarginine, SDMA) [4,5] (Figure 1). MMA, ADMA, and SDMA inhibit the activity of the three NOS isoforms by distinctly different mechanisms and inhibitory potency [6,7]. Like their parent molecule Arg, MMA, ADMA, and SDMA exist in two forms: as residues of certain proteins, and as free acids produced by regular proteolysis of those N^G-methylated proteins. The free guanidine group of Arg moieties in proteins undergo posttranslational methylation, which is catalyzed by the family of the protein arginine methyltransferases (PRMT; EC 2.1.1.125) [8–10]; the methyl group for this reaction is provided by the universal cofactor *S*-adenosylmethionine (SAM) (Figure 1).

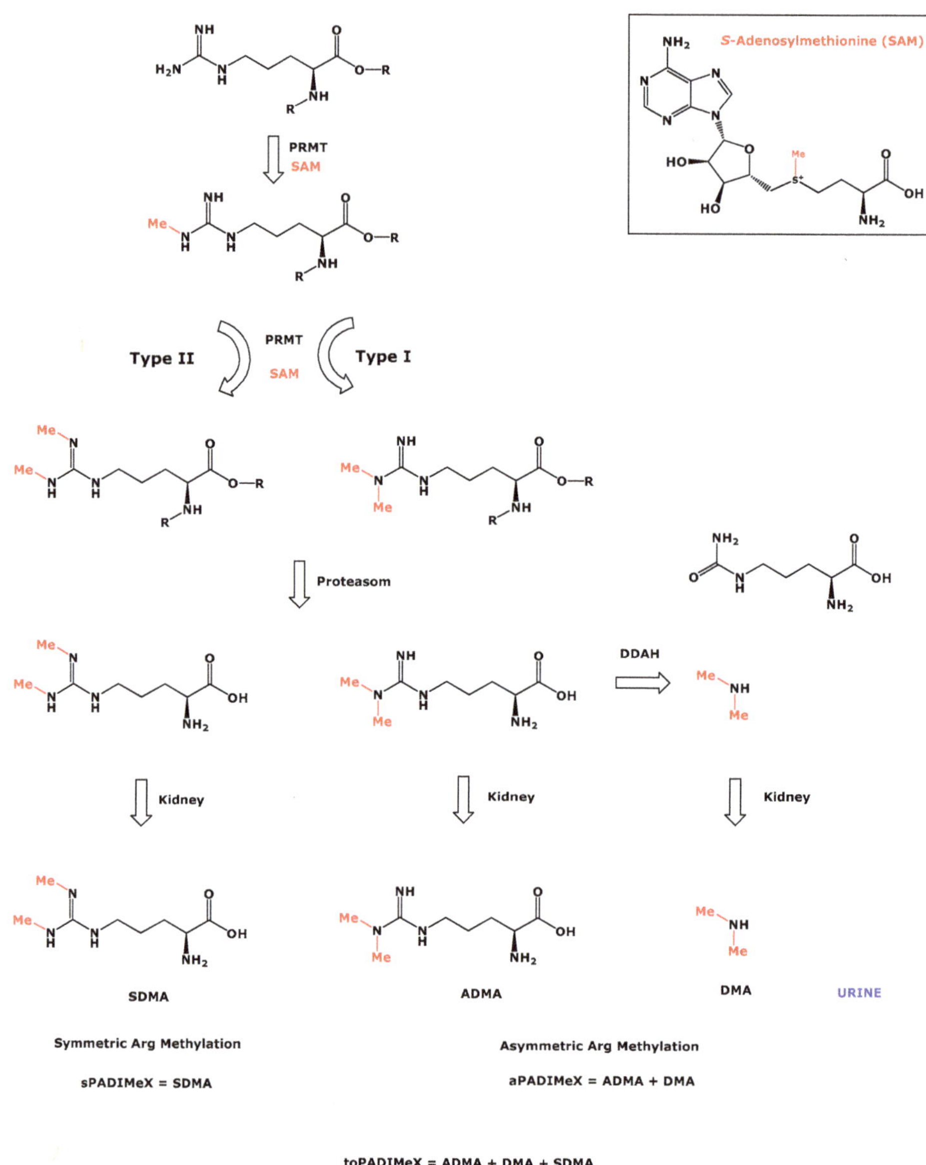

Figure 1. Simplified schematic of the asymmetric and symmetric methylation (ADMA and SDMA respectively) of arginine residues in proteins, their proteolysis to free ADMA and SDMA, metabolism of ADMA by dimethylarginine dimethylaminohydrolase (DDAH) to dimethylamine (DMA), and their excretion in the urine. Proposal of the protein arginine dimethylation index (PADiMeX).

High circulating ADMA and SDMA concentrations are considered risk factors in the renal and cardiovascular systems [11–16]. In these systems, ADMA is thought to exert its detrimental effects by inhibiting NOS activity in the endothelium. Yet, there is an increasing indication that free and/or proteinic ADMA and SDMA exert NO-independent biological effects that have not yet been fully elucidated [6,7]. As an example, mass proteomic studies identified mono- and dimethylated Arg residues in the cardiac sodium channel, suggesting a potential role of proteinic Arg methylation in the regulation of the cardiac voltage-gated Na^+ channel, presumably via mutual Arg methylation–phosphorylation crosstalk [17–19]. It has also been discussed that methylation of arginine residues in proteins are essential for proper regeneration of skeletal muscles, presumably by regulating muscle stem cell function [9].

A major fraction (about 90%) of endogenously produced ADMA is hydrolyzed to dimethylamine (DMA) by dimethylarginine dimethylaminohydrolase (DDAH; EC 3.5.3.18); only a minor fraction of about 10% of the daily produced ADMA is excreted, unchanged, in the urine [4,20]. Unlike MMA and ADMA, SDMA is not hydrolyzed by DDAH and is excreted almost unchanged in the urine [21]. The urinary concentrations of ADMA, DMA, and SDMA can be considered markers of whole-body asymmetric (ADMA + DMA), symmetric (SDMA), and total (ADMA + DMA + SDMA) dimethylation of Arg residues in proteins. Therefore, this could be a practical way to assess whole-body Arg dimethylation, and its variations in vivo. Previously, we and others have used Arg methylation indices for symmetric [22] and asymmetric Arg dimethylation, including the DMA/ADMA molar ratio in urine of healthy and diseased children and adults [23,24]. In the present work, we further build on the rationale of indices for assessing protein Arg dimethylation by defining and using the following terms: aPADiMeX for asymmetric dimethylation; sPADiMeX for symmetric dimethylation; toPADiMeX (ADMA + DMA + SDMA) for total Arg dimethylation; and a/sPADiMeX for the molar ratio of asymmetric-to-symmetric Arg dimethylation, i.e., (ADMA + DMA)/SDMA (Figure 1). In 24 h-collected urine samples, the amounts excreted within a day can be used in these terms. In urine samples collected by spontaneous micturition, the creatinine-corrected concentrations of ADMA, DMA, and SDMA are considered.

MMA, SDMA, ADMA, and DMA are natural compounds, and the ingestion of vegetable and meaty food may contribute to endogenously produced SDMA, ADMA, and most notably to DMA [25–27]. Application of the above described protein Arg dimethylation concept in health and disease requires taking proper measures to minimize exogenous contributors to urinary SDMA, ADMA, and DMA. Previously, we found that dietary fat ingestion increased the plasma concentration of ADMA marginally (by 6%) in lean and obese healthy subjects [28]. In 10 overweight men, we previously found that high-fat protein meals (HFMs) acutely increased plasma ADMA concentrations [29]. Yet, in those studies, we did not measure ADMA, DMA, and SDMA in urine samples.

The aim of the present study was to apply the above proposed indices of Arg methylation to a human study, and to test the hypothesis that dimethylarginine methylation in proteins would increase after a HFM. We measured, by fully validated and previously reported gas chromatography–mass spectrometry (GC–MS) methods, the concentration of ADMA, DMA, SDMA, and creatinine in spot urine samples collected in previous study [30] before, during, and after HFM meals consumed on three occasions by the 10 healthy overweight volunteers.

2. Materials and Methods

2.1. Ingestion of High-Fat Protein Meals by Healthy Overweight Men

The urine samples analyzed in the present study had been collected in a previous study, reported by us in detail [30]. The study was conducted in accordance with the Declaration of Helsinki, approved by the Ethics Committee of Saint-Germain-en-Laye Hospital (Reference #08001), and authorized by the French Ministry for Health (Reference 2007-A01296-47). All participants gave their written informed consent prior to enrolment. The study recruited eleven healthy overweight (body max index (BMI)> 25 kg/m^2) men aged 21–50 years, with enlarged waist circumference (>94 cm), and without any established illnesses. This sample size was set taking into account 10% attrition, and considering that 10 individuals were necessary to detect medium effect size (Cohen's d = 0.5) of the treatment on the primary outcome of the clinical trial (postprandial endothelial dysfunction). Volunteers had no regular use of medication or nutritional supplements, were not heavy smokers or alcohol drinkers, and had no moderate/high level of physical activity. They had blood hemoglobin >130 g/L, and no hypertension. The volunteers had the following characteristics: age, 34 ± 9 years; height, 178 ± 3 cm; weight, 96 ± 6 kg; BMI, 30.2 ± 1.5; body fat, 24.3 ± 2.0%; waist circumference, 96 ± 3 cm. Three HFMs of the same nutritional composition, but differing in the protein source, were tested in a randomized crossover design. Each period consisted of a postprandial study separated by at least two weeks. The test meals consisted of a mixture of 233 g cream containing 40% fat, 45 g sucrose, 45 g protein as protein isolates, and 160 mL water. The composition of the meals was as follows: energy, 1200 kcal; fat, 93 g (70% energy); carbohydrates, 45 g (15% energy); crude protein, 45 g (15% energy). After the overnight fasting (9–12 h), the subjects ingested the meal, and spot urine samples were collected before the meal (0 h, T0) and 2 h (T2), 4 h (T4), and 6 h (T6) after the meal. One subject chose to withdraw from the study during the first session because he felt nauseated after the meal. The urine samples collected during the meal were also analyzed and considered in statistics.

2.2. Measurement of Urinary ADMA, DMA, SDMA, Creatinine, and Quality Control

Creatinine, ADMA, DMA, and SDMA were measured by previously reported fully validated methods based on gas chromatography–mass spectrometry (GC–MS) methods [22,24,31,32]. Urine donated by a healthy volunteer served as a quality control (QC) sample, and was analyzed alongside the study samples within 8 runs. The following analyte concentrations were measured in the QC samples (mean ± SD): 10.7 ± 0.02 mmol/L (RSD, 2%) for creatinine, 21.3 ± 0.4 µmol/L (RSD, 2.0%) for ADMA, 36.2 ± 1.27 µmol/L (RSD, 2%) for SDMA, and 243 ± 18 µmol/L (RSD, 7.4%) for DMA. These results underline the reliability of the GC–MS methods in the measurements of the study samples.

2.3. Measurement of Inflammation and Cardiovascular Biomarkers

The biochemical parameters apolipoprotein B 48 (apoB48), monocytes chemoattractant protein-1 (MCP-1), myeloperoxidase (MPO), non-esterified fatty acids (NEFA), reflexion index (RI), soluble intracellular adhesion molecule-1 (sICAM-1), soluble vascular cell adhesion molecule-1 (sVCAM-1), triacylglyceride (TAG), tumor necrosis factor-alpha (TNF-α), tissue plasminogen activator inhibitor-1 (tPAI-1), and the physiological parameters reflexion index (RI) from pulse wave analysis were measured as described elsewhere [30].

2.4. Statistical Analyses

Statistical analyses were performed, and graphs were constructed using Origin 7.5G, GraphPad Prism 7 (GraphPad Prism Software Inc. San Diego, CA, USA). Distribution of variables was tested by D'Agostino and Pearson omnibus K2 test. Normally distributed parameters are presented as mean ± SD or mean ± SEM. Non-normally distributed parameters are presented as median and interquartile range (25th–75th percentile). Correlations between variables were assessed by Pearson (parametric) or Spearman (non-parametric) statistical tests. Repeated measures one-way ANOVA with Tukey's multiple comparisons test was used to test the effect of postprandial time on the urinary parameters. p-values < 0.05 were considered as statistically significant.

3. Results

Considering all volunteers, meals, and time points (n = 124 values), the urinary analyte concentrations were 40.2 (22–57) µmol/L for ADMA, 55.2 (31–76) µmol/L for SDMA, 348 (180–536) µmol/L for DMA, and 11.9 (6.7–19.6) mmol/L for creatinine. The creatinine-corrected excretion rates (µmol/mmol) were 3.39 (2.50–4.15) for ADMA, 4.48 (3.33–5.78) for SDMA, and 29 (21.7–37.0) for DMA. The other values were 32.8 (24.6–41.1) µmol/mmol for aPADiMeX, 37.9 (28.5–47.0) µmol/mmol for toPADiMeX, and 7.26 (6.27–8.20) for a/sPADiMeX. The concentrations of ADMA, SDMA, DMA, and creatinine correlated strongly with each other (Table 1a). The creatinine-corrected concentrations of ADMA, SDMA, and DMA also correlated with each other (Table 1b).

Table 1. Spearman correlation coefficients between the concentrations (**a**) and creatinine-corrected excretion rates (**b**) of the analytes in the urine samples for all volunteers, meals, and time points (n = 124 values). toPADiMeX is ADMA + DMA + SDMA for total Arg dimethylation.

Table 1a	ADMA	SDMA	DMA
SDMA (µmol/L)	0.939, $p < 0.0001$		
DMA (µmol/L)	0.900, $p < 0.0001$	0.923, $p < 0.0001$	
Creatinine (mmol/L)	0.842, $p < 0.0001$	0.886, $p < 0.0001$	0.883, $p < 0.0001$
Table 1b	**ADMA**	**SDMA**	**DMA**
SDMA (µmol/mmol)	0.811, $p < 0.0001$		
DMA (µmol/mmol)	0.652, $p < 0.0001$	0.765, $p < 0.0001$	
toPADiMeX (µmol/mmol)	0.755, $p < 0.0001$	0.842, $p < 0.0001$	0.981, $p < 0.0001$

At baseline, many of the plasma clinical chemistry biochemical parameters correlated moderately-to-strongly with each other (Table 2). All of the found statistically significant correlations were positive, except for NEFA and tPAI-1. The strongest correlation was observed between IL-6 and TNF-α ($r = 0.847$, $p < 0.0001$).

The correlations found between plasma clinical chemistry biochemical parameters and the creatinine-uncorrected urinary concentrations of ADMA, SDMA, DMA, and toPADiMeX at baseline are summarized in Table 3. These parameters correlated with TNF-α and IL-6, with SDMA showing the strongest correlation. The BMI value of the volunteers (range 26.9–33.4 kg/m^2) was found to correlate with insulin, sICAM-1, sVCAM-1, and E-selectin (Table 3). At baseline, the urinary concentrations (µmol/L) of ADMA, SDMA, DMA, and toPADiMeX, or with a/sPADiMeX, did not correlate with the BMI values. The creatinine-corrected excretion rates of DMA ($r = -0.354$, $p = 0.051$) and toPADiMeX ($r = -0.324$, $p = 0.076$) only tended to correlate with the BMI (not shown in Table 3).

We did not observe any postprandial changes in the indices for the individual HFM (data not shown), and so the dataset of the 3 individual meals ($n = 10$) were collapsed to a single dataset ($n = 30$). The urinary creatinine concentration, the creatinine-corrected excretion of ADMA, DMA, and SDMA, and their indices aPADiMeX, toPADiMeX, and a/sPADiMeX are summarized in Table 4 for all three meals. Statistically significant overall time effects (ANOVA) were obtained for creatinine ($p = 0.0021$) and DMA ($p = 0.019$). Statistically significant time effects were obtained for creatinine (T0 vs. T4, $p = 0.0007$; T4 vs. T6 $p = 0.0011$), ADMA (T2 vs. T6, $p = 0.0006$), DMA (T0 vs. T4, $p = 0.0027$), SDMA (T4 vs. T6, $p = 0.0499$), and aPADiMeX (T0 vs. T4, $p = 0.0290$).

Table 2. Spearman correlation coefficients between plasma clinical laboratory parameters at baseline.

	Glucose	TAG	apoB48	NEFA	Insulin	TNF-α	IL-6	MCP-1	sICAM-1	sVCAM-1	MPO	E-Selectin	tPAI-1	RI
TAG	0.086													
apoB48	0.034	0.785 [c]												
NEFA	−0.265	0.099	0.055											
Insulin	0.249	−0.225	−0.062	−0.312										
TNF-α	0.395 [a]	−0.082	−0.010	−0.154	0.372 [a]									
IL-6	0.489 [a]	−0.026	0.023	−0.158	0.446 [a]	0.847 [c]								
MCP-1	0.109	−0.116	0.077	−0.301	0.557 [b]	0.690 [c]	0.754 [c]							
sICAM-1	−0.235	0.076	0.121	−0.218	0.495 [a]	0.198	0.164	0.283						
sVCAM-1	−0.231	−0.048	0.090	−0.471 [a]	0.303	−0.012	−0.059	0.131	0.726 [c]					
MPO	−0.187	−0.050	−0.108	−0.164	0.073	−0.212	−0.108	0.118	−0.010	0.063				
E-Selectin	−0.242	−0.281	−0.224	−0.403 [a]	0.454 [a]	0.216	−0.028	0.225	0.630 [c]	0.598 [c]	0.203			
tPAI-1	0.126	−0.171	−0.153	−0.061	0.089	0.004	0.168	0.016	0.059	−0.073	0.039	0.168		
RI	−0.104	−0.163	−0.169	−0.467 [a]	0.197	0.055	−0.074	0.033	0.348	0.485 [a]	−0.092	0.565 [b]	0.111	
SI	−0.087	0.436 [a]	0.317	0.296	−0.022	0.017	0.058	0.118	−0.025	−0.156	0.136	−0.226	−0.673	−0.236

[a], $p < 0.05$; [b], $p < 0.001$; [c], $p < 0.0001$.

Table 3. Spearman correlation coefficients between baseline urinary concentrations (µmol/L; not corrected for creatinine) of ADMA, SDMA, DMA, toPADiMeX, and a/sPADiMeX, BMI, and plasma clinical chemistry biochemical parameters. Data are not shown for correlations among the plasma parameters (i.e., from Glucose to Basal_SI). (*), $0.05 < p < 0.1$; *, $p < 0.05$; **, $p < 0.001$; ***, $p < 0.0001$.

	ADMA	SDMA	DMA	toPADiMeX	a/sPADiMeX	BMI
BMI	0.072	0.168	0.057	0.064	−0.279	
Glucose	0.249	0.328 (*)	0.312 (*)	0.332 (*)	−0.068	−0.191
TAG	−0.143	−0.141	−0.164	−0.180	−0.125	−0.235
apoB48	−0.285	−0.260	−0.236	−0.264	−0.017	−0.076
NEFA	0.171	0.167	0.156	0.150	−0.107	−0.117
Insulin	0.197	0.240	0.160	0.189	−0.205	0.458 *
TNF-α	0.360 *	0.526 **	0.392 *	0.422 *	−0.241	0.310 (*)
IL-6	0.349 (*)	0.496 **	0.370 *	0.411	−0.298	0.200
MCP-1	0.116	0.279	0.172	0.198	−0.212	0.339 (*)
sICAM-1	0.134	0.126	−0.021	−0.013	−0.376 *	0.627 ***
sVCAM-1	−0.207	−0.208	−0.291	−0.302	−0.208	0.415 *
MPO	−0.214	−0.187	−0.217	−0.223	−0.233	0.113
E-Selectin	−0.103	−0.034	−0.150	−0.147	−0.257	0.774 ***
tPAI-1	−0.001	0.004	0.080	0.071	0.094	0.295
Basal_RI	0.066	−0.012	0.056	0.053	0.186	0.217
Basal_SI	0.005	0.008	−0.126	−0.103	−0.317 (*)	−0.211

Table 4. Urinary creatinine (mmol/L), creatinine-corrected excretion of ADMA, DMA, SDMA, aPADiMeX, toPADiMeX (μmol/mmol), and a/sPADiMeX (median (25th–75th) or mean $\pm$ SD) at the indicated time points.

Measure	T0	T2	T4	T6	ANOVA
Creatinine	10.9 (5.5–17.0)	12.0 (8.6–21.9)	17.4 ± 9.5 T4 vs. T0: $p = 0.0007$	8.9 (5.2–15.4) T6 vs. T4: $p = 0.0011$	$p = 0.0021$
ADMA	3.59 ± 1.38	3.62 ± 1.3	3.43 ± 1.23	3.17 ± 1.09 T6 vs. T2: $p = 0.0006$	$p = 0.0718$
DMA	26.9 (19.9–35.1)	27.5 (20.3–36.1)	33.7 (22.2–37.3) T4 vs. T0: $p = 0.0027$	30.8 ± 10.1	$p = 0.0190$
SDMA	4.45 (3.48–5.59)	4.78 (3.1–6.3)	4.52 (3.35–5.86)	4.05 (3.22–5.64) T6 vs. T2: $p = 0.0499$	$p = 0.2687$
aPADiMeX	29.9 (24.3–39)	30.7 (22.2–40.4)	37.4 (24.8–41.9) T4 vs. T0: $p = 0.0290$	33.9 ± 10.9	$p = 0.0677$
toPADiMeX	33.5 (29.9–44.0)	33.8 (28.2–45.4)	37.6 (26.1–43.0)	38.3 ± 11.4	$p = 0.2980$
a/sPADiMeX	7.21 (5.25–8.63)	7.57 ± 3.15	7.90 ± 2.81	8.46 ± 3.19	$p = 0.2213$

4. Discussion

In blood and urine, MMA is present at much lower concentrations than ADMA and SDMA. For instance, in healthy humans, the mean creatinine-corrected excretion rate of MMA was reported to be 0.017 µmol/mmol, and the ratio of the mean clearance rates of MMA, ADMA, and SDMA were reported to be 1:69:71 [33]. This may be an indicator that MMA, the first PRMT-catalyzed product of Arg-methylation in proteins, is immediately methylated to form ADMA and SDMA proteins. Other ADMA and SDMA metabolites from N^{α}-acetylation and N^{α}-oxidation pathways occur in urine, yet at much lower concentrations than ADMA and SDMA, such as 0.013 µmol/mmol creatinine, and in the range 0.011–1.03 µmol/mmol creatinine for the ADMA metabolites, respectively [34,35]. Consequently, the urinary concentrations of ADMA, DMA, and SDMA are useful for the determination of the whole-body N^{G}-dimethylation of Arg residues in proteins.

In the present work, we propose the use of the Protein Arginine Dimethylation indeX: aPADiMeX for asymmetric, sPADiMeX for symmetric, toPADiMeX for total dimethylation, and a/sPADiMeX for the asymmetric-to-symmetric (a/s) dimethylation state (Figure 1). We applied this proposal to (1) investigate potential postprandial effects of HFMs on protein Arg dimethylation; and (2) to test potential correlations of the indices with clinical chemistry laboratory biomarkers of inflammation and vascular functions, such as IL-6 and TNF-α.

The ADMA/SDMA molar ratio in our study is close to 1, and is almost identical with that reported for healthy subjects [33]. However, this ratio does not mean that asymmetric and symmetric protein Arg dimethylation rates are equal. This is because ADMA is metabolized to DMA, of which the excretion is about 10 times higher than non-metabolized ADMA in healthy adults [24,33].

The creatinine-corrected excretion rates of the ADMA, DMA, and SDMA measured in the present study at baseline, are within ranges reported by us and others for healthy and diseased adults [24,33]. In our healthy overweight men, the average baseline creatinine-corrected urinary excretion rates (µmol/mmol) were 3.59 for ADMA, 4.48 for SDMA, and 26.9 for DMA. The baseline indices were calculated to be 30 µmol/mmol for aPADiMeX, 37.1 µmol/mmol for toPADiMeX, and 6.7 for a/sPADiMeX. These data indicate that DMA is the strongest quantitative contributor to the proposed indices. The whole-body asymmetric dimethylation of proteinic Arg is about 7 times higher than the symmetric in the healthy overweight men of the present study.

In urine samples from 14 healthy non-overweight men (age, 41 ± 11 years; range, 26–60 years; BMI, 23.9 ± 3.3 kg/m^2) from previous work [32], we measured creatinine-corrected excretion rates of 29.6 ± 3.9 µmol/mmol DMA, 2.74 ± 0.51 µmol/mmol ADMA, and 2.99 ± 0.44 µmol/mmol SDMA. The Arg dimethylation indices were calculated to be 35.0 ± 5.2 µmol/mmol for aPADiMeX, 38.0 ± 5.5 µmol/mmol for toPADiMeX, and 11.8 ± 1.6 for a/sPADiMeX. In urine samples from 5 healthy non-overweight women (age, 42 ± 10 years; range, 28–56 years; BMI, 26.9 ± 7.8 kg/m^2) from the same study [32], we measured creatinine-corrected excretion rates of 42.5 ± 5.8 µmol/mmol DMA, 4.2 ± 1.2 µmol/mmol ADMA, and 3.91 ± 0.6 µmol/mmol SDMA. The Arg dimethylation indices were calculated to be 46.7 ± 6.6 µmol/mmol for aPADiMeX, 50.6 ± 7.2 µmol/mmol for toPADiMeX, and 10.8 ± 0.8 for a/sPADiMeX. Statistically significant differences between men and women were found for DMA ($p = 0.01$), ADMA ($p = 0.003$), SDMA ($p = 0.022$), aPADiMeX ($p = 0.007$), and toPADiMeX ($p = 0.0046$), but not for a/sPADiMeX ($p = 0.95$), suggesting potential effects of gender on whole-body asymmetric and symmetric proteinic Arg dimethylation, yet not on their balance.

In previous studies, we found that HFM taken by the same healthy overweight men is associated with considerable postprandial changes in many circulating biochemical biomarkers, including Arg, L-homoarginine (hArg), and ADMA [29,30,36]. The present study indicates that HFM has no appreciable postprandial effects on total asymmetric and symmetric protein arginine dimethylation (toPADiMeX), and asymmetric-to-symmetric protein arginine dimethylation (a/sPADiMeX). However, we found some significant temporary changes on ADMA (decrease, between T2 and T6), DMA, SDMA, and aPADiMeX (increases, all between T0 and T4). At baseline, circulating TNF-α and IL-6 correlated with urinary creatinine-corrected SDMA excretion. Previously, we found no statistically significant changes in circulating TNF-α and IL-6 upon meal ingestion [30], suggesting that these factors may not be responsible for the observed changes in SDMA excretion. The results of the present study may suggest that the HFM themselves did not contain appreciable amounts of ADMA, DMA, and SDMA (not investigated), and did not exert appreciable effects on dimethylation of proteinic Arg. As creatinine excretion changed relatively strongly, an effect on the glomerular filtration rate (GFR) of the kidney on the excretion rates of ADMA, DMA, and SDMA cannot be excluded. In renal transplant recipients (median estimated GFR of 43.5 mL/min/1.73 m^2), the mean ADMA-to-SDMA molar ratio was found to be only 0.6 [37].

The non-invasive measurement of ADMA, DMA, and SDMA in human urine provides a relevant approach to estimate the extent of proteinic Arg dimethylation and the relative contribution of the asymmetric and symmetric dimethylation, which can be translated into the individual PRMTs. The proposed indices may provide valuable information of the status of protein arginine dimethylation in health and disease. In healthy overweight men, HFM ingestion caused temporary changes in creatinine excretion, and creatinine-corrected excretion rates of ADMA, DMA, and SDMA. However, these changes do not indicate changes in the total protein arginine dimethylation, and the balance between asymmetric and symmetric protein arginine dimethylation. Of note, asymmetric dimethylation tended to increase after HFMs.

A limitation of our study is the small number of participants. Strengths of the study are the closely controlled, repeated postprandial testing on the same volunteers, controlled meal composition, and validated methods for measuring arginine metabolites in urine. Urinary DMA is by far the greatest term of the indices proposed in this work. Further studies on larger cohorts are warranted to assess potential differences in methylation profiles of proteinic arginine in various conditions in health and disease, and to assess the effects of gender and age. In studies addressing the in vivo protein arginine dimethylation in non-closely controlled studies, subjects must abstain from ingestion of DMA-rich food, notably fish [24–27].

Author Contributions: All authors have contributed to the manuscript and have approved this final version of the work.

Funding: This research received no external funding.

Conflicts of Interest: The authors declare no conflict of interest.

References

1. Wu, G.; Bazer, F.W.; Davis, T.A.; Kim, S.W.; Li, P.; Marc Rhoads, J.; Carey Satterfield, M.; Smith, S.B.; Spencer, T.E.; Yin, Y. Arginine metabolism and nutrition in growth, health and disease. *Amino Acids* **2009**, *37*, 153–168. [CrossRef] [PubMed]
2. Wu, G. Functional amino acids in nutrition and health. *Amino Acids* **2013**, *45*, 407–411. [CrossRef] [PubMed]
3. Moncada, S.; Higgs, A. The L-arginine-nitric oxide pathway. *N. Engl. J. Med.* **1993**, *329*, 2002–2012. [CrossRef] [PubMed]
4. Leiper, J.; Vallance, P. Biological significance of endogenous methylarginines that inhibit nitric oxide synthases. *Cardiovasc. Res.* **1999**, *43*, 542–548. [CrossRef]
5. Tsikas, D.; Böger, R.H.; Sandmann, J.; Bode-Böger, S.M.; Frölich, J.C. Endogenous nitric oxide synthase inhibitors are responsible for the L-arginine paradox. *FEBS Lett.* **2000**, *478*, 1–3. [CrossRef]
6. Tsikas, D. Does the inhibitory action of asymmetric dimethylarginine (ADMA) on the endothelial nitric oxide synthase activity explain its importance in the cardiovascular system? The ADMA paradox. *J. Controversies Biomed. Res.* **2017**, *3*, 16–22. [CrossRef]
7. Tsikas, D.; Bollenbach, A.; Hanff, E.; Kayacelebi, A.A. Asymmetric dimethylarginine (ADMA), symmetric dimethylarginine (SDMA) and homoarginine (hArg): The ADMA, SDMA and hArg paradoxes. *Cardiovasc. Diabetol.* **2018**, *17*, 1. [CrossRef]
8. Blanc, R.S.; Richard, S. Arginine Methylation: The Coming of Age. *Mol. Cell* **2017**, *65*, 8–24. [CrossRef]

9.   Blanc, R.S.; Richard, S. Regenerating muscle with arginine methylation. *Transcription* **2017**, *8*, 175–178. [CrossRef]

10. Peng, C.; Wong, C.C. The story of protein arginine methylation: Characterization, regulation, and function. *Expert Rev. Proteomics* **2017**, *14*, 157–170. [CrossRef]

11. Zoccali, C.; Bode-Böger, S.; Mallamaci, F.; Benedetto, F.; Tripepi, G.; Malatino, L.; Cataliotti, A.; Bellanuova, I.; Fermo, I.; Frölich, J.; et al. Plasma concentration of asymmetrical dimethylarginine and mortality in patients with end-stage renal disease: A prospective study. *Lancet* **2001**, *358*, 2113–2117. [CrossRef]

12. Zoccali, C.; Benedetto, F.A.; Maas, R.; Mallamaci, F.; Tripepi, G.; Malatino, L.S.; Böger, R. Asymmetric dimethylarginine, C-reactive protein, and carotid intima-media thickness in end-stage renal disease. *JASN* **2002**, *13*, 490–496. [PubMed]

13. Frenay, A.R.; van den Berg, E.; de Borst, M.H.; Beckmann, B.; Tsikas, D.; Feelisch, M.; Navis, G.; Bakker, S.J.; van Goor, H. Plasma ADMA associates with all-cause mortality in renal transplant recipients. *Amino Acids* **2015**, *47*, 1941–1949. [CrossRef] [PubMed]

14. Schlesinger, S.; Sonntag, S.R.; Lieb, W.; Maas, R. Asymmetric and symmetric dimethylarginine as risk markers for total mortality and cardiovascular outcomes: A systematic review and meta-analysis of prospective studies. *PLoS ONE* **2016**, *11*, e0165811. [CrossRef] [PubMed]

15. Emrich, I.E.; Zawada, A.M.; Martens-Lobenhoffer, J.; Fliser, D.; Wagenpfeil, S.; Heine, G.H.; Bode-Böger, S.M. Symmetric dimethylarginine (SDMA) outperforms asymmetric dimethylarginine (ADMA) and other methylarginines as predictor of renal and cardiovascular outcome in non-dialysis chronic kidney disease. *Clin. Res. Cardiol.* **2018**, *107*, 201–213. [CrossRef] [PubMed]

16. Zobel, E.H.; von Scholten, B.J.; Reinhard, H.; Persson, F.; Teerlink, T.; Hansen, T.W.; Parving, H.H.; Jacobsen, P.K.; Rossing, P. Symmetric and asymmetric dimethylarginine as risk markers of cardiovascular disease, all-cause mortality and deterioration in kidney function in persons with type 2 diabetes and microalbuminuria. *Cardiovasc. Diabetol.* **2017**, *16*, 88. [CrossRef]

17. Beltran-Alvarez, P.; Pagans, S.; Brugada, R. The cardiac sodium channel is post-translationally modified by arginine methylation. *J. Proteom Res.* **2011**, *10*, 3712–3719. [CrossRef]

18. Beltran-Alvarez, P.; Tarradas, A.; Chiva, C.; Perez-Serra, A.; Batlle, M.; Perez-Villa, F.; Schulte, U.; Sabido, E.; Brugada, R.; Pagans, S. Identification of N-terminal protein acetylation and arginine methylation of the voltage-gated sodium channel in end-stage heart failure human heart. *J. Mol. Cell. Cardiol.* **2014**, *76*, 126–129. [CrossRef]

19. Beltran-Alvarez, P.; Feixas, F.; Osuna, S.; Diaz-Hernandez, R.; Brugada, R.; Pagans, S. Interplay between R513 methylation and S516 phosphorylation of the cardiac voltage-gated sodium channel. *Amino Acids* **2015**, *47*, 429–434. [CrossRef]

20. Achan, V.; Broadhead, M.; Malaki, M.; Whitley, G.; Leiper, J.; MacAllister, R.; Vallance, P. Asymmetric dimethylarginine causes hypertension and cardiac dysfunction in humans and is actively metabolized by dimethylarginine dimethylaminohydrolase. *Arterioscl. Thromb. Vascular Biol.* **2003**, *23*, 1455–1459. [CrossRef]

21. Nijveldt, R.J.; Van Leeuwen, P.A.; Van Guldener, C.; Stehouwer, C.D.; Rauwerda, J.A.; Teerlink, T. Net renal extraction of asymmetrical (ADMA) and symmetrical (SDMA) dimethylarginine in fasting humans. *Nephrology Dialysis Transplantation* **2002**, *17*, 1999–2002. [CrossRef] [PubMed]

22. Bollenbach, A.; Hanff, E.; Beckmann, B.; Kruger, R.; Tsikas, D. GC-MS quantification of urinary symmetric dimethylarginine (SDMA), a whole-body symmetric l-arginine methylation index. *Anal. Biochem.* **2018**, *556*, 40–44. [CrossRef] [PubMed]

23. Kuo, H.C.; Hsu, C.N.; Huang, C.F.; Lo, M.H.; Chien, S.J.; Tain, Y.L. Urinary arginine methylation index associated with ambulatory blood pressure abnormalities in children with chronic kidney disease. *JASH* **2012**, *6*, 385–392. [CrossRef]

24. Tsikas, D.; Thum, T.; Becker, T.; Pham, V.V.; Chobanyan, K.; Mitschke, A.; Beckmann, B.; Gutzki, F.M.; Bauersachs, J.; Stichtenoth, D.O. Accurate quantification of dimethylamine (DMA) in human urine by gas chromatography-mass spectrometry as pentafluorobenzamide derivative: Evaluation of the relationship between DMA and its precursor asymmetric dimethylarginine (ADMA) in health and disease. *J. Chromatogr. B* **2007**, *851*, 229–239. [CrossRef] [PubMed]

25. Asatoor, A.M.; Simenhoff, M.L. The origin of urinary dimethylamine. *Biochim. Biophys. Acta* **1965**, *111*, 384–392. [CrossRef]

26. Mitchell, S.C.; Zhang, A.Q.; Smith, R.L. Dimethylamine and diet. *Food Chem. Toxicol.* **2008**, *46*, 1734–1738. [CrossRef] [PubMed]

27. Servillo, L.; Giovane, A.; Cautela, D.; Castaldo, D.; Balestrieri, M.L. The methylarginines NMMA, ADMA, and SDMA are ubiquitous constituents of the main vegetables of human nutrition. *Nitric Oxide* **2013**, *30*, 43–48. [CrossRef] [PubMed]

28. Engeli, S.; Tsikas, D.; Lehmann, A.C.; Bohnke, J.; Haas, V.; Strauss, A.; Janke, J.; Gorzelniak, K.; Luft, F.C.; Jordan, J. Influence of dietary fat ingestion on asymmetrical dimethylarginine in lean and obese human subjects. *Nutr. Metab. Cardiovasc. Dis.* **2012**, *22*, 720–726. [CrossRef] [PubMed]

29. Kayacelebi, A.A.; Langen, J.; Weigt-Usinger, K.; Chobanyan-Jurgens, K.; Mariotti, F.; Schneider, J.Y.; Rothmann, S.; Frolich, J.C.; Atzler, D.; Choe, C.U.; et al. Biosynthesis of homoarginine (hArg) and asymmetric dimethylarginine (ADMA) from acutely and chronically administered free L-arginine in humans. *Amino Acids* **2015**, *47*, 1893–1908. [CrossRef] [PubMed]

30. Mariotti, F.; Valette, M.; Lopez, C.; Fouillet, H.; Famelart, M.H.; Mathe, V.; Airinei, G.; Benamouzig, R.; Gaudichon, C.; Tome, D.; et al. Casein compared with whey proteins affects the organization of dietary fat during digestion and attenuates the postprandial triglyceride response to a mixed high-fat meal in healthy, overweight men. *J. Nutr.* **2015**, *145*, 2657–2664. [CrossRef] [PubMed]

31. Tsikas, D.; Schubert, B.; Gutzki, F.M.; Sandmann, J.; Frölich, J.C. Quantitative determination of circulating and urinary asymmetric dimethylarginine (ADMA) in humans by gas chromatography-tandem mass spectrometry as methyl ester tri(N-pentafluoropropionyl) derivative. *J. Chromatogr. B* **2003**, *798*, 87–99. [CrossRef]

32. Tsikas, D.; Beckmann, B.; Gutzki, F.M.; Jordan, J. Simultaneous gas chromatography-tandem mass spectrometry quantification of symmetric and asymmetric dimethylarginine in human urine. *Anal. Biochem.* **2011**, *413*, 60–62. [CrossRef]

33. Torremans, A.; Marescau, B.; Vanholder, R.; De Smet, R.; Billiouw, J.M.; De Deyn, P.P. The low nanomolar levels of N G-monomethylarginine in serum and urine of patients with chronic renal insufficiency are not significantly different from control levels. *Amino Acids* **2003**, *24*, 375–381. [CrossRef] [PubMed]

34. Martens-Lobenhoffer, J.; Rodionov, R.N.; Drust, A.; Bode-Boger, S.M. Detection and quantification of alpha-keto-delta-(N(G),N(G)-dimethylguanidino)valeric acid: A metabolite of asymmetric dimethylarginine. *Anal. Biochem.* **2011**, *419*, 234–240. [CrossRef] [PubMed]

35. Martens-Lobenhoffer, J.; Rodionov, R.N.; Bode-Böger, S.M. Determination of asymmetric Nalpha-acetyldimethylarginine in humans: A phase II metabolite of asymmetric dimethylarginine. *Anal. Biochem.* **2014**, *452*, 25–30. [CrossRef] [PubMed]

36. Schneider, J.Y.; Rothmann, S.; Schröder, F.; Langen, J.; Lücke, T.; Mariotti, F.; Huneau, J.F.; Frölich, J.C.; Tsikas, D. Effects of chronic oral L-arginine administration on the L-arginine/NO pathway in patients with peripheral arterial occlusive disease or coronary artery disease: L-Arginine prevents renal loss of nitrite, the major NO reservoir. *Amino Acids* **2015**, *47*, 1961–1974. [CrossRef]

37. Said, M.Y.; Bollenbach, A.; Minovic, I.; van Londen, M.; Frenay, A.R.; de Borst, M.H.; van den Berg, E.; Kayacelebi, A.A.; Tsikas, D.; van Goor, H.; et al. Plasma ADMA, urinary ADMA excretion, and late mortality in renal transplant recipients. *Amino Acids* **2019**. [CrossRef] [PubMed]

Article

Peripheral Blood Mononuclear Cell Metabolism Acutely Adapted to Postprandial Transition and Mainly Reflected Metabolic Adipose Tissue Adaptations to a High-Fat Diet in Minipigs

Yuchun Zeng, Jérémie David, Didier Rémond, Dominique Dardevet, Isabelle Savary-Auzeloux and Sergio Polakof *

INRA, UNH, Unité de Nutrition Humaine, CRNH Auvergne, Université Clermont Auvergne, F-63000 Clermont-Ferrand, France; zengyuchunzlh@gmail.com (Y.Z.); jeremie.david@inra.fr (J.D.); didier.remond@inra.fr (D.R.); dominique.dardevet@inra.fr (D.D.); isabelle.savary-auzeloux@inra.fr (I.S.-A.)
* Correspondence: sergio.polakof@inra.fr; Tel.: +33-(0)-473-624-895; Fax: +33-(0)-473-624-638

Received: 29 September 2018; Accepted: 15 November 2018; Published: 21 November 2018

Abstract: Although peripheral blood mononuclear cells (PBMCs) are widely used as a valuable tool able to provide biomarkers of health and diseases, little is known about PBMC functional (biochemistry-based) metabolism, particularly following short-term nutritional challenges. In the present study, the metabolic capacity of minipig PBMCs to respond to nutritional challenges was explored at the biochemical and molecular levels. The changes observed in enzyme activities following a control test meal revealed that PBMC metabolism is highly reactive to the arrival of nutrients and hormones in the circulation. The consumption, for the first time, of a high fat–high sucrose (HFHS) meal delayed or sharply reduced most of the observed postprandial metabolic features. In a second experiment, minipigs were subjected to two-month HFHS feeding. The time-course follow-up of metabolic changes in PBMCs showed that most of the adaptations to the new diet took place during the first week. By comparing metabolic (biochemical and molecular) PMBC profiles to those of the liver, skeletal muscle, and adipose tissue, we concluded that although PBMCs conserved common features with all of them, their response to the HFHS diet was closely related to that of the adipose tissue. As a whole, our results show that PBMC metabolism, particularly during short-term (postprandial) challenges, could be used to evaluate the whole-body metabolic status of an individual. This could be particularly interesting for early diagnosis of metabolic disease installation, when fasting clinical analyses fail to diagnose the path towards the pathology.

Keywords: peripheral blood mononuclear cells; postprandial metabolism; high fat–high sugar diet; minipig; adipose tissue; biomarkers

---

1. Introduction

Peripheral blood mononuclear cells (PBMCs) have been widely used for more than 10 years now as a valuable tool to provide reliable biomarkers of health and diseases [1–3]. Although very recent studies have approached PBMC metabolism from a more functional point of view, such as metabolomics studies [4,5], most of them are based on transcriptomics analyses only [1,6,7]. Very little is actually known about the functional, i.e., biochemical-level, metabolism of PBMCs following nutrition-related interventions. Previous studies showed for specific individual white cell populations (such as lymphocytes and macrophages), these rapidly dividing cells had all the necessary biochemical machinery to utilise the most abundant energy substrates, such as glucose, lipids, and some amino acids, particularly glutamine [8]. More recent studies based on enzyme activities carried out on dogs, cats, and cattle confirm that PBMCs have an active metabolism that could be similar to that of other organs [9–11]. However, very little is known about PBMC metabolism in minipigs, as most of the studies are focused on immunological approaches. Only a few papers have explored the pig PBMC transcriptomic response to diverse conditions, such as mitogenic stimulation, [12], growth performance [13], stress [14], or leptin administration [15]. Thus, functional metabolic studies are lacking in minipigs, especially with respect to nutritionally-related metabolic diseases.

An aspect of interest in PBMCs is that they circulate permanently and through all body parts in the blood stream, being subjected to any variation of this fluid composition, including those related to fluctuations in circulating nutrients, substrates, and hormones [16]. All these molecules can then potentially impact PBMC metabolism and deeply modify their gene expression profile. On chronic exposure to a modified nutritional environment, PBMC transcriptomics have been used to discover valuable biomarkers of many metabolism-related diseases, such as diabetes [17], obesity [3], or insulin resistance [18]. Despite the current knowledge about the capacity of PBMCs to be imprinted at the long-term molecular level by altered nutritional conditions, it is not clear whether PBMC metabolism could be representative of nutritionally-induced short-term and rapid modifications in blood or tissue metabolites and metabolic activity. The postprandial phase (i.e., the period that follows meal intake) represents one of the most challenging phenomena in whole-body metabolism taking place in healthy conditions; it occurs every day, several times a day. Following meal intake, body metabolism must adapt to major changes in blood composition: the increased levels of circulating nutrients could be harmful remain elevated for a long time, and on the other hand, levels of several hormones increase for just few minutes after the nutrient arrival. As do all the other organs, PBMCs must adapt to these changes [19]. More importantly, while the contact of most of the organs with circulating nutrients and hormones will depend on vascularisation and changes in blood circulation, PBMCs are directly and permanently affected by the blood, as they are surrounded by the fluid. Changes in blood composition may be then particularly sharp for PBMCs, forcing them to rapidly adapt to these changes as suggested in transcriptomic-based postprandial studies [20,21]. For example, this has been demonstrated in studies on humans [22] and rats [16] during fasting, at least at the gene expression level. Whether this flexibility to adapt to a meal is compromised in PBMCs as has been shown for the whole body and individual organs during the onset of many metabolic diseases, such as diabetes or obesity [23,24], remains to be elucidated.

The main objective of the present study was to determine if PBMC metabolism at the biochemical and molecular levels could adapt to different nutritional conditions known to alter the profile of circulating metabolites and hormones, and to induce major changes in whole body metabolism. We submitted Yucatan minipigs to different nutritional challenges, including a meal test (a meal regular or high in fat and sugar, HFHS) and long-term HFHS feeding. In the first case, we aimed at determining whether PBMCs were able to adapt rapidly to changes occurring after meal intake and if their metabolic adaptive capacity might provide information about the capacity of the individual to handle the meal. In the second case we aimed at studying the long-term metabolic footprint of HFHS feeding on PBMC metabolism and to compare it to the metabolism of other tissues, including the liver, skeletal muscle, and adipose tissue (AT). For the postprandial challenges, the biochemical approach (enzyme activities) was chosen to explore the short-term adaptation (hours). For the long-term trial (months), both enzyme activities and mRNA levels were assessed to explore the metabolic adaptations installed in the fasting state during HFHS feeding.

2. Materials and Methods

2.1. Animals

The study involved 10 female adult (6 month-old) Yucatan mini-pigs (30 $\pm$ 1 kg). They were housed in subject pens (1 $\times$ 1.5 m) in a ventilated room with controlled temperature (21 $^\circ$C) and regular light cycle (L12:D12). They were fed once daily with 400 g/day of a concentrate feed containing 17.5% proteins, 3.2% fat, 4.3% cellulose, and 5.2% ash (Porcyprima; Sanders Centre Auvergne, Aigueperse,

France) and had free access to tap water. All procedures were in accordance with the guidelines formulated by the European Community for the use of experimental animals (L358-86/609/EEC, Council Directive, 1986).

2.2. Experimental Procedure

2.2.1. Postprandial Meal Test

The postprandial meal test trial consisted in the ingestion of 400 g of a regular (control) or a high fat-high sugar (HFHS) meal after an overnight fasting period. Five animals were involved in the trial, consuming first the control diet, and after one week of wash-out, the HFHS diet. The HFHS diet consisted of a regular pig diet enriched with fat (12% butter) and sugar (10% sucrose). Animals ingested the whole mixture in no more than 10 min. Blood was collected though an arterial catheter before (T0) and then 2, 4, and 7 h after the meal.

2.2.2. High Fat–High Sucrose Long-Term Trial

The HFHS trial consisted in the ingestion of an obesogenic diet during two months. Five animals were involved in the trial and were fed twice a day (500 g). Arterial blood was sampled through a permanent catheter after a fast of 24 h (day 0) and after 7, 14, 30, and 60 days of diet consumption also at the fasting state. Liver, skeletal muscle and AT samples were collected during the catheter surgery procedure (before the trial) and by the end of the 60 days. In both cases, minipigs were under deep anaesthesia and samples were immediately frozen at $-80\,^{\circ}\text{C}$ for further analyses.

2.3. Analytical Procedures

2.3.1. Blood Treatment and PBMC Collection

At each point, 10 mL of blood were collected on a dry syringe (without anticoagulant) and then gently transferred to a tube CPT™ tube (BD vacutainer) containing a polyester gel and a density gradient liquid (FICOLL™ Hypaque™ solution). This filter allows cell separation during a single centrifugation step. Briefly, blood was centrifuged at 1650 g during 20 min at 20 °C. After centrifugation, the plasma was collected and quickly stored at $-80\,^{\circ}\text{C}$ until further analyses. The layer containing the PBMC was collected (buffy coat), washed twice with PBS solution and centrifuged at 300 g during 15 min at 20 °C. After washing, the PBMC were pelleted, the PBS discarded and cells suspended on RNLater™ (for further mRNA analysis) or a lysis solution.

2.3.2. Plasma Insulin and Metabolite Determination

Glucose, triacylglycerol (TG), lactate, and urea levels were enzymatically measured using commercial kits on an automotive ABX Pentra 400 (Horiba Medical, Grabels, France) test system. Plasma insulin levels were assessed using a commercial ELISA kit (Mercodia, Uppsala, Sweden). Branched-chain amino acids (BCAA) were assessed enzymatically as described in Polakof et al. 2017 [25].

2.3.3. Measurements of Enzyme Activities

For enzyme activities, PBMC were immediately homogenized on the lysis solution (20 mM Tris, pH 7.4, 250 mM sucrose, 2 mM EDTA, 10 mM β-mercaptoethanol, 100 mM NaF, 0.5 mM EDTA). The homogenate was centrifuged for 20 min at $10{,}000\times g$ and the supernatant immediately used. Enzyme activities were determined by spectrophotometry using a microplate reader (Infinite® 200 PRO NanoQuant, Tecan, Grödig, Austria), based on mini-pigs methods adapted to PBMC [23,25,26], including hexokinase (HK), pyruvate kinase (PK), glucose-6-phosphate dehydrogenase (G6PD), malate dehydrogenase (MDH), lactate dehydrogenase (LDH), glutamate dehydrogenase (GDH), glutaminase

P-dependent, branched-chain amino acid aminotransferase (BCAT) and aspartate aminotransferase (AspAT).

2.3.4. Western Blot Analyses

Samples were homogenized as described above and for each sample, total protein lysates (4 μg) were subjected to SDS-PAGE, electrotransferred on a PVDF membrane and probed with the indicated antibodies: total serine/threonine kinase protein kinase B (AKT) and phospho AKT at serine 473 (pAKT) and total eEF2-α and phosphor eEF2-α at threonine 56 (peEF2-α) (Cell Signaling Technology, Ozyme, St Quentin-en-Yvelines, France). After washing, membranes were incubated with an IRDye infrared secondary antibody (LI-COR Biotechnology, Lincoln, NE, USA). Bands were visualized by infrared fluorescence using the Odyssey imaging system (LI-COR Inc. Biotechnology, Lincoln, NE, USA) and quantified by Odyssey infrared imaging system software (version 1.2).

2.3.5. PCR Analyses

Total RNA was extracted using RNEasy Mini Kit® (Qiagen) and mRNA levels were determined by RT-PCR. RNA quality was verified on 1% agarose ethidium-bromide stained gel. cDNA was generated from 500 ng RNA using the High Capacity cDNA Reverse Transcription Kit (Life Technologies, Villebon-sur-Yvette, France). Real-time PCR was performed in the CFX96 Touch™ Real-Time PCR Detection System (BIO-RAD, Hercules, CA, USA) as in [23]. Primers were designed so that they are overlapping an intron (Primer3 software; Whitehead Institute for Biomedical Research/MIT Center, Cambridge, MA, USA) using known sequences in nucleotide databases. Primers sequences are available in [26].

2.4. Statistical Analyses

Data from the postprandial challenges were analysed using a repeated measures two-way ANOVA test (time and diet as variables). The PBMC and plasma parameters from the HFHS trial were analysed using a one-way ANOVA test followed by post-hoc Holm-Sidak. The p-value significance threshold for all factors was set to 0.05.

The differences between the D0 and D60 HFHS trial in liver, skeletal muscle and AT enzyme activities and mRNA levels were analysed using a Mann–Whitney nonparametric test using for the post hoc analysis the Student–Newman–Keuls test (SigmaPlot 12, Systat Software, San Jose, CA, USA). The p-value significance threshold for all factors and ions was set to 0.05. p-values between 0.05.

3. Results

3.1. Short-Term Control vs HFHS Postprandial Trial

Glucose (insulin, glucose, lactate), lipid (TG), and nitrogen (urea, BCAA) metabolism were evaluated before and after the control and the HFHS meal based on plasma hormone and metabolite assessment in order to evaluate the postprandial metabolic phenotype of minipig followings the meal (Figure 1). Plasma glucose levels increased (between 2 and 4 h after the meal) after the control meal intake, recovering to basal level 7 h after the meal. No changes with time were observed following the HFHS meal, although plasma glucose levels were lower between 2 and 4 h than after the control meal. Plasma insulin levels showed a very important variability among individuals to put in evidence a time effect. However, 2 h after the meal, a tendency ($p < 0.081$) to have higher levels than before the meal was observed, which is in agreement with the changes observed in the plasma glucose profile. No differences were found among the animals consuming the different meals. Lactate plasma levels, a good indicator of glucose utilisation at the whole body level, increased significantly after both meal tests. However, the increase observed in the control group was more important, at least 2 h after the meal. Plasma TG levels increased also after the control and HFHS meals, but the maximum values were different among the groups. For the control group the peak was observed 2 h after the meal (+355%), while for the HFHS group it was delayed up to 4 h (+281%), most likely reflecting differences in the meal digestion and absorption. Postprandial urea plasma levels (indicative amino acid catabolism at the whole body level) showed different patterns: while in the control group levels steadily increased up to 151% at 4 h when compared to the fasting point, in the HFHS group no changes were observed. While BCAA levels after the control meal slightly increased (+12%), following the HFHS meal, their concentration was reduced up to 25% 7 h after the meal intake, suggesting their withdrawal from the blood compartment exceeded their appearance following the digestion process.

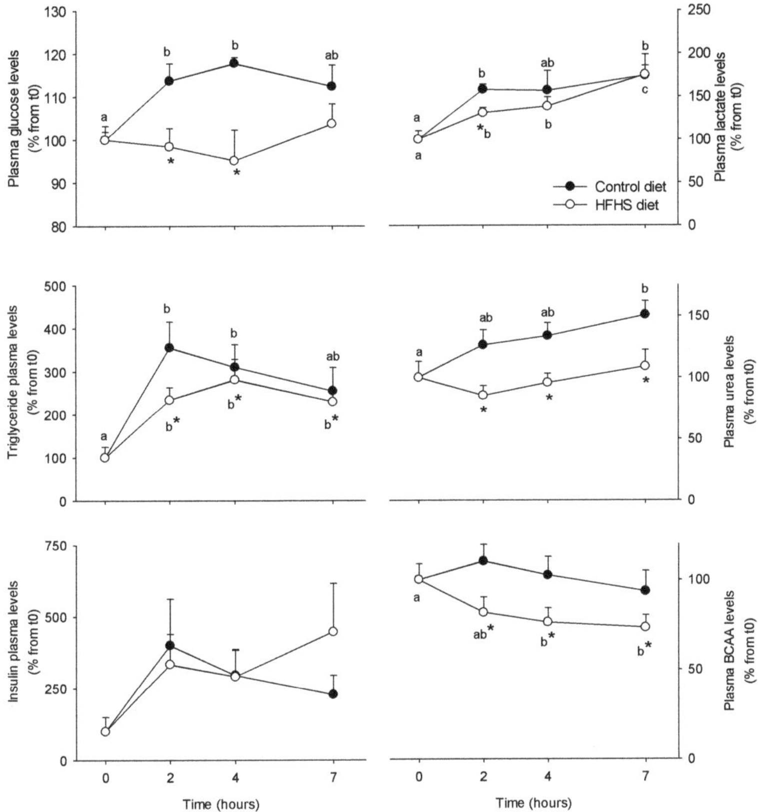

Figure 1. Plasma parameters in Yucatan mini-pigs fed either a control or a high fat–high sucrose (HFHS) test meals. Results are expressed as means + SEM (n = 5) and were analysed using a repeated measures 2-way ANOVA test. Different letters indicate significant differences between sampling points for a given meal test. * Significant different for a given sampling point between the control and the HFHS meals ($p < 0.05$). BCAA: branched-chain amino acids.

Enzyme activities involved in glucose and amino acids metabolism in PBMC (Figure 2) were assessed in order to determine if they were affected by the meal (postprandial period) and its composition (regular vs. HFHS). We aimed also to explore if such changes could be related to the circulating metabolites and hormones described above. Two opposite profiles were observed: enzymes which activity increased after the control meal, including those participating in the glucose phosphorylation (HK, +300%), glycolysis (PK +400-600%, LDH, +150%), pentose phosphate pathway (G6PD, +200%), and amino acid catabolism (BCAT +125%, GDH, +140%); and enzymes for which activity was inhibited by the meal, such as those participating in amino acids metabolism, including glutaminase (−50%) and AspAT (−50%). For all of them, the response to the meal was further blunted when the HFHS test was performed. MDH activity (enzyme involved in the Krebs cycle) was not modified by the regular meal, but it was inhibited (−75% at 2 h) following the HFHS test, suggesting that those changes were meal-specific. Finally, we also explored the insulin (Akt) and protein synthesis (eEF2α) signalling pathways in order to evaluate the capacity of PBMC to transduce the nutritional stimuli from the meal into intracellular information able to regulate the glucose and protein metabolisms (Figure 3). In both cases, their phosphorylation statuses increased following the control meal (about +200% for both proteins), but no changes were observed after the HFHS test.

3.2. Long-Term HFHS Trial

After two months of HFHS feeding minipigs developed an obesity-like phenotype, with a significant increase in body weight (from 31.5 ± 1.4 kg to 44.7 ± 1.7 kg), most likely as the consequence of fat deposition at the visceral and subcutaneous adipose tissue [23]. The main metabolic features associated to this phenotype are described in Table 1. No changes were observed in glucose and lactate levels, while TG concentrations slightly increased between day 7 and 14. Insulin levels increased significantly from day 7 (about three times from day 0), and remained elevated up to the end of trial. In contrast, urea plasma levels were reduced from day 7 onwards (about 1.5 times) up to day 60. Finally, the inflammation status of minipigs was evaluated by assessing the CRP circulating levels at the beginning and at the end of the trial. CRP levels increased significantly from 0.52 ± 0.10 up to 0.99 ± 0.21 mg/L following the two months of HFHS feeding.

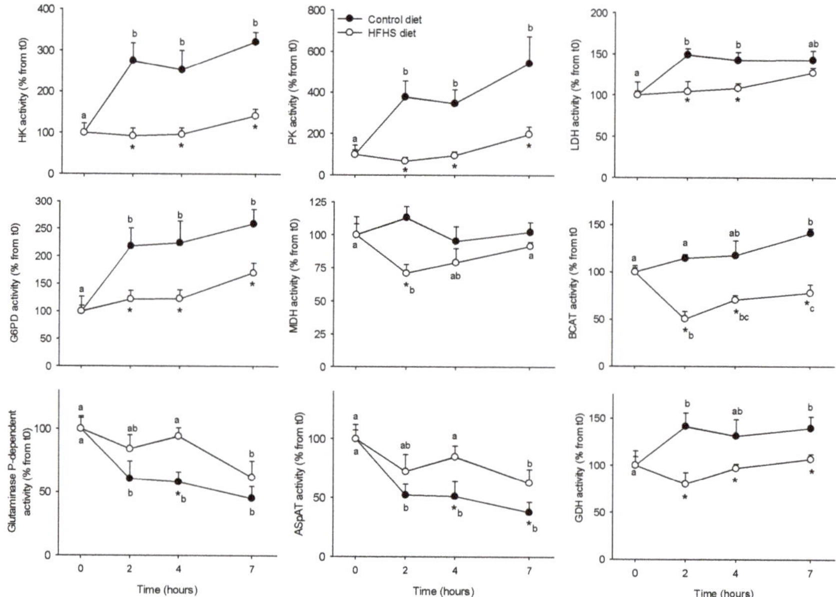

Figure 2. Hexokinase (HK), pyruvate kinase (PK), lactate dehydrogenase (LDH), glucose 6-phosphate dehydrogenase (G6PD), malate dehydrogenase (MDH), branched chain amino-acid transaminase (BCAT), glutaminase P-dependent, aspartate aminotransferase (AspAT) and glutamate dehydrogenase (GDH) alanine transaminase activities in PBMCs from Yucatan mini-pigs fed either a control or a high fat–high sucrose (HFHS) test meals. Enzyme activity units (mU) are defined as nmol of substrate converted to product, per min, at 37 °C and per mg protein. Results are expressed as means + SEM (*n* = 5) and were analysed using a repeated measures two-way ANOVA test. Different letters indicate significant differences between sampling points for a given meal test. * Significantly different for a given sampling point between the control and the HFHS meals ($p < 0.05$). PMBC: peripheral blood mononuclear cell.

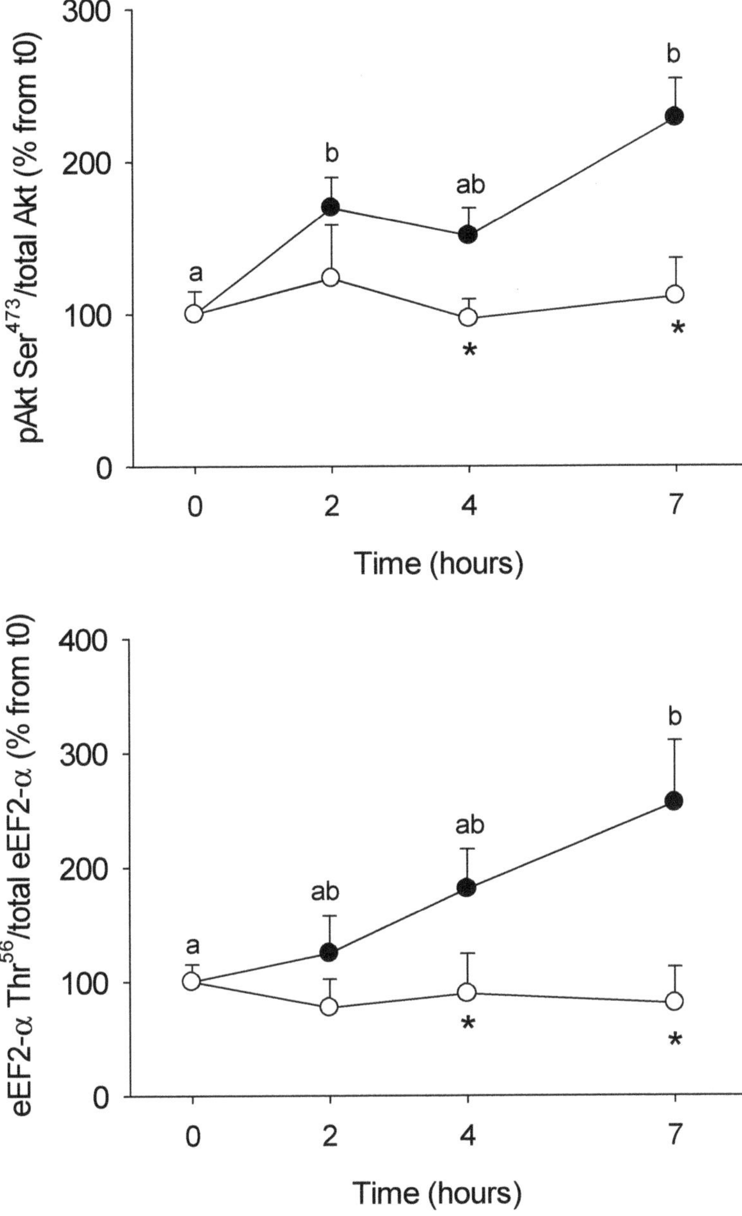

Figure 3. Phosphorylation levels of Akt Ser^{473} and eEF2-α Thr^{56} in PBMCs and S6 Ser$^{235/236}$ from Yucatan mini-pigs fed either a control or a high fat–high sucrose (HFHS) test meals. Analysis was made by Western blot and levels of phosphorylated protein were normalized to the levels of the respective total protein (Akt Ser473 and eEF2-α). Results are expressed as means + SEM ($n = 5$) and were analysed using a repeated measures two-way ANOVA test. Different letters indicate significant differences between sampling points for a given meal test. *Significant different for a given sampling point between the control and the HFHS meals ($p < 0.05$).

Table 1. Plasma fasting metabolites in Yucatan minipigs fed a high fat–high sucrose diet (HFHS) over 2 months.

	Time (Days)				
	1	*7*	*14*	*30*	*60*
Glucose (mM)	3.78 ± 0.28	3.54 ± 0.11	3.42 ± 0.12	3.56 ± 0.15	3.54 ± 0.07
Lactate (mM)	0.53 ± 0.07	0.51 ± 0.03	0.49 ± 0.03	0.48 ± 0.04	0.58 ± 0.06
Triglycerides (mM)	$0.19 \pm 0.03a$	$0.35 \pm 0.06b$	$0.33 \pm 0.07b$	$0.21 \pm 0.03a$	$0.32 \pm 0.05ab$
Insulin (ng/mL)	$0.05 \pm 0.03a$	$0.17 \pm 0.04b$	$0.16 \pm 0.04b$	$0.17 \pm 0.05b$	$0.14 \pm 0.03b$
Urea (mM)	$5.38 \pm 0.43a$	$3.41 \pm 0.32b$	$3.79 \pm 0.15b$	$3.92 \pm 0.31b$	$4.06 \pm 0.31b$

Results are expressed as means + SEM ($n = 5$) and were analysed using a one-way ANOVA test followed by post-hoc Holm-Sidak. Different letters indicate significant differences between the sampling points ($p < 0.05$).

Once the metabolic phenotype of minipigs evaluated during the 2 months of HFHS feeding, we aimed at determining if the PBMC metabolism was able to respond to the modifications induced by the diet at the circulating level. We explored the PBMC metabolism at both, the enzyme activity and mRNA levels, as shown in Figures 4 and 5, respectively. Overall, most of the explored enzymes increased their activities from day 7, recovering progressively their basal levels after 2 months of feeding. Only glutaminase and AspAT activities were reduced by the HFHS feeding (-50%), also from day 7. The tendency was conserved up to the end of the trial. Concerning the molecular analyses, several genes resulted up-regulated by the long-term HFHS feeding, including *acox, fasn, bckdhb* and *bckdk*, for which levels increased progressively with time. Only *dld* mRNA levels increased at day 7 and remained stable up to the end of the trial. *Bcat* and *bckdhb* mRNA levels were down-regulated by the HFHS consumption, while no changes were observed on *srebpf2* levels. Finally, *srebpf1* levels were transiently reduced between day 7 and 30 (-40%) but recovered the initial levels at day 60.

The third objective of the present study was to compare PBMC metabolic potential to those of other organs, such as the liver, the skeletal muscle and the AT. We therefore explored the metabolism of those tissues before and 2 months after HFHS feeding. As for the PBMCs, we choose a double approach, including biochemical (glycogen levels, enzyme activities) and molecular (mRNA levels) targets as presented in Figures 6 and 7, as well as in Table 2, respectively. At the hepatic level, the HFHS feeding resulted in increased glycogen levels ($+24$-fold), and in enzymes participating at the gluconeogenesis (FBPase, $+3$-fold), lipogenesis (FAS, $+5$-fold), lipid oxidation (Acox, $+2.5$-fold) and amino acids catabolism (GDH, $+3$-fold). Other enzyme activities were inhibited by the HFHS feeding, including those participating at the glycolysis (HK, PK, about -4-fold), the pentose phosphate pathway (G6PD, -1.8-fod), and amino acid transamination (AspAT, -1.3-fod). Except for the AspAT activity (which was inhibited (-2-fold)), all the explored enzymes were up-regulated in the adipose tissue, including HK ($+5$-fold), G6PD ($+3$-fold) and GDH ($+2$-fold). At the muscle level, glycogen levels increased by 2-fold after 2 months of HFHS feeding, while HK, GDH, and AspAT activities were reduced.

Concerning the mRNA levels of proteins involved in glucose, lipids and amino acids metabolism are at the hepatic level, the general tendency was the reduction of expression induced by the HFHS feeding including those genes involved in glucose metabolism such as *hk1* (-1.5-fold) and *g6pc* (-2-fold), or in lipid oxidation like *cptl* (-5-fold) or lipogenesis (*acly*; -1.3-fod). The only exception was another gene involved in lipogenesis, *fasn*, which mRNA levels were dramatically ($+23$-fold) increased by the HFHS diet intake. Only minor changes were observed in the skeletal muscle, with a 50% reduction it the mRNA levels of the *cptm*, involved in β-oxidation. Several genes were up-regulated by the HFHS feeding in the AT, including those participating at the glucose transport, like *slc2a4* ($+6$-fold), *hk1* ($+1.4$-fold), or lipogenesis, like *acly* ($+10$-fold) and *fasn* ($+9$-fold). In contrast, the hepatic isoform of CPT (*cptl*) was down-regulated (-5-fold) the HFHS feeding.

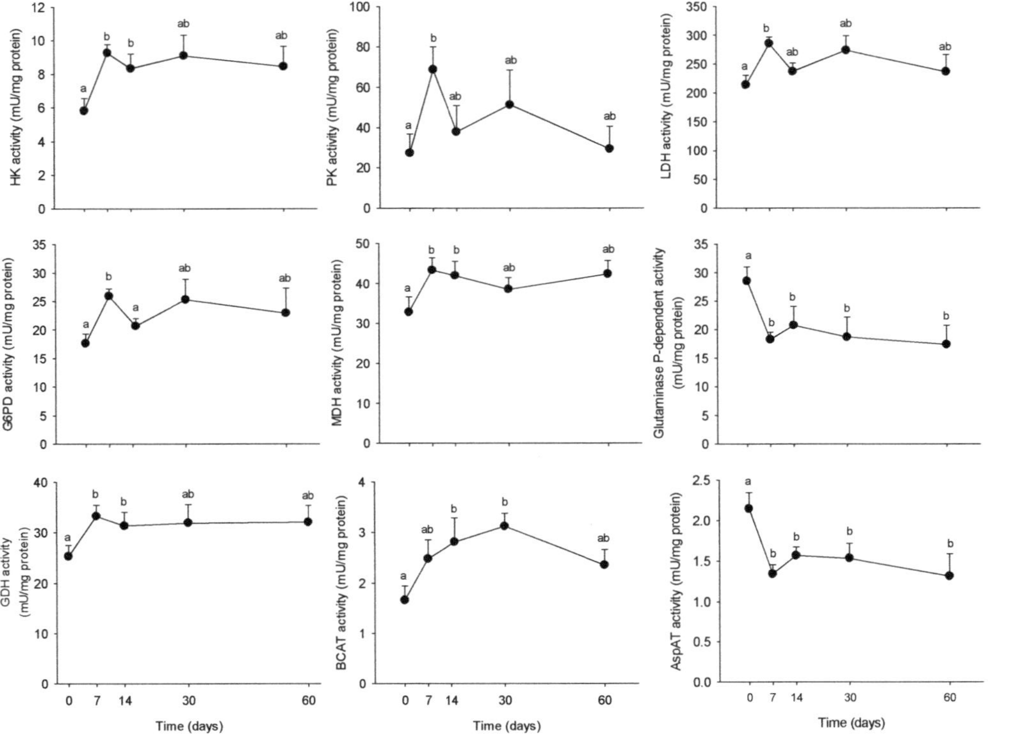

Figure 4. Hexokinase (HK), pyruvate kinase (PK), lactate dehydrogenase (LDH), glucose 6-phosphate dehydrogenase (G6PD), malate dehydrogenase (MDH), branched chain amino-acid transaminase (BCAT), glutaminase P-dependent, aspartate aminotransferase (AspAT), and glutamate dehydrogenase (GDH) alanine transaminase activities in PBMC from Yucatan mini-pigs fed a high fat-high sucrose (HFHS) diet for 2 months. Samples were obtained after an overnight fast period before (0) and 7, 14, 30, and 60 days after HFHS feeding. Enzyme activity units (mU) are defined as nmol of substrate converted to product, per min, at 37 °C and per mg protein. Results are expressed as means + SEM ($n = 5$) and were analysed using a one-way ANOVA test followed by post-hoc Holm-Sidak. Different letters indicate significant differences between the sampling points ($p < 0.05$).

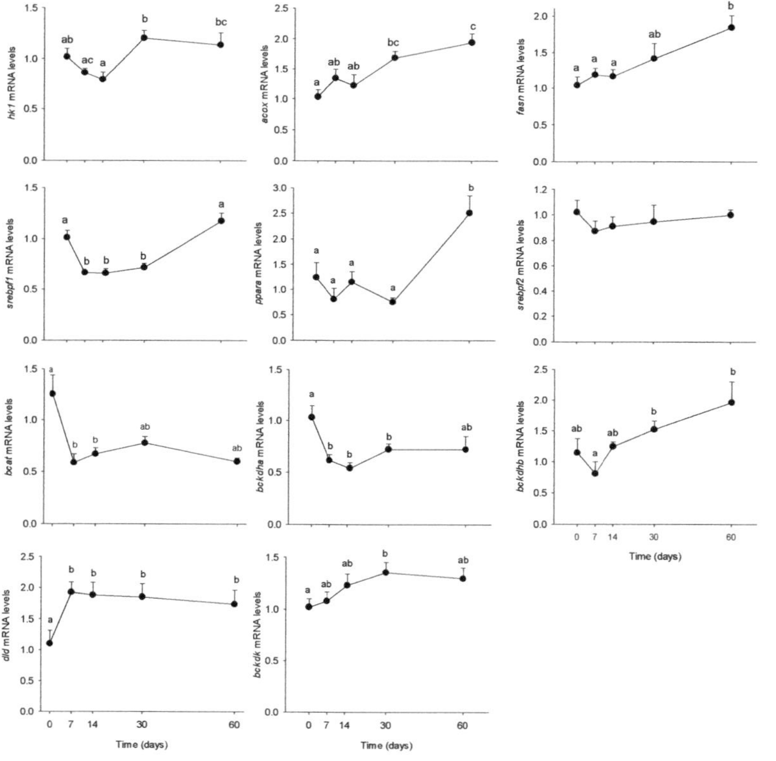

Figure 5. Hexokinase (*hk1*), acyl-coenzyme A oxidase (*acox*), fatty acid synthase (*fasn*), sterol regulatory element binding protein-1c (*srebf1c*), peroxisome proliferator-activated receptor alpha (*ppara*), sterol regulatory element binding protein-2 (*srebf2*), branched-chain alpha-keto acid dehydrogenase alpha/beta (*bckdha/b*), dihydrolipoyl dehydrogenase (*dld*), and branched chain ketoacid dehydrogenase kinase (*bckdk*) mRNA levels in PBMCs from Yucatan mini-pigs fed on high fat-high sucrose (HFHS) diet for 2 months. Samples were obtained after an overnight fast period before (0) and 7, 14, 30, and 60 days after HFHS feeding. Results are expressed as means + SEM (*n* = 5) and were analysed using a one-way ANOVA test followed by post-hoc Holm-Sidak. Different letters indicate significant differences between the sampling points (*p* < 0.05).

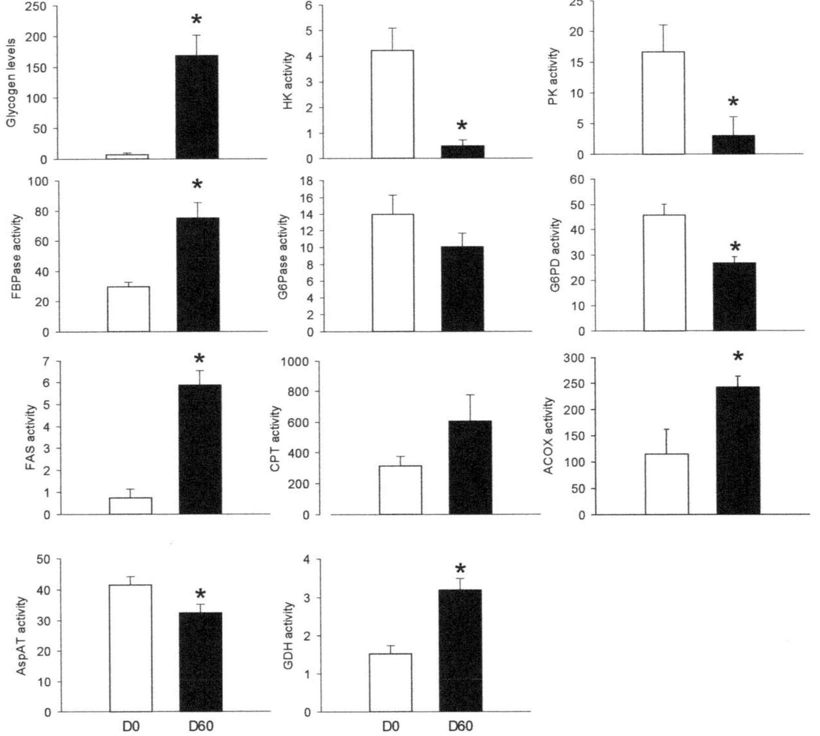
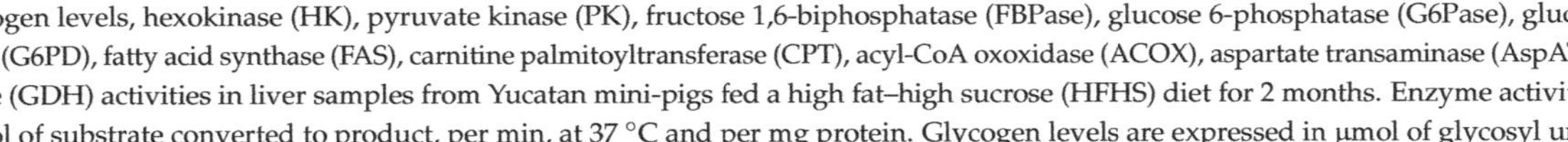

Figure 6. Glycogen levels, hexokinase (HK), pyruvate kinase (PK), fructose 1,6-biphosphatase (FBPase), glucose 6-phosphatase (G6Pase), glucose 6-phosphate dehydrogenase (G6PD), fatty acid synthase (FAS), carnitine palmitoyltransferase (CPT), acyl-CoA oxoxidase (ACOX), aspartate transaminase (AspAT), and glutamate dehydrogenase (GDH) activities in liver samples from Yucatan mini-pigs fed a high fat–high sucrose (HFHS) diet for 2 months. Enzyme activity units (mU) are defined as nmol of substrate converted to product, per min, at 37 °C and per mg protein. Glycogen levels are expressed in μmol of glycosyl units/g wet tissue. Results are expressed as means + SEM (n = 5) and were analysed using a t-student test. *, significantly different from the D0 ($p < 0.05$).

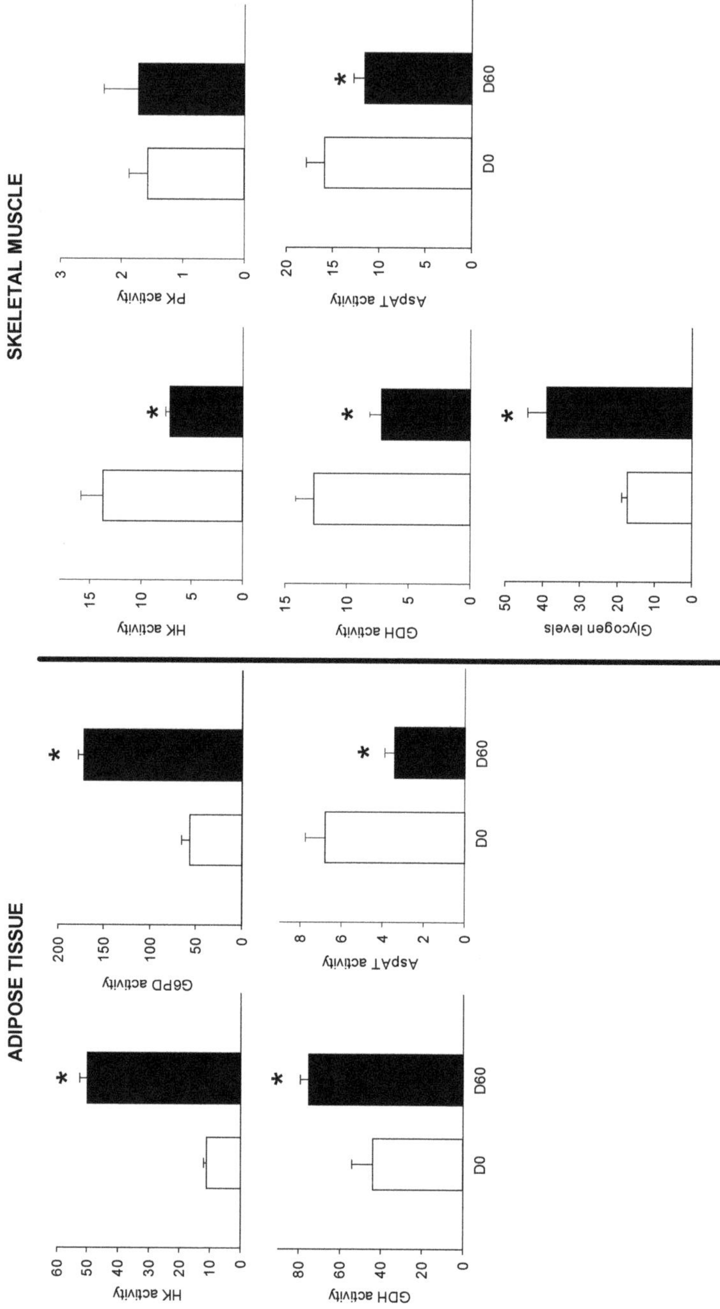

Figure 7. Glycogen levels, hexokinase (HK), pyruvate kinase (PK), glucose 6-phosphate dehydrogenase (G6PD), aspartate transaminase (AspAT), glutamate dehydrogenase (GDH) activities in adipose tissue and skeletal muscle samples from Yucatan mini-pigs fed a high fat–high sucrose (HFHS) diet for 2 months. Enzyme activity units (mU) are defined as nmol of substrate converted to product, per min, at 37 °C and per mg protein. Glycogen levels are expressed in µmol of glycosyl units/g wet tissue. Results are expressed as means + SEM ($n = 5$) and were analysed using a *t*-student test. *, significantly different from the D0 ($p < 0.05$).

Table 2. mRNA levels of proteins involved in glucose and lipid metabolism in the liver, adipose tissue and skeletal muscle samples from Yucatan minipigs fed on a high fat–high sucrose diet (HFHS) over 2 months.

		Liver			Skeletal Muscle			Adipose Tissue	
Metabolic Pathway	Gene	D0	D60	Gene	D0	D60	Gene	D0	D60
Glucose metabolism	gck	0.63 ± 0.20	1.57 ± 0.54	slc2a4	1.03 ± 0.12	1.42 ± 0.14	slc2a4	1.03 ± 0.11	6.40 ± 1.69 *
	g6pc	1.51 ± 0.21	0.78 ± 0.15 *						
	pck1	1.09 ± 0.09	0.94 ± 0.07	hk-1	0.70 ± 0.08	1.50 ± 0.37	hk-1	1.01 ± 0.06	1.45 ± 0.16 *
	hk1	1.23 ± 0.10	0.84 ± 0.05 *						
Lipid oxidation	cpt1-l	2.43 ± 0.30	0.44 ± 0.04 *	cpt1-l	1.03 ± 0.11	0.50 ± 0.06 *	cpt1-l	1.27 ± 0.26	0.27 ± 0.04 *
	acox	1.01 ± 0.06	0.86 ± 0.05	acox	0.87 ± 0.04	0.82 ± 0.09	acox	1.03 ± 0.11	0.79 ± 0.17
	ppara	1.25 ± 0.34	1.03 ± 0.09	ppara	1.02 ± 0.09	0.73 ± 0.08 [t]	ppara	1.04 ± 0.14	1.46 ± 0.18 [t]
Lipogenesis	acly	1.01 ± 0.08	0.75 ± 0.04 *				acly	1.16 ± 0.34	11.88 ± 1.29 *
	fasn	0.23 ± 0.03	5.37 ± 1.19 *		-		fasn	1.19 ± 0.36	10.34 ± 2.57 *
	srebf1c	0.92 ± 0.16	1.27 ± 0.21				srebf1c	1.29 ± 0.47	0.61 ± 0.05

Each value is the mean $\pm$ SEM of $n = 5$ animals. mRNA levels were estimated using real-time RT-PCR. mRNA expression values were normalized with β-actin expressed transcripts and are indicated as fold variation of the control group. * significantly different from D0 ($p < 0.05$) among groups. *gck*, glucokinase; *pck1*, phosphoenolpyruvate kinase; *g6pc*, glucose 6-phosphatase; *acly*, ATP-citrate lyase; *fasn*, fatty acid synthase; *srebf1c*, sterol regulatory element binding protein-1c; *cpt1-l/m*, carnitine palmitoyltransferase-liver/muscle isoform; *acox*, acyl-coenzyme A oxidase; *ppara*, peroxisome proliferator-activated receptor alpha; *hk-1*, hexokinase 1.

4. Discussion

In the present study we show for the first time that the metabolism of mini-pig PBMC can be altered at the biochemical and molecular levels and that it was closely related to the AT profile, at least in the context of a long-term HFHS feeding. Further, we showed that the PBMC metabolism, particularly at the biochemical level, can be modulated at the postprandial level.

4.1. PBMC Metabolism Is Highly Reactive to the Meal, But Its Response Is Blunted by the Intake of a HFHS Meal

After meal intake, the body must adapt to the greater influx of nutrients in a very short period of time (from minutes to a couple of hours). Although several studies have previously explored the postprandial response of PBMC, they were focussed on transcriptomics analyses only [20,21,27]. Given the very short-term response needed after the meal, we choose to explore the postprandial metabolism at the biochemical and signalling levels rather than at the molecular level, which as far as we are aware has never been done before. The objective was therefore to elucidate if PBMC metabolism could respond in the very short-term to a nutritional stimulus and provide valuable information about changes at the whole body metabolism and metabolic flexibility that could not be visible at the fasting state [16,28]. In our study, PBMC metabolism was indeed highly responsive to the control meal challenge. The glucose phosphorylation potential and further channelling through the glycolysis was strongly increased, highlighting the importance of glucose for fuelling PBMC metabolism [29]. This was further supported by the meal-stimulated phosphorylation of Akt, a major actor on the insulin signalling pathway. Other metabolic processes were also stimulated by the meal, as the NADPH production to support lipogenesis or nucleotide biosynthesis through the pentose phosphate pathway.

PBMCs not only rely on glucose as energy source, as glutamine utilization in these cells can be even greater than that of glucose [8]. In our study glutaminase activity was significantly reduced following the meal, in line with its increased activity in lymphocytes from starved rats [29]. The fact that the AspAT activity followed the same profile than the glutaminase, further confirms that this enzyme rather than GDH is mainly responsible of the glutamate to α-ketoglutarate conversion in PBMCs. However, PBMCs do not only use glutamine for energy purposes, as the production of non-essential amino acids (such as alanine and aspartate) via the transaminases for biosynthetic purposes is of great importance for this kind of highly proliferating cell [8]. Following the meal, when circulating amino acid levels increase, PBMCs would rather rely on them for biosynthesis processes, reducing the demand from transaminases and resulting in an increase in the glutamate channelling through the GDH reaction [30], explaining the increased GDH activity in our study. Finally, the fact that the glucose utilisation was enhanced in response to the meal, whereas glutamine utilisation was reduced, strongly suggests that during the fasted-to-fed transition a metabolic shift allows PBMC to switch from glutamine to glucose utilisation [31].

During the HFHS meal test, we surprisingly found that the PBMC response was often blunted or in some cases delayed when compared to the regular test meal. The postprandial trial aimed at challenging the metabolic response of PBMC, which has been achieved with the control meal. By providing for the first time an HFHS meal to the same animals, our goal was to further alter the PBMC environment in terms of composition and time-course appearance of nutrients and hormones. Thus, the kinetics of most of the plasma parameters were altered following the HFHS meal, most likely due to the impact of the diet on gut physiology (motility, gastric emptying) and hormone release (CCK, PYY, GLP-1) [32]. Thus, despite a higher content of sugar than in the regular meal, the HFHS consumption did not result in a significant postprandial increase in blood glucose levels. Accordingly, the same lack of responsiveness was observed in the insulin signalling pathway, resulting (together with the low glucose available) in a non-stimulated glucose utilisation, and supporting the idea that PBMC strongly depend on circulating substrates, particularly glucose. This could explained in part by a delayed and reduced excursion of glucose and insulin mediated by a slower gastric emptying and reduced gut hormone secretion, as previously observed in humans subjected to high fat diet tests [33–35].

Unlike after the control meal, following the HFHS intake we did not longer observe the fasting-to-fed transition between glucose and glutamine utilisation. This suggests that PBMC metabolism has adapted to the reduced glucose levels and continue to rely on glutamine, such as during the post-absorptive period. In contrast to other circulating metabolites, BCAA levels following the HFHS meal showed reduced levels during the postprandial period, which resulted in a lower transamination potential as well, likely reducing the flux of ketoacids up to the Krebs cycle. In agreement with this, the phosphorylation status of the protein eEF2α, responsible for determine the end of the translation and sensitive to the leucine signal role, remained unchanged after the HFHS meal. This may represent another sign of the altered response to the meal stimulus. Another symptom of impaired normal metabolic activity after the HFHS meal was the reduction on MDH activity. Since MDH plays crucial roles in the malate–aspartate shuttle, essential for coupling activation of mitochondrial and energy production [36], a decrease in MDH activities may suggest a depression of the transfer of NADH, and less ATP production in mitochondria.

4.2. Time-Course Changes in PBMC Metabolism during Long-Term High Fat Feeding

The second goal of our study was to evaluate if the metabolism of PBMC could be modified by a long-term nutritional change, such as HFHS feeding-induced obesity. As in our previous study focused on tissue/organ metabolism [23], we observed in the present study that the same major metabolic changes occurred also in PBMC after only one week of HFHS feeding. However, unlike during the short-term stimulation (postprandial period) PBMC do not seem to rely on available circulating metabolites during the long-term HFHS imprinting. Thus, despite the lack of changes in post-absorptive blood glucose levels, PBMC glucose metabolism was enhanced, as previously shown in a PBMC transcriptome-based study comparing lean vs. obese subjects [20]. It is known that alterations in postprandial glucose kinetic precedes those at the fasting state in the insulin resistance installation [37]. The increased glucose metabolism at the fasting state observed in the circulating PBMC may be the result of a long-term important glucose postprandial excursions (as supported by the hyperinsulinemia), exposing the circulating cells to elevated postprandial glucose levels. If our results are further investigated in future studies, alterations in these enzyme activities could constitute early events signalling the beginning of glucose alterations leading to insulin resistance, even when fasting blood glucose levels remain unchanged.

PBMCs seem to also have very active lipid metabolism [22]. Thus, genes involved in lipid catabolism, such as *acox* and *ppara*, were up-regulated by the long-term HFHS feeding, suggesting that minipig PBMC metabolism adapted to the increased dietary lipids, as previously observed in PBMCs from obese rats fed on cafeteria diets [2,38]. On the other hand, markers of lipogenesis (*fasn, srebpf1*) also resulted up-regulated. A recent study showed that lipogenesis was active in PBMCs and that was highly dependent on glucose [39], a strong activator of lipogenic genes [40]. This effect of glucose on de novo lipid synthesis represents a common feature for lipogenic organs, such as the liver or the AT from HFHS-fed animals [28,41,42].

As far as we are aware, the metabolism of amino acids has never been explored on PBMC from obese animals. After two months of HFHS feeding, both glutaminase and AspAT activities resulted rapidly (one week) down-regulated, suggesting that amino acids use for energy was reduced in detriment of other alternative fuels like glucose and lipids [23]. BCAA related-enzyme activities and mRNA levels were also reduced by the end of the trial, supporting the idea of a progressively reduction of BCAA catabolism from obese minipigs. We and other have shown that in obese and insulin resistant animal models, BCAA circulating levels increased with time [25,43]. Interestingly, the increased transamination potential during the first weeks of HFHS feeding could be the consequence of a metabolic adaptation to the progressively increase in BCAA levels. However, by the end of the trial, the potential to catabolise BCAA was blunted, in agreement with a defective oxidative deamination observed in the adipose tissue of HFHS-fed minipigs [25].

4.3. The PBMC Metabolism Rather Resembled that of the Adipose Tissue

As explained above, PBMC metabolism is different from any other organ, as these cells are actually permanently stimulated by changes in circulating blood metabolites and hormones. Despite this, we demonstrated that their metabolism has many common features with several major organs of the body. The second objective of the present study was therefore to compare PBMC response to the long-term HFHS meal to that of the liver, skeletal muscle and the AT.

We observed that an adaptive strategy was settled at the hepatic level in order to handle the excess of nutrients and energy brought by the HFHS diet. Thus, those pathways involved in glucose utilisation, such as glycolysis and pentose phosphate pathway were reduced, while lipid oxidation potential increased, illustrating a metabolic switch aiming at giving the priority to the more abundant energy substrates available. In accordance with this, we observed that the liver increased its capacity to store the excess of energy in the form of lipids (lipogenesis) and glycogen. Of note, the increased FBPase activity and reduced G6Pase potential strongly suggests that the glucose produced via the gluconeogenesis was further stored as glycogen rather than being exported into the blood stream. Unlike the liver, the overall metabolic features observed in the skeletal muscle and the AT suggest that both organs enhanced the uptake, utilisation, and storage of glucose. Two other catabolic features were also different from those of the liver, including lipids and amino acids oxidation pathways, that were reduced in the muscle and the AT. As for the liver, lipogenesis potential was strongly up-regulated in the AT. The integrative view of the HFHS minipig physiology suggests therefore that at the whole-body level, inflammation was settled. Despite this, the metabolism seemed flexible enough to adapt and kept blood glucose levels into within the physiological range. This was achieved through a coordinated increased in glucose uptake, utilisation and storage from all the explored organs in order to remove as much as possible glucose from the blood circulation. Further, the liver and the AT increased also their potential of other highly consuming glucose pathways, such as *de novo* lipogenesis. Finally, the lipid handling seemed to be organ-dependent, as the liver increased the oxidation potential of lipids, while the muscle and the adipose tissue showed an opposite trend.

When compared to the results discussed above, the metabolic response observed in the PBMCs in the obese minipig at the fasting state seems to be coherent with the adaptations reported in the major organs involved in the intermediary metabolism. As in a previous study in hamsters [44] our results show that PBMC metabolism does not reflect that of the liver. In contrast, the metabolic profile of PBMC evaluated at the molecular and biochemical levels was particularly close to the one observed in the AT. This is in line with similar conclusions draws in several studies in humans [45] and laboratory animals [16,44], specially concerning lipid metabolism [46]. In contrast to this, other metabolic features, like pentose phosphate potential seems to have a PBMC-specific behaviour, with a strong dependence on glucose uptake and phosphorylation and a lower induction capacity when compared to the AT. The responsiveness of PBMCs, particularly at the biochemical level could be then potentially used to explore the whole body metabolism in response to different dietary interventions or to evaluate the functional metabolic status of an individual with a single blood sample.

5. Conclusions

As a whole, the evaluation of the metabolic response to a control test meal of PBMC showed for the first time at the biochemical level that these cells were able to modify and adapt in a very short period (often < 3 h) their metabolic activity under the influence of circulating nutrient and hormones. Most of the changes recorded showed that PBMCs relayed glutamine metabolism during the post-absorptive period, and that they switched to glucose when quantity of carbohydrates from the diet increased in the circulation. Interestingly, we showed that major and fast changes in the PBMC environment induced by the HFHS during the postprandial period were able to trigger in PBMCs the necessary metabolic changes to adapt to this challenge. Finally, the exploration of the PBMC metabolism in long-term HFHS feeding confirmed that the regulation pattern observed in PBMCs fits with that expected in tissues involved in energy balance. It would be therefore reasonable to use them to evaluate whole-body metabolism without the need to perform invasive organs biopsies, as they can act as metabolic sentinels of nutritional-related changes. The common features observed between the PBMC and AT metabolism in HFHS minipigs make these cells particularly attractive for exploring metabolism and looking for biomarkers of dysregulation in obesity-related conditions.

Author Contributions: For research articles with several authors, a short paragraph specifying their individual contributions must be provided. The following statements should be used "Conceptualization, S.P. and I.S.A.; Methodology, Y.Z., D.R. and J.D.; Software, S.P.; Validation, S.P. and Y.Z.; Formal Analysis, Y.Z. and J.D.; Investigation, Y.Z., I.S.A. and S.P.; Resources, S.P.; Data Curation, S.P.; Writing-Original Draft Preparation, S.P.; Writing-Review & Editing, D.D., D.R.; Visualization, Y.Z. and S.P.; Supervision, S.P.; Project Administration, S.P.; Funding Acquisition, S.P.", please turn to the CRediT taxonomy for the term explanation. Authorship must be limited to those who have contributed substantially to the work reported.

Funding: No external funding received.

Acknowledgments: The authors acknowledge D. Durand and the personnel of the Animal Facility (C. de L'Homme, B. Cohade) for technical assistance.

Conflicts of Interest: The authors declare no conflict of interest.

References

1. Liew, C.C.; Ma, J.; Tang, H.C.; Zheng, R.; Dempsey, A.A. The peripheral blood transcriptome dynamically reflects system wide biology: A potential diagnostic tool. *J. Lab. Clin. Med.* **2006**, *147*, 126–132. [CrossRef] [PubMed]

2. Reynes, B.; Diaz-Rua, R.; Cifre, M.; Oliver, P.; Palou, A. Peripheral blood mononuclear cells as a potential source of biomarkers to test the efficacy of weight-loss strategies. *Obesity* **2015**, *23*, 28–31. [CrossRef] [PubMed]

3. Ghosh, S.; Dent, R.; Harper, M.E.; Gorman, S.A.; Stuart, J.S.; McPherson, R. Gene expression profiling in whole blood identifies distinct biological pathways associated with obesity. *BMC Med. Genom.* **2010**, *3*, 56. [CrossRef] [PubMed]

4. Kim, M.; Kim, M.; Han, J.Y.; Lee, S.H.; Jee, S.H.; Lee, J.H. The metabolites in peripheral blood mononuclear cells showed greater differences between patients with impaired fasting glucose or type 2 diabetes and healthy controls than those in plasma. *Diabetes Vasc. Dis. Res.* **2017**, *14*, 130–138. [CrossRef] [PubMed]

5. Stepien, M.; Nugent, A.P.; Brennan, L. Metabolic profiling of human peripheral blood mononuclear cells: Influence of vitamin D status and gender. *Metabolites* **2014**, *4*, 248–259. [CrossRef] [PubMed]

6. Mohr, S.; Liew, C.C. The peripheral-blood transcriptome: New insights into disease and risk assessment. *Trends Mol. Med.* **2007**, *13*, 422–432. [CrossRef] [PubMed]

7. De Boever, P.; Wens, B.; Forcheh, A.C.; Reynders, H.; Nelen, V.; Kleinjans, J.; Van Larebeke, N.; Verbeke, G.; Valkenborg, D.; Schoeters, G. Characterization of the peripheral blood transcriptome in a repeated measures design using a panel of healthy individuals. *Genomics* **2014**, *103*, 31–39. [CrossRef] [PubMed]

8. Newsholme, E.A.; Crabtree, B.; Ardawi, M.S. Glutamine metabolism in lymphocytes: Its biochemical, physiological and clinical importance. *Q. J. Exp. Physiol.* **1985**, *70*, 473–489. [CrossRef] [PubMed]

9. Washizu, T.; Takahashi, M.; Azakami, D.; Ikeda, M.; Arai, T. Activities of Enzymes in the Malate–Aspartate Shuttle in the Peripheral Leukocytes of Dogs and Cats. *Vet. Res. Commun.* **2001**, *25*, 623–629. [CrossRef] [PubMed]

10. Mori, A.; Kenyon, P.R.; Mori, N.; Yamamoto, I.; Tanaka, Y.; Suzuki, N.; Tazaki, H.; Ozawa, T.; Hayashi, T.; Hickson, R.E.; et al. Changes in metabolite, energy metabolism related enzyme activities and peripheral blood mononuclear cell (PBMC) populations in beef heifers with two differing liveweight change profiles in New Zealand. *Vet. Res. Commun.* **2008**, *32*, 159–166. [CrossRef] [PubMed]

11. Washizu, T.; Tanaka, A.; Sako, T.; Washizu, M.; Arai, T. Comparison of the activities of enzymes related to glycolysis and gluconeogenesis in the liver of dogs and cats. *Res. Vet. Sci.* **1999**, *67*, 205–206. [CrossRef] [PubMed]

12. Wilkinson, J.M.; Dyck, M.K.; Dixon, W.T.; Foxcroft, G.R.; Dhakal, S.; Harding, J.C. Transcriptomic analysis identifies candidate genes and functional networks controlling the response of porcine peripheral blood mononuclear cells to mitogenic stimulation. *J. Anim. Sci.* **2012**, *90*, 3337–3352. [CrossRef] [PubMed]

13. Adler, M.; Murani, E.; Ponsuksili, S.; Wimmers, K. PBMC transcription profiles of pigs with divergent humoral immune responses and lean growth performance. *Int. J. Biol. Sci.* **2013**, *9*, 907–916. [CrossRef] [PubMed]

14. Oster, M.; Murani, E.; Ponsuksili, S.; D'Eath, R.B.; Turner, S.P.; Evans, G.; Tholking, L.; Kurt, E.; Klont, R.; Foury, A.; et al. Transcriptional responses of PBMC in psychosocially stressed animals indicate an alerting of the immune system in female but not in castrated male pigs. *BMC Genom.* **2014**, *15*, 967. [CrossRef] [PubMed]

15. Weber, T.E.; Spurlock, M.E. Leptin alters antibody isotype in the pig in vivo, but does not regulate cytokine expression or stimulate STAT3 signaling in peripheral blood monocytes in vitro. *J. Anim. Sci.* **2004**, *82*, 1630–1640. [CrossRef] [PubMed]

16. Caimari, A.; Oliver, P.; Keijer, J.; Palou, A. Peripheral blood mononuclear cells as a model to study the response of energy homeostasis-related genes to acute changes in feeding conditions. *OMICS J. Integr. Biol.* **2010**, *14*, 129–141. [CrossRef] [PubMed]

17. Manoel-Caetano, F.S.; Xavier, D.J.; Evangelista, A.F.; Takahashi, P.; Collares, C.V.; Puthier, D.; Foss-Freitas, M.C.; Foss, M.C.; Donadi, E.A.; Passos, G.A.; et al. Gene expression profiles displayed by peripheral blood mononuclear cells from patients with type 2 diabetes mellitus focusing on biological processes implicated on the pathogenesis of the disease. *Gene* **2012**, *511*, 151–160. [CrossRef] [PubMed]

18. Tangen, S.E.; Tsinajinnie, D.; Nunez, M.; Shaibi, G.Q.; Mandarino, L.J.; Coletta, D.K. Whole blood gene expression profiles in insulin resistant Latinos with the metabolic syndrome. *PLoS ONE* **2013**, *8*, e84002. [CrossRef] [PubMed]

19. Drew, J.E.; Farquharson, A.J.; Horgan, G.W.; Duthie, S.J.; Duthie, G.G. Postprandial cell defense system responses to meal formulations: Stratification through gene expression profiling. *Mol. Nutr. Food Res.* **2014**, *58*, 2066–2079. [CrossRef] [PubMed]

20. Fazelzadeh, P.; Hangelbroek, R.W.J.; Joris, P.J.; Schalkwijk, C.G.; Esser, D.; Afman, L.; Hankemeier, T.; Jacobs, D.M.; Mihaleva, V.V.; Kersten, S.; et al. Weight loss moderately affects the mixed meal challenge response of the plasma metabolome and transcriptome of peripheral blood mononuclear cells in abdominally obese subjects. *Metabolomics* **2018**, *14*, 46. [CrossRef] [PubMed]

21. Bouwens, M.; Grootte Bromhaar, M.; Jansen, J.; Muller, M.; Afman, L.A. Postprandial dietary lipid-specific effects on human peripheral blood mononuclear cell gene expression profiles. *Am. J. Clin. Nutr.* **2010**, *91*, 208–217. [CrossRef] [PubMed]

22. Bouwens, M.; Afman, L.A.; Muller, M. Fasting induces changes in peripheral blood mononuclear cell gene expression profiles related to increases in fatty acid beta-oxidation: Functional role of peroxisome proliferator activated receptor alpha in human peripheral blood mononuclear cells. *Am. J. Clin. Nutr.* **2007**, *86*, 1515–1523. [CrossRef] [PubMed]

23. Polakof, S.; Remond, D.; Bernalier-Donadille, A.; Rambeau, M.; Pujos-Guillot, E.; Comte, B.; Dardevet, D.; Savary-Auzeloux, I. Metabolic adaptations to HFHS overfeeding: How whole body and tissues postprandial metabolic flexibility adapt in Yucatan mini-pigs. *Eur. J. Nutr.* **2018**, *57*, 119–135. [CrossRef] [PubMed]

24. Polakof, S.; Dardevet, D.; Lyan, B.; Mosoni, L.; Gatineau, E.; Martin, J.F.; Pujos-Guillot, E.; Mazur, A.; Comte, B. Time course of molecular and metabolic events in the development of insulin resistance in fructose-fed rats. *J. Proteome Res.* **2016**, *15*, 1862–1874. [CrossRef] [PubMed]

25. Polakof, S.; Remond, D.; David, J.; Dardevet, D.; Savary-Auzeloux, I. Time-course changes in circulating branched-chain amino acid levels and metabolism in obese Yucatan minipig. *Nutrition* **2017**, *50*, 66–73. [CrossRef] [PubMed]

26. Polakof, S.; Rémond, D.; Rambeau, M.; Pujos-Guillot, E.; Sébédio, J.-L.; Dardevet, D.; Comte, B.; Savary-Auzeloux, I. Postprandial metabolic events in mini-pigs: New insights from a combined approach using plasma metabolomics, tissue gene expression, and enzyme activity. *Metabolomics* **2015**, *11*, 964–979. [CrossRef]

27. Sagaya, F.M.; Hurrell, R.F.; Vergeres, G. Postprandial blood cell transcriptomics in response to the ingestion of dairy products by healthy individuals. *J. Nutr. Biochem.* **2012**, *23*, 1701–1715. [CrossRef] [PubMed]

28. Castro, H.; Pomar, C.A.; Pico, C.; Sanchez, J.; Palou, A. Cafeteria diet overfeeding in young male rats impairs the adaptive response to fed/fasted conditions and increases adiposity independent of body weight. *Int. J. Obes.* **2014**. [CrossRef] [PubMed]

29. Ardawi, M.S.; Newsholme, E.A. Maximum activities of some enzymes of glycolysis, the tricarboxylic acid cycle and ketone-body and glutamine utilization pathways in lymphocytes of the rat. *Biochem. J.* **1982**, *208*, 743–748. [CrossRef] [PubMed]

30. Coloff, J.L.; Murphy, J.P.; Braun, C.R.; Harris, I.S.; Shelton, L.M.; Kami, K.; Gygi, S.P.; Selfors, L.M.; Brugge, J.S. Differential Glutamate Metabolism in Proliferating and Quiescent Mammary Epithelial Cells. *Cell Metab.* **2016**, *23*, 867–880. [CrossRef] [PubMed]

31. Brand, K.; Fekl, W.; von Hintzenstern, J.; Langer, K.; Luppa, P.; Schoerner, C. Metabolism of glutamine in lymphocytes. *Metabolism* **1989**, *38*, 29–33. [CrossRef]

32. Little, T.J.; Horowitz, M.; Feinle-Bisset, C. Modulation by high-fat diets of gastrointestinal function and hormones associated with the regulation of energy intake: Implications for the pathophysiology of obesity. *Am. J. Clin. Nutr.* **2007**, *86*, 531–541. [CrossRef] [PubMed]

33. Gentilcore, D.; Chaikomin, R.; Jones, K.L.; Russo, A.; Feinle-Bisset, C.; Wishart, J.M.; Rayner, C.K.; Horowitz, M. Effects of Fat on Gastric Emptying of and the Glycemic, Insulin, and Incretin Responses to a Carbohydrate Meal in Type 2 Diabetes. *J. Clin. Endocrinol. Metab.* **2006**, *91*, 2062–2067. [CrossRef] [PubMed]

34. Cunningham, K.M.; Read, N.W. The effect of incorporating fat into different components of a meal on gastric emptying and postprandial blood glucose and insulin responses. *Br. J. Nutr.* **2007**, *61*, 285–290. [CrossRef]

35. Welch, I.M.; Bruce, C.; Hill, S.E.; Read, N.W. Duodenal and ileal lipid suppresses postprandial blood glucose and insulin responses in man: Possible implications for the dietary management of diabetes mellitus. *Clin. Sci.* **1987**, *72*, 209–216. [CrossRef] [PubMed]

36. Eto, K.; Tsubamoto, Y.; Terauchi, Y.; Sugiyama, T.; Kishimoto, T.; Takahashi, N.; Yamauchi, N.; Kubota, N.; Murayama, S.; Aizawa, T.; et al. Role of NADH Shuttle System in Glucose-Induced Activation of Mitochondrial Metabolism and Insulin Secretion. *Science* **1999**, *283*, 981–985. [CrossRef] [PubMed]

37. American Diabetes Association. 2. Classification and diagnosis of diabetes. *Diabetes Care* **2018**, *40*, S11–S24.

38. Oliver, P.; Reynes, B.; Caimari, A.; Palou, A. Peripheral blood mononuclear cells: A potential source of homeostatic imbalance markers associated with obesity development. *Pflug. Arch.* **2013**, *465*, 459–468. [CrossRef] [PubMed]

39. Dufort, F.J.; Gumina, M.R.; Ta, N.L.; Tao, Y.; Heyse, S.A.; Scott, D.A.; Richardson, A.D.; Seyfried, T.N.; Chiles, T.C. Glucose-dependent de novo lipogenesis in B lymphocytes: A requirement for ATP-citrate lyase in lipopolysaccharide-induced differentiation. *J. Biol. Chem.* **2014**, *289*, 7011–7024. [CrossRef] [PubMed]

40. Ferramosca, A.; Conte, A.; Damiano, F.; Siculella, L.; Zara, V. Differential effects of high-carbohydrate and high-fat diets on hepatic lipogenesis in rats. *Eur. J. Nutr.* **2014**, *53*, 1103–1114. [CrossRef] [PubMed]

41. Oosterveer, M.H.; van Dijk, T.H.; Tietge, U.J.F.; Boer, T.; Havinga, R.; Stellaard, F.; Groen, A.K.; Kuipers, F.; Reijngoud, D.-J. High Fat Feeding Induces Hepatic Fatty Acid Elongation in Mice. *PLoS ONE* **2009**, *4*, e6066. [CrossRef] [PubMed]

42. Polakof, S.; Diaz-Rubio, M.E.; Dardevet, D.; Martin, J.F.; Pujos-Guillot, E.; Scalbert, A.; Sebedio, J.L.; Mazur, A.; Comte, B. Resistant starch intake partly restores metabolic and inflammatory alterations in the liver of high-fat-diet-fed rats. *J. Nutr. Biochem.* **2013**, *24*, 1920–1930. [CrossRef] [PubMed]

43. Lynch, C.J.; Adams, S.H. Branched-chain amino acids in metabolic signalling and insulin resistance. *Nat. Rev. Endocrinol.* **2014**, *10*, 723–736. [CrossRef] [PubMed]

44. Caimari, A.; Crescenti, A.; Puiggròs, F.; Boqué, N.; Arola, L.; del Bas, J.M. The intake of a high-fat diet and grape seed procyanidins induces gene expression changes in peripheral blood mononuclear cells of hamsters: Capturing alterations in lipid and cholesterol metabolisms. *Genes Nutr.* **2014**, *10*, 438. [CrossRef] [PubMed]

45. O'Grada, C.M.; Morine, M.J.; Morris, C.; Ryan, M.; Dillon, E.T.; Walsh, M.; Gibney, E.R.; Brennan, L.; Gibney, M.J.; Roche, H.M. PBMCs reflect the immune component of the WAT transcriptome—Implications as biomarkers of metabolic health in the postprandial state. *Mol. Nutr. Food Res.* **2014**, *58*, 808–820. [CrossRef] [PubMed]

46. Konieczna, J.; Sánchez, J.; van Schothorst, E.M.; Torrens, J.M.; Bunschoten, A.; Palou, M.; Picó, C.; Keijer, J.; Palou, A. Identification of early transcriptome-based biomarkers related to lipid metabolism in peripheral blood mononuclear cells of rats nutritionally programmed for improved metabolic health. *Genes Nutr.* **2013**, *9*, 366. [CrossRef] [PubMed]

Article

Profound Changes in Net Energy and Nitrogen Metabolites Fluxes within the Splanchnic Area during Overfeeding of Yucatan Mini Pigs That Remain Euglycemic

Isabelle Savary-Auzeloux [1,*], Ahmed-Ben Mohamed [1], Benoit Cohade [1], Dominique Dardevet [1], Jérémie David [1], Noureddine Hafnaoui [1], Carole Migné [2], Estelle Pujos-Guillot [2], Didier Rémond [1] and Sergio Polakof [1]

[1] Université Clermont Auvergne, INRA, UNH, Unité de Nutrition Humaine, CRNH Auvergne, F-63000 Clermont Ferrand, France; ahmed-ben.mohamed@inra.fr (A.-B.M.); benoit.cohade@inra.fr (B.C.); dominique.dardevet@clermont.inra.fr (D.D.); jeremie.david@inra.fr (J.D.); noureddine.hafnaoui@inra.fr (N.H.); didier.remond@inra.fr (D.R.); sergio.polakof@inra.fr (S.P.)
[2] Université Clermont Auvergne, INRA, UNH, Unité de Nutrition Humaine, PFEM, Metabo-Hub Clermont, CRNH Auvergne, F-63000 Clermont Ferrand, France; carole.migne@inra.fr (C.M.); estelle.pujos-guillot@inra.fr (E.P.-G.)
* Correspondence: isabelle.savary-auzeloux@inra.fr; Tel.: +33-(0)-4-73-62-47-32

Received: 15 January 2019; Accepted: 6 February 2019; Published: 19 February 2019

Abstract: A dysregulation of nutrient exchange between tissues (gut, liver, muscles, adipose) occurs during overnutrition and could induce obesity and metabolic diseases. We aimed to evaluate how, in overfed mini pigs, nutrients use and partition were regulated in the gut and liver. Net nutrients fluxes were assessed in the fed (PP) and post absorptive (PA) states at 1, 14 and 60 days of adaptation to overfeeding in five adult Yucatan female multicatheterized minipigs. Pigs PA glycaemia and PP-induced hyperglycemia remained unchanged over the experimental period, suggesting that the management of the excess of energy intake allowed the maintenance of glucose levels. This was associated with (1) an increased PA plasma insulin, (2) an increased gut lactate production (increased lactate net release +89%, 1 h PP, D1 vs. D60) probably from an increased glucose oxidation, (3) a shift in utilization of gluconeogenic precursor (lactate, propionate) in the liver, and (4) a reduced gut utilization of nitrogen moieties for energy purposes (glutamine), a nitrogen sparing effect at the whole body level (decreased plasma urea in PA (−24% D1 vs. D60) and PP states) and a specific increased level of AA involved in lipids handling and bile recycling in the gut lumen (taurine and glycine).

Keywords: liver; gut; obesity; amino acid; glucose; lactate; nutrient flux; short chain fatty acid; minipig

1. Introduction

Prevalence of overweight and obesity has increased dramatically over the last 35 years, as well as the associated chronic diseases (diabetes, cardiovascular diseases, cancers) [1]. Obesity results from an imbalance between energy supply and expenditure due to increased food intake and reduced physical activity. This imbalance, when occurring over a long period of time, induces metabolic dysregulations leading firstly to metabolic adaptations capable to maintain normal homeostatic state and later to the development of metabolic disturbances such as low grade inflammation [2], insulin resistance [3] and, ultimately, pathologies. These late perturbations occur when tissues and organs are no longer capable to deal with the nutrient oversupply and maintain homeostasis compatible with a healthy status.

The most deleterious effects of over-nutrition on tissues and organs are observed on liver (development of non-alcoholic fatty liver disease) [4], subcutaneous and visceral adipose tissues [5], muscle [6], inflammatory cells [7], etc. An altered gut profile of bacteria species [8] and richness [9] have also been demonstrated in obese or overfed or diabetic humans and animals, which explains the increasing interest on the impact of the gut (and its tightly connected microbiota) on the development of insulin resistance and associated pathologies. Consequently, the gut-liver axis has recently regained increasing interest [10]. Interestingly, it has been shown that the splanchnic area (and particularly the liver) was metabolically disturbed early (i.e., within the first days) in streptozotocin treated rats [11,12]. These disturbances at the splanchnic level occur before dysfunctions of peripheral tissues are demonstrated. However, the interaction between gut and liver, although considered to be one of the keys in the development of hepatic and whole body obesity-related pathologies [13–15], has been relatively scarcely studied due to a lack of access to metabolites/factors present the portal vein which is the only vascular link from the gut to the liver. In addition, the investigation of the gut-liver axis is complex due to the combination of various metabolically intertwined tissues and organs in this area (microbiota, gut, liver, visceral adipose tissues). On the top of this, the investigation of the adaptation mechanisms stimulated to deal with excess of nutrients is also complex because gut and liver, but also nutrients and metabolites are at the crossroad of various metabolic pathways leading to a competition between tissues for metabolites utilization [14,16,17].

Consequently, from this analysis of the data obtained so far in the field, we focused on the investigation of (1) the gut-liver nutrients utilization within the first steps of obesity/IR development in a situation of overfeeding and (2) the major routes of exchange of nutrients between gut, liver, and the periphery during obesity development. This approach will help to decipher what the gut actually releases (as nutrients and signals) to the liver, what is taken up/released to the liver, and what is finally available to peripheral tissues in the early phases of adaptation to overfeeding and obesity development.

To address the point concerning the early steps of obesity development, we have recently investigated the whole body and tissues metabolic adaptations by analyzing genes expressions and activities of enzymes (in the fed state) involved in the major metabolic pathways of nutrients utilization/synthesis in gut, liver, and adipose and muscle tissues in Yucatan mini pigs overfed for two months with a high fat, high sucrose diet [18–20]. We have confirmed specific inter-organ metabolic adaptations developed by the pigs to handle the sudden and massive dietary supply of nutrients and we have shown the development of an obese phenotype (rapid weight gain, increased cholesterol, and insulin plasma levels in the fasted state), as well as a reduced potential to phosphorylate glucose, a decreased capacity for de novo lipogenesis and an increased arterial insulin and branched chain plasma amino acids. The present work aims at investigating the adaptive dynamics of the net fluxes of energy and nitrogenous nutrients (i.e., the consequence of the genes regulation) between tissues in the splanchnic area both in the fasted and the fed states. The main objective is to understand how the gut and the liver net uptake and release of major nutrients/metabolites in a situation of overfeeding affects nutrients available to peripheral tissues.

To address this issue, multicatheterized (in artery, portal vein, and hepatic vein) Yucatan minipigs were overfed for a period of two months as previously published [18,19]. Net fluxes of energy and nitrogenous metabolites were measured across the gut and the liver in the fasted and fed states over the several weeks of overfeeding.

2. Materials and Methods

2.1. Animals and Experimental Procedure

The study involved five female adult Yucatan mini-pigs (30 $\pm$ 1 kg). Three weeks before the experimentation, the mini-pigs were surgically fitted with permanent catheters (polyvinyl chloride; 1.1 mm i.d., 1.9 mm o.d.) in the abdominal aorta, the portal vein, and the sus hepatic vein for blood sampling and in mesenteric vein for infusions. The animals were housed in subject pens (1×1.5 m) in a ventilated room with controlled temperature (21 °C) and regular light cycle (L12:D12). They were fed once daily with 400 g/d of a concentrate feed containing 17.5% protein, 3.2% fat, 4.3% cellulose, and 5.2% ash (Porcyprima; Sanders Centre Auvergne, Aigueperse, France) and had free access to tap water. Catheters were flushed and filled with a saline solution containing heparin (1/10) three times a week over the experimental period to avoid the formation of blood clots in the catheters. All procedures were in accordance with the guidelines formulated by the European Community for the use of experimental animals (L358-86/609/EEC, Council Directive, 1986; authorization 02090.01).

At least two weeks after surgery, the mini-pigs were fed a High Fat High Sugar (HFHS) diet consisting in a regular pig diet enriched with fat (13% palm oil) and sugar (10% sucrose) (1 kg/day, 13.3 kJ/day) for two months. Lipids represent 27% of energy supply in the diet. This diet was offered in two meals (500g each meal) at 8:00 am and 16:00 pm. Over the entire experimental period, the animals ingested their meal in no more than 10 min and no refusal was observed. Samplings were performed on D1 (1st day of HFHS exposure), D14 and D60. After an overnight fast, blood samples were simultaneously withdrawn from the aorta, the portal vein and the sus-hepatic vein on heparinized or EDTA treated tubes, before the meal, 30, 60, 120, 180, 240, 330, 420 and 510 min post-meal ingestion. Blood was centrifuged at $4500 \times g$ for 10 min, plasma rapidly collected and stored at −80 °C until further analyses. Body weight was determined weekly.

For measurement of plasma flow, a solution of 0.185 M of sodium p-aminohippurate (PAH) (pH 7.4) was infused in the mesenteric vein at a rate of 12 mL/h. The infusion started 1 h before feeding and lasted over the entire post prandial sampling period. The plasma flows in each vessel was calculated according to the Fick principle [21].

After the two-month experimental period, the mini-pigs were euthanized after an overnight fast by intravenous administration of Dolethal® (pentobarbitone sodium 200 mg/L, Vetoquinol®, Magny-Vernois, France).

2.2. Analytical Procedures

Glucose, lactate, urea, triacylglycerol (TG), High Density Lipoprotein (HDL)-cholesterol, Low Density Lipoprotein (LDL)-cholesterol, total cholesterol, albumin, aspartate aminotransferase (AST), alanine aminotransferase (ALT), and fructosamine concentrations were enzymatically measured using commercial kits on an automotive ABX Pentra 400 (Horiba Medical, Montpellier, France) test system. Plasma insulin levels were assayed using a commercial ELISA kit (Mercodia, Uppsala, Sweden). Data at D1 and D60 for glucose, urea, triacylglycerol (TG), HDL-cholesterol, LDL-cholesterol, total cholesterol, and insulin in artery have already been published [18]. Data for these parameters at D14 have been added to evaluate the short-term vs. longer term kinetics of adaptation to overfeeding.

For amino acid measurements, detailed procedures are described elsewhere [22]. In brief, plasma samples were deproteinised with sulphosalicylic acid after adding norleucine as an internal standard. The supernatant was diluted (2/3) with a lithium injection buffer containing glucosaminic acid as an injection standard and amino acids concentrations were determined with an amino acid analyzer (Hitachi L8900, Sciencetec, Villebon/Yvette, France) by ion exchange chromatography with postcolumn derivatisation with ninhydrine.

PAH concentration in plasma was measured according to [23], 250 µL of plasma samples were deproteinized with sulphosalicylic acid acid, thoroughly mixed and centrifuged at $10,000 \times g$, 4 °C, for 15 min. 80 µL of the supernanant was deacetylated by adding 20 µL of 5 M HCL, followed by an incubation at 90 °C for 1 h [24]. Sodium nitrite (625 mg/L) was then added manually. The samples were then inserted into the automotive ABX Pentra 400 (Horiba Medical, Montpellier, France) which added successively ammonium sulfamate (0.64 g/L) and N-(1-Naphtyl) ethylenediamine dihydrochloride (1 mg/mL). Concentrations were determined by comparison with PAH standard and read out at 600 nm.

Short chain fatty acids (SCFA) in plasma were measured using 1-(tert-butyldimethylsilyl) imidazole (MTBSTFA) derivatization and analysis by gas chromatography (GC) according to [25]. Shortly, to 500 µL of plasma was added 50 µL of a mixture of 13C labelled 1-^{13}C-acetate (4 mM), 1-^{13}C-propionate (1.5 mM), 1-^{13}C-butyrate (0.6 mM) (Cortecnec, Voisins Le Bretonneux, France), and 10 µL of 37% (*v/v*) HCL solution. A total of 2 mL of diethyl ether was added to plasma and the mixture was centrifuged (10 min, 2000 rpm). A total of 50 µL of MTBSTFA (Tokyo Chemical Industry, Tokyo, Japan) was added to supernatant for SCFA derivatization. The mixture was injected in GC–MS system 7890A (Agilent Technologies California, USA) using the splitless mode equipped with a quadrupole detector (5975C) and autoinjector (7683). The ionisation mode was operated in electron impact (electron energy 70 eV). The GC system was fitted with a nonpolar capillary column DB-5 MS (J&W Scientific, Folsom, CA, USA, 30 m × 0.25 mm i.d. × 0.25 µm film thickness) for chromatographic separation. Quantification of the SCFA was performed using the Selected Ion Monitoring acquisition Mode by measurement of the m/z ratios of the specific ^{13}C and ^{12}C ions of each quantified SCFA and comparison to a standard curve: 117/118 (^{12}C acetate/1-^{13}C-acétate), 131/132 (^{12}C propionate/1-^{13}C-propionate), 145/146 (^{12}C butyrate/1-^{13}C-butyrate). SCFA content in the feces was determined by NMR following a water extraction of feces samples.

2.3. Calculations

The HOMA2-IR (Homeostasis Model Assessment 2, Insulin-Resistance) was calculated from arterial insulin and glucose levels by the program HOMA Calculator v2.2.3 (The Oxford Centre for Diabetes, Endocrinology and Metabolism. Diabetes Trials Unit, Oxford University, Oxford, United Kingdom; http://www.dtu.ox.ac.uk/ToolsSoftware/).

Net nutrient fluxes through the gut (viscera drained by the portal vein), the liver and total splanchnic tissues (gut + liver) were calculated as described by Katz et al. [21]. The net nutrient fluxes were calculated as differences between the efferent flux and the afferent flux. Consequently, a positive net flux indicates a net release whereas a negative net flux indicates a net uptake.

Metabolite (MET) net flux across the portal-drained viscera was calculated as follows: ([MET]$_{PV}$-[MET]$_A$) × PF$_{PV}$ where PF$_{PV}$ is the portal plasma flow and [MET]$_{PV,A}$ the plasma concentrations of the metabolite in the portal vein and the aorta, respectively. The net hepatic flux of metabolites was calculated as follows: ([MET]$_{HV}$ × PF$_{HV}$ − ([MET]$_{PV}$ × PF$_{PV}$ + [MET]$_A$ × PF$_{AH}$) where [MET]$_{HV, PV \text{ and } A}$ are the plasma concentrations of the metabolite in the hepatic vein, portal vein and aorta, respectively and PF $_{HV, PV \text{ and } A}$ are the plasma flows in the hepatic vein, portal vein and artery, respectively. Lastly, the net flux of AA and urea across overall splanchnic tissues was calculated as follows: ([MET]$_{HV}$-[MET]$_A$) × PF$_{HV}$ where PF$_{HV}$ is the plasma flow in the hepatic vein and [MET]$_{HV \text{ and } A}$, the metabolite concentrations in the hepatic vein and artery. Indispensable amino acids (IAA) included histidine, isoleucine, leucine, lysine, methionine, phenylalanine, threonine, tryptophane, and valine; non-indispensable amino acids (NIAA) included alanine, glutamate, glutamine, glycine, serine, tyrosine, cysteine, citrulline, ornithine, arginine, proline, 3-methylhistidine; and branched-chain amino acids (BCAA) included isoleucine, leucine, and valine. Total amino acids (TAA) were the sum of IAA and NIAA.

2.4. Statistics

All data are expressed as means ± SEM. Comparisons of data between D0, D14, and D60 in the fasted state were performed using a one way repeated measures ANOVA (SigmaPlot 12, Systat software, San Jose, CA, USA) followed by a post hoc analysis using the Holm Sidak test. Comparison of data between D1, D14, and D60 (adaptation to the diet) and between hours post test meal ingestion in the fed state (t0 (fasted), t30, t60, t120, t180, t240, t330, t420, t510 min post meal ingestion) were performed using a two way repeated measures ANOVA and followed by post hoc analysis using the Holm Sidak test. Differences were considered significant if $p < 0.05$ and as a tendency (t) for $0.05 < p < 0.1$.

3. Results

3.1. Overfeeding, Impact on Pigs' Weight, Insulin Levels, Alteration of Energy Nutrients Concentrations and Net Splanchnic Uptake in the Fasted and Fed States: Data Presented in Tables 1–5 and Figure 6 (for Insulin)

3.1.1. Energy Nutrients

Overfeeding led to a 43% increase of pigs' weight over the 60 days of the experimental period ($p < 0.05$). At the plasma level and as already observed in a similar model of overfed animals [18], lactate, total plasma cholesterol, HDL cholesterol, and LDL cholesterol were significantly ($p < 0.05$) or tended to be increased over the experimental period (+54%, +44%, +58%, D60 vs. D1 for lactate, total and HDL cholesterol, respectively; +34% D14 vs. D1 for LDL cholesterol) whereas no effect on glucose, triglycerides, albumin, alanine aminotransferase, and aspartate aminotransferase was observed in the fasted state at D14 and D60 relatively to D1 (Table 1). The stable glucose arterial levels observed between D1 and D60, suggests that animals were still capable to maintain their glycaemia although glucose content in the diet (supplied as saccharose or starch) and net plasma gut release of glucose is higher at D60 vs. D14 and D1 in the fasted state after adaptation to overfeeding ($4.9\times$ between D1 and D60, $p = 0.06$, Table 4). This active glucose utilization by the liver (net splanchnic release in the fasted state not significantly altered between D1, D14 and D60, Table 4) can be associated with the increased insulin levels ($p < 0.05$), as well as HOMA index (Figure 6) in the artery at D14 and D60.

Table 1. Animals' weight, arterial concentration in the fasted state of glucose, lactate, urea, triglycerides, total cholesterol, Low Density Lipoprotein (LDL)-cholesterol, High Density Lipoprotein (HDL)-cholesterol, albumin, alanine aminotransferase (ALT), aspartate aminotransferase (AST) before (D1) and after 14 days (D14) and 60 days (D60) of adaptation to High Fat High Sugar (HFHS) diet. Average plasma flow in portal vein (PV), hepatic vein (HV), and hepatic artery (HA) from 0 to 8.5 h post meal ingestion before (D1) and after 14 days (D14) and 60 days (D60) of adaptation to HFHS diet.

Parameter	Vessel	D1	D14	D60	ANOVA
Pig's weight Kg		31.52 ± 1.49 a	35.70 ± 1.51 b	45.16 ± 1.56 c	<0.001
Glucose mmol/L	A	5.20 ± 0.35	5.46 ± 0.29	5.26 ± 0.11	0.85
Lactate mmol/L	A	0.55 ± 0.05 a	0.86 ± 0.94 b	0.85 ± 0.05 b(t)	0.03
Urea mmol/L	A	5.34 ± 0.43 a	3.83 ± 0.31 b	4.04 ± 0.31 b(t)	0.03
Triglycerides mmol/L	A	0.19 ± 0.01	0.25 ± 0.04	0.24 ± 0.05	0.15
Total Cholesterol mmol/L	A	1.76 ± 0.10 a	2.79 ± 0.09 b	2.53 ± 0.16 b	<0.001
LDL Cholesterol mmol/L	A	0.97 ± 0.03 a	1.30 ± 0.02 b	1.00 ± 0.08 a	<0.001
HDL Cholesterol mmol/L	A	0.67 ± 0.03 a	1.28 ± 0.09 b	1.06 ± 0.08 c	<0.001
Albumin µmol/L	A	536 ± 20	544 ± 14	537 ± 17	0.70
ALT U/L	A	29.7 ± 0.8	27.1 ± 1.8	25.4 ± 1.7	0.15
AST U/L	A	36.2 ± 4.2	37.9 ± 5.2	28.8 ± 2.3	0.36
Plasma flow L/min	PV	0.60 ± 0.05	0.69 ± 0.05	0.66 ± 0.03	0.37

Table 1. *Cont.*

Parameter	Vessel	D1	D14	D60	ANOVA
Plasma flow L/min	HV	0.78 ± 0.03	0.82 ± 0.08	0.84 ± 0.01	0.80
Plasma flow L/min	HA	0.17 ± 0.02	0.13 ± 0.03	0.18 ± 0.03	0.34

Part of the data (D1 and D60) for glucose, lactate, triglycerides, cholesterol, and urea are published elsewhere [18]. Values are means ± SEM; for details on the statistical treatment of the data see Materials and Methods.

Lactate levels were severely altered in the fasted state, as previously shown in a study in a similar model [18]: arterial, portal and hepatic vein concentrations increased significantly at D14 and D60 vs. D1 (+34% and +89% at D60 vs. D1 in the portal and hepatic vein, respectively, $p < 0.05$, data not shown). Only an increased net hepatic lactate release between D1 and D60 was observed ($p = 0.03$) (Table 4).

Fasting arterial short chain fatty acids levels (acetate + propionate + butyrate: C2 + C3 + C4) were increased (+37%, D1 vs. D60, $p < 0.05$), essentially due to an increased acetate level (+38%, D1 vs. D60, $p < 0.05$), which is the only major SCFA significantly altered at the arterial level (Table 2). Minor SCFA (isobutyrate and isovalerate) were also significantly modified over the experimental period: their concentration decreased between D1 and D14 and significantly re-increased between D14 and D60 (+32% for isobutyrate and +25% for isovalerate between D14 and D60) (Table 2). Except for fasting propionate in the hepatic vein that tend to be increased at D60 (+23% D60 vs. D1 $p = 0.06$, Table 2), the SCFA levels in the hepatic vein are not significantly modified. However, they tended to be increased in the portal vein (+18%, D60 vs. D1 for C2 + C3 + C4, $p = 0.09$, Table 2). Concerning the net overall splanchnic release of SCFA (Table 4), it was not modified by overfeeding even if the net gut release of butyrate (+43%, D60 vs. D1, $p = 0.05$) and to a lesser extend propionate (+33%, D60 vs. D1, $p = 0.09$) increased over the experimental period. Quantitatively, it also should be noted that except for acetate, the net splanchnic release represents only a small percentage of what is released by the gut: 3%, 27%, 7%, 11% at D1 for propionate, butyrate, isobutyrate and isovalerate, respectively, implying an important utilization of these molecules by the liver, whatever the diet (Table 4).

Table 2. Acetate (C2), propionate (C3), butyrate (C4), isobutyrate, isovalerate and C2+C3+C4 concentrations in the fasted state in artery (A), portal vein (PV), and hepatic vein (HV) before (D1) and after 14 days (D14) and 60 days (D60) of adaptation to HFHS diet.

Metabolite	Vessel	D1	D14	D60	ANOVA
Acetate (C2) µmol/L	A	552.3 ± 14.5 a	572.1 ± 34.6 a	761.8 ± 50.0 b	0.02
	PV	844.8 ± 53.8	897.0 ± 54.7	1000.2 ± 25.5	0.07
	HV	744.5 ± 53.1	861.3 ± 81.2	896.2 ± 41.0	0.33
Propionate (C3) µmol/L	A	22.7 ± 0.5	26.8 ± 4.4	27.1 ± 4.4	0.50
	PV	199.8 ± 22.6	232.0 ± 30.2	229.5 ± 16.7	0.50
	HV	26.1 ± 1.2 a	28.0 ± 2.7 ab	32.0 ± 4.4 b(t)	0.05
Butyrate (C4) µmol/L	A	2.0 ± 0.6	5.0 ± 1.4	2.8 ± 1.1	0.30
	PV	63.0 ± 6.7	80.7 ± 11.1	81.0 ± 4.3	0.09
	HV	14.4 ± 1.7	19.9 ± 6.5	15.4 ± 2.1	0.62
Isobutyrate µmol/L	A	2.3 ± 0.1 ab	1.9 ± 0.19 b	2.5 ± 0.2 a	0.03
	PV	15.3 ± 2.1 a	11.1 ± 1.2 b(t)	12.8 ± 0.2 ab	0.08
	HV	3.0 ± 0.2	1.8 ± 0.6	2.8 ± 0.3	0.20
Isovalerate µmol/L	A	1.8 ± 0.1 a	1.6 ± 0.1 b	2.0 ± 0.1 c	0.002
	PV	8.7 ± 1.4	7.0 ± 0.6	8.3 ± 0.2	0.12
	HV	2.3 ± 0.3	2.5 ± 0.5	2.6 ± 0.2	0.37
C2+C3+C4 µmol/L	A	577 ± 17 a	603 ± 40 a	792 ± 55 b	0.02
	PV	1108 ± 81	1210 ± 91	1311 ± 32	0.09
	HV	851 ± 50	909 ± 88	944 ± 37	0.46

Values are means ± SEM; for details on the statistical treatment of the data see Materials and Methods.

3.1.2. Nitrogenous Nutrients

Urea levels were strongly decreased in the fasted state in all blood vessels at D14 (-39%, -26% and -27% at D14 vs. D0 in artery, portal vein and hepatic vein respectively, $p < 0.05$), and to a lesser extent at D60 (Table 1). The small differences in concentrations between vessels did not allow us to determine significant variations in net gut or hepatic uptake/release of urea throughout the experimental period (Table 4) although sites or urea recycling and synthesis are mainly located at the splanchnic level. Arterial BCAA, and particularly leucine levels, were significantly increased between D1 and D60 ($+14\%$ and $+18\%$ for BCAA and leucine, respectively, Table 3), as already shown in a previous paper, using other methodologies for assessment of total BCAA levels [19]. Other AA, such as tryptophan, tended to increase ($+28\%$ D60 vs. D1, $p = 0.09$) as well as non-indispensable AA, such as glycine ($+26\%$, D60 vs. D1, $p = 0.01$), proline ($+76\%$, D60 vs. D1, $p = 0.003$), and serine ($+24\%$, D60 vs. D1, $p = 0.09$) (Table 3). Taken together, arterial NIAA and TAA (but not IAA) were increased at D60 vs. D1 ($+26\%$ and $+17\%$ for arterial NIAA and TAA, $p = 0.03$ and $p = 0.07$, respectively). On the contrary, methionine, and to a lesser extent phenylalanine, decreased over the same period (-28% ($p = 0.02$) and -17% ($p = 0.06$) between D1 and D14 for methionine and phenylalanine respectively). Except for methionine and phenylalanine, the significant increased arterial concentration observed for leucine, isoleucine, valine, tryptophane, glycine, serine, proline, NIAA, and TAA was (or tended to be) present also in the portal and hepatic veins (Table 3).

Table 3. Amino acids concentrations (μmol/L) in the fasted state in artery (A), portal vein (PV) and hepatic vein (HV) before (D1) and after 14 days (D14) and 60 days (D60) of adaptation to HFHS diet.

Metabolite	Vessel	D1	D14	D60	ANOVA
Leucine	A	145.2 ± 8.3 a	146.9 ± 5.8 a	172.0 ± 9.6 b	0.03
μmol/L	PV	149.0 ± 10.5 a	151.1 ± 7.3 a	204.0 ± 7.2 b	0.01
	HV	150.2 ± 8.9 a	161.0 ± 9.8 a	190.8 ± 12.5 b(t)	0.06
Isoleucine	A	102.9 ± 8.0 a	126.3 ± 2.6 b	120.8 ± 2.2 b(t)	0.03
μmol/L	PV	103.8 ± 10.7 a	127.4 ± 2.5 ab	139.3 ± 5.6 b(t)	0.06
	HV	104.2 ± 9.5 a	137.8 ± 7.5 b(t)	134.0 ± 7.0 b(t)	0.05
Valine	A	275.2 ± 14.5 a	276.7 ± 12.0 a	302.8 ± 16.9 b(t)	0.06
μmol/L	PV	276.0 ± 15.0 a	275.8 ± 11.8 a	336.5 ± 5.2 b	0.02
	HV	279.8 ± 14.4 a	285.3 ± 10.0 a	323.4 ± 16.5 b(t)	0.06
Lysine	A	139.5 ± 14.4	157.7 ± 18.5	166.4 ± 5.0	0.37
μmol/L	PV	143.4 ± 18.5	164.2 ± 12.3	193.5 ± 10.0	0.24
	HV	134.0 ± 15.7	167.8 ± 9.8	171.8 ± 9.9	0.08
Phenylalanine	A	56.8 ± 2.1 a	47.2 ± 2.0 b(t)	53.0 ± 2.0 ab	0.06
μmol/L	PV	58.8 ± 3.3	52.4 ± 3.7	66.0 ± 2.8	0.12
	HV	54.4 ± 2.8	51.0 ± 3.0	53.2 ± 1.8	0.68
Methionine	A	26.4 ± 2.3 a	19.0 ± 0.8 b	19.0 ± 0.4 b	0.02
μmol/L	PV	26.0 ± 3.1	19.8 ± 0.9	22.8 ± 1.0	0.25
	HV	22.2 ± 2.2	21.0 ± 1.5	19.8 ± 0.9	0.65
Threonine	A	132.8 ± 13.2	123.2 ± 7.8	135.3 ± 4.9	0.43
μmol/L	PV	129.8 ± 12.2	123.6 ± 5.3	151.3 ± 6.9	0.25
	HV	129.4 ± 9.4	123.0 ± 5.4	130.8 ± 6.6	0.97
Tryptophane	A	28.0 ± 2.0 a	30.8 ± 1.4 a	35.8 ± 1.9 b(t)	0.09
μmol/L	PV	27.8 ± 2.2 a	27.6 ± 1.4 a	36.0 ± 2.8 b(t)	0.07
	HV	27.0 ± 2.0 a	35.5 ± 1.0 b(t)	32.2 ± 2.7 ab	0.05
Histidine	A	74.0 ± 4.2	69.4 ± 2.5	76.3 ± 1.4	0.34
μmol/L	PV	75.0 ± 5.9	72.0 ± 2.2	84.8 ± 1.4	0.28
	HV	71.0 ± 5.2	70.5 ± 1.0	78.0 ± 2.0	0.39
Alanine	A	202.8 ± 31.5	216.4 ± 24.4	222.3 ± 19.6	0.93
μmol/L	PV	233.0 ± 42.7	268.8 ± 26.4	286.0 ± 19.1	0.68
	HV	158.0 ± 32.2	224.3 ± 21.5	215.4 ± 12.2	0.20

Table 3. *Cont.*

Metabolite	Vessel	D1	D14	D60	ANOVA
Glutamate	A	129.6 ± 15.4	156.8 ± 12.3	140.8 ± 15.7	0.48
μmol/L	PV	120.2 ± 16.9	130.4 ± 9.5	131.0 ± 19.3	0.90
	HV	280.8 ± 44.0	279.8 ± 53.2	357.4 ± 42.2	0.41
Glutamine	A	244.5 ± 13.8	215.3 ± 24.2	213.5 ± 16.2	0.46
μmol/L	PV	200.4 ± 12.6	179.4 ± 12.4	183.3 ± 15.6	0.50
	HV	196.4 ± 11.7	190.3 ± 9.9	181.3 ± 15.3	0.75
Glycine	A	624.4 ± 72.2 a	864.6 ± 51.9 b	785.3 ± 39.2 b	0.01
μmol/L	PV	650.2 ± 84.0 a	977.2 ± 59.5	953.3 ± 19.5	0.007
	HV	620.0 ± 78.2 a	894.3 ± 57.3 b	824.0 ± 41.5	0.001
Serine	A	133.8 ± 7.2 a	149.8 ± 4.3 ab	165.8 ± 15.4 b(t)	0.09
μmol/L	PV	130.4 ± 10.8 a	152.6 ± 5.7 b(t)	182.8 ± 13.0 c	0.009
	HV	132.4 ± 11.1	156.5 ± 10.6	158.4 ± 11.5	0.11
Tyrosine	A	66.4 ± 7.7	62.0 ± 7.5	76.5 ± 12.2	0.28
μmol/L	PV	66.4 ± 7.0 a	66.2 ± 9.3 ab	79.0 ± 14.2 b(t)	0.09
	HV	60.0 ± 7.2	69.3 ± 12.0	68.2 ± 9.0	0.17
Cystine	A	24.4 ± 2.6	22.2 ± 2.1	26.0 ± 0.9	0.66
μmol/L	PV	26.0 ± 3.8	25.2 ± 1.2	29.5 ± 0.5	0.74
	HV	27.0 ± 3.3	23.0 ± 0.9	24.4 ± 2.3	0.75
Citrulline	A	69.6 ± 5.9	77.4 ± 4.9	70.0 ± 4.3	0.38
μmol/L	PV	86.2 ± 8.3	92.8 ± 4.7	97.5 ± 5.3	0.52
	HV	78.6 ± 6.6	92.3 ± 6.4	89.8 ± 5.1	0.13
Ornithine	A	54.6 ± 4.7 a	76.2 ± 7.6 b(t)	68.0 ± 4.3 ab	0.06
μmol/L	PV	61.8 ± 6.3	83.2 ± 8.9	77.8 ± 5.1	0.17
	HV	59.6 ± 6.1 a	86.5 ± 6.1 b	75.6 ± 4.2 ab	0.04
Arginine	A	87.8 ± 6.0	87.4 ± 9.2	100.8 ± 4.6	0.60
μmol/L	PV	86.0 ± 12.0	93.4 ± 8.7	114.5 ± 4.5	0.32
	HV	80.6 ± 9.4	98.8 ± 7.0	102.0 ± 4.2	0.20
Taurine	A	72.7 ± 2.3	75.1 ± 4.5	73.8 ± 2.4	0.85
μmol/L	PV	76.3 ± 2.7	76.0 ± 4.3	74.0 ± 4.8	0.81
	HV	77.1 ± 3.0	80.7 ± 5.6	74.8 ± 3.8	0.78
Proline	A	215.8 ± 33.3 a	379.2 ± 10.7 b	379.0 ± 9.8 b	0.003
μmol/L	PV	208.4 ± 26.5 a	381.4 ± 6.6 b	379.3 ± 40.6 b	0.006
	HV	229.8 ± 35.7 a	394.8 ± 15.5 b	402.6 ± 7.7 b	0.002
3-methyl histidine	A	24.2 ± 0.6 ab	21.2 ± 2.5 a	28.5 ± 2.2 b(t)	0.09
μmol/L	PV	24.2 ± 0.8	24.2 ± 1.2	25.3 ± 5.0	0.99
	HV	26.0 ± 2.2	22.0 ± 2.4	29.6 ± 2.1	0.13
Carnosine	A	16.8 ± 1.9	20.2 ± 2.3	18.5 ± 2.2	0.17
μmol/L	PV	17.8 ± 2.2	19.4 ± 1.5	18.0 ± 1.2	0.72
	HV	20.6 ± 2.1	22.0 ± 2.9	20.4 ± 2.2	0.98
BCAA	A	523.3 ± 29.2 a	549.8 ± 19.4 ab	595.6 ± 28.5 b(t)	0.07
μmol/L	PV	528.8 ± 35.3 a	554.3 ± 20.0 a	679.8 ± 16.2 b	0.03
	HV	534.2 ± 31.6 a	581.0 ± 24.3 ab	648.2 ± 35.0 b(t)	0.06
IAA	A	985.9 ± 57.8	997.1 ± 41.8	1081.2 ± 32.9	0.22
μmol/L	PV	989.6 ± 73.7	1009.9 ± 43.85	1234.0 ± 33.4	0.09
	HV	972 ± 58.8	1044.5 ± 38.6	1107.8 ± 48.9	0.19
NIAA	A	1877.9 ± 102.3 a	2330.3 ± 120.7 b	2276.2 ± 14.3 b(t)	0.03
μmol/L	PV	1873.2 ± 145.9 a	2474.8 ± 123.8 b	2539.1 ± 85.7 b	0.01
	HV	1929.2 ± 165.4 a	2530.5 ± 132.7 b	2528.7 ± 95.3 b	0.01
TAA	A	2742.4 ± 147.0 a	3198.4 ± 157.4 a	3210.9 ± 44.8 b(t)	0.07
μmol/L	PV	2771.0 ± 205.9 a	3352.3 ± 161.7 b(t)	3609.0 ± 88.2 b (t)	0.04
	HV	2789.8 ± 207.2 a	3440.7 ± 160.5 b	3498.7 ± 139.7 b	0.03

Branched chain amino acids (BCAA): leucine + isoleucine + valine; Indispensable amino acids (IAA): leucine + isoleucine + valine + lysine + phenylalanine + methionine + threonine + histidine + tryptophane; Non indispensable amino acids (NIAA): alanine + glutamate + glutamine + glycine + tyrosine + citrulline + cystine + 3 methyl-histidine + ornithine + arginine + proline + serine; Total amino acids (TAA): NIAA + IAA. Values are means ± SEM; for details on the statistical treatment of the data see Materials and Methods.

Table 4. Net gut, liver and splanchnic (gut + liver) fluxes of acetate (C2), propionate (C3), butyrate (C4), isobutyrate, isovalerate and C2+C3+C4 in the fasted state before (D1), 14 days (D14) and 60 days (D60) of adaptation to the HFHS diet.

Metabolite		D1	D14	D60	ANOVA
Glucose	Gut	6.84 ± 6.67 a	22.45 ± 6.31 a	33.41 ± 5.78 b(t)	0.05
mmol/h	Liver	29.78 ± 7.32	31.60 ± 2.65	36.31 ± 11.17	0.51
	Splanchnic	34.72 ± 12.18	50.81 ± 10.23	59.54 ± 15.16	0.56
Lactate	Gut	4.85 ± 1.49	1.45 ± 1.14	2.04 ± 0.73	0.17
mmol/h	Liver	−1.78 ± 3.08 a	2.90 ± 4.15 a	8.53 ± 0.63 b	0.03
	Splanchnic	2.85 ± 2.36	3.63 ± 4.36	10.57 ± 0.77	0.09
Urea	Gut	1.78 ± 0.89	5.74 ± 1.92	2.24 ± 3.49	0.47
mmol/h	Liver	7.11 ± 2.45	3.36 ± 1.47	6.60 ± 2.57	0.49
	Splanchnic	7.10 ± 1.55	9.69 ± 0.94	9.31 ± 2.29	0.32
Acetate	Gut	10,645 ± 1736	12,921 ± 1738	8717 ± 1563	0.47
μmol/h	Liver	−722 ± 2416	689 ± 1607	−418 ± 1205	0.42
	Splanchnic	9984 ± 1916	14,635 ± 3153	8299 ± 2739	0.26
Propionate	Gut	6193 ± 688	8210 ± 1038	8227 ± 664	0.09
μmol/h	Liver	−6447 ± 817	−8822 ± 622	−8191 ± 1205	0.11
	Splanchnic	164 ± 55	252 ± 119	35 ± 130	0.57
Butyrate	Gut	2176 ± 296 a	3048 ± 295 b(t)	3121 ± 189 b(t)	0.05
μmol/h	Liver	−1583 ± 357	−2362 ± 280	−2515 ± 187	0.23
	Splanchnic	581 ± 84	747 ± 216	607 ± 122	0.79
Isobutyrate	Gut	465 ± 86	375 ± 46	414 ± 29	0.37
μmol/h	Liver	−428 ± 99	−386 ± 32	−409 ± 27	0.64
	Splanchnic	31 ± 16	4 ± 29	5 ± 16	0.71
Isovalerate	Gut	252 ± 51	218 ± 18	253 ± 13	0.56
μmol/h	Liver	−227 ± 53	−177 ± 15	−229 ± 20	0.54
	Splanchnic	27 ± 15	44 ± 17	24 ± 9	0.85
C2+C3+C4	Gut	19,014 ± 2608	24,179 ± 2961	20,065 ± 816	0.50
μmol/h	Liver	−8476 ± 4498	−10,494 ± 961	−11,124 ± 1948	0.87
	Splanchnic	12,173 ± 1550	15,634 ± 3346	8942 ± 2720	0.40

Values are means ± SEM; for details on the calculations and statistical treatment of the data see Materials and Methods.

Although concentrations of several AA were modified between D1 and D60, an absence of impact of adaptation to overfeeding on nearly all the net gut, hepatic and splanchnic amino acids fluxes in the fasted state (Table 5) were observed. Only an increased glutamate net gut uptake at D14 (+ 180%, $p < 0.05$), a tendency for an increased glycine release by the gut at D14 and D60 (+339% at D14, $p < 0.1$) and alterations of tryptophan net plasma gut release and liver uptake ($p < 0.05$) occurred.

Table 5. Net gut, liver and splanchnic (gut + liver) fluxes of amino acids in the fasted state before (D1) and after 14 days (D14) and 60 days (D60) of adaptation to HFHS diet.

Metabolite	Area	D1	D14	D60	ANOVA
Leucine	Gut	147 ± 227	188 ± 133	793 ± 37	0.28
μmol/h	Liver	152 ± 84	343 ± 566	−71 ± 167	0.93
	Splanchnic	246 ± 182	620 ± 531	586 ± 194	0.86
Isoleucine	Gut	37 ± 143	46 ± 83	423 ± 21	0.27
μmol/h	Liver	71 ± 61	290 ± 367	−13 ± 141	0.84
	Splanchnic	70 ± 116	388 ± 357	343 ± 120	0.76
Valine	Gut	46 ± 265	−38 ± 71	619 ± 159	0.20
μmol/h	Liver	291 ± 75	446 ± 625	54 ± 237	0.97
	Splanchnic	238 ± 254	427 ± 592	623 ± 156	0.90
Lysine	Gut	154 ± 218	248 ± 110	816 ± 47	0.22
μmol/h	Liver	−341 ± 90	−31 ± 407	−948 ± 111	0.24
	Splanchnic	−238 ± 186	280 ± 391	−160 ± 101	0.55

Table 5. *Cont.*

Metabolite	Area	D1	D14	D60	ANOVA
Phenylalanine	Gut	81 ± 142	230 ± 131	563 ± 104	0.37
µmol/h	Liver	-161 ± 89	-145 ± 207	-512 ± 27	0.36
	Splanchnic	-101 ± 110	-157 ± 125	-1 ± 75	0.51
Methionine	Gut	-9 ± 61	2 ± 24	121 ± 26	0.61
µmol/h	Liver	-157 ± 92	84 ± 71	-105 ± 62	0.53
	Splanchnic	-176 ± 108	110 ± 44	-1 ± 35	0.1
Threonine	Gut	-64 ± 198	-8 ± 119	359 ± 37	0.55
µmol/h	Liver	20 ± 268	-6 ± 274	-639 ± 240	0.3
	Splanchnic	-97 ± 423	43 ± 328	-307 ± 106	0.76
Tryptophane	Gut	-3 ± 32 ab	-123 ± 30 b	111 ± 68 a	0.03
µmol/h	Liver	-29 ± 10 ab	305 ± 136 b	-328 ± 218 a(t)	0.05
	Splanchnic	-43 ± 31	166 ± 139	-189 ± 163	0.1
Histidine	Gut	42 ± 114	105 ± 42	319 ± 37	0.34
µmol/h	Liver	-145 ± 30	-35 ± 129	-288 ± 90	0.38
	Splanchnic	-132 ± 110	$97 \pm 110\,103$	$36 \pm 110\,58$	0.47
Alanine	Gut	1128 ± 583	2141 ± 447	2917 ± 719	0.23
µmol/h	Liver	-3106 ± 680	-2776 ± 1204	-2989 ± 1263	0.93
	Splanchnic	-2065 ± 837	-318 ± 965	-810.4 ± 863	0.33
Glutamate	Gut	-357 ± 268 a	-1023 ± 228 b(t)	-567 ± 72 a	0.04
µmol/h	Liver	6387 ± 1315	7692 ± 2556	9063 ± 273	0.68
	Splanchnic	5976 ± 1502	6531 ± 2759	8926 ± 434	0.61
Glutamine	Gut	-1545 ± 265	-1612 ± 547	-1628 ± 248	0.90
µmol/h	Liver	-468 ± 673	157 ± 413	-440 ± 395	0.53
	Splanchnic	-2106 ± 757	-1460 ± 1079	-1888 ± 208	0.89
Glycine	Gut	1049 ± 884 a	4605 ± 978 b(t)	5072 ± 1916 ab	0.08
µmol/h	Liver	-1003 ± 373	-4675 ± 1412	-4549 ± 2028	0.31
	Splanchnic	-170 ± 669	744 ± 977	562 ± 232	0.74
Serine	Gut	-110 ± 211	70 ± 246	428 ± 64	0.52
µmol/h	Liver	107 ± 84	-99 ± 421	-810 ± 193	0.32
	Splanchnic	-52 ± 288	103 ± 509	-430 ± 104	0.58
Tyrosine	Gut	7 ± 150	151 ± 89	371 ± 62	0.29
µmol/h	Liver	-259 ± 96	-167 ± 238	-605 ± 160	0.51
	Splanchnic	-279 ± 141	21 ± 250	-251 ± 79	0.64
Cystine	Gut	49 ± 123	120 ± 35	168 ± 47	0.79
µmol/h	Liver	57 ± 137	-141 ± 101	-350 ± 135	0.39
	Splanchnic	98 ± 208	-46 ± 114	-124 ± 93	0.94
Citrulline	Gut	580 ± 107	662 ± 146	1029 ± 44	0.19
µmol/h	Liver	-161 ± 202	-88 ± 221	-167 ± 130	0.82
	Splanchnic	396 ± 184	581 ± 405	920 ± 108	0.38
Ornithine	Gut	230 ± 84	293 ± 77	366 ± 48	0.79
µmol/h	Liver	5 ± 61	-99 ± 115	-68 ± 62	0.54
	Splanchnic	261 ± 112	225 ± 61	315 ± 38	0.36
Arginine	Gut	-75 ± 231	234 ± 41	638 ± 74	0.14
µmol/h	Liver	-218 ± 120	-3 ± 181	-623 ± 96	0.18
	Splanchnic	-324 ± 192	226 ± 184	38 ± 119	0.34
Taurine	Gut	149 ± 80	24 ± 44	132 ± 162	0.65
µmol/h	Liver	144 ± 67	185 ± 170	22 ± 181	0.75
	Splanchnic	223 ± 75	117 ± 121	225 ± 105	0.25
Proline	Gut	-237 ± 360	-14 ± 370	108 ± 2154	0.97
µmol/h	Liver	955 ± 1483	555 ± 1110	695 ± 2464	0.21
	Splanchnic	653 ± 1598	472 ± 1434	811 ± 376	0.95
3-methyl	Gut	1 ± 20	47 ± 31	28 ± 14	0.27
histidine	Liver	75 ± 97	-102 ± 108	22 ± 18	0.21
µmol/h	Splanchnic	67 ± 94	-73 ± 73	38 ± 24	0.31
Carnosine	Gut	39 ± 17	-17 ± 67	40 ± 79	0.63
µmol/h	Liver	162 ± 144	101 ± 96	25 ± 73	0.77
	Splanchnic	194 ± 147	44 ± 49	144 ± 57	0.47

Table 5. *Cont.*

Metabolite	Area	D1	D14	D60	ANOVA
BCAA	Gut	230 ± 629	196 ± 279	1834 ± 147	0.23
µmol/h	Liver	553 ± 203	1079 ± 1556	−29 ± 544	0.93
	Splanchnic	533 ± 543	1435 ± 1474	1552 ± 448	0.88
IAA	Gut	432 ± 1166	549 ± 586	4123 ± 299	0.17
µmol/h	Liver	−300 ± 618	1120 ± 2610	−5185 ± 1781	0.28
	Splanchnic	−235 ± 1239	2026 ± 2115	−816 ± 1210	0.61
NIAA	Gut	720 ± 2330	5674 ± 1476	8962 ± 5024	0.14
µmol/h	Liver	2371 ± 2231	257 ± 4121	−821 ± 6378	0.93
	Splanchnic	2409 ± 3985	7005 ± 3836	8118 ± 1275	0.64
TAA	Gut	1307 ± 3243	6141 ± 1766	12,704 ± 5186	0.15
µmol/h	Liver	2099 ± 1873	1542 ± 6094	−5203 ± 6534	0.80
	Splanchnic	2400 ± 4524	8945 ± 5525	7822 ± 1020	0.77

Branched chain amino acids (BCAA): leucine + isoleucine + valine; Indispensable amino acids (IAA): leucine + isoleucine + valine + lysine + phenylalanine + methionine + threonine + histidine + tryptophane; Non indispensable amino acids (NIAA): alanine + glutamate + glutamine + glycine + tyrosine + citrulline + cystine + 3 methyl-histidine + ornithine + arginine + proline + serine; Total amino acids (TAA): NIAA + IAA. Values are means ± SEM; for details on the calculations and statistical treatment of the data see Materials and Methods.

3.2. Overfeeding, Impact on Pigs' Weight, Insulin Levels, Alteration of Energy Nutrient Concentrations, and Net Splanchnic Uptake in the Fed State. Data Presented in Figures 1–6

3.2.1. Energy Nutrients

Even if arterial, portal, and sus-hepatic vein glucose concentrations were increased after meal ingestion (PP time < 0.05), the duration of adaptation to overfeeding (D1 vs. D14 vs. D60) did not modify the plasma postprandial glucose profile in artery, portal, or hepatic veins (Figure 1). These data are consistent with what found in the fasted state. No significant "day effect" was found concerning postprandial glucose net gut and splanchnic release (Figure 1).

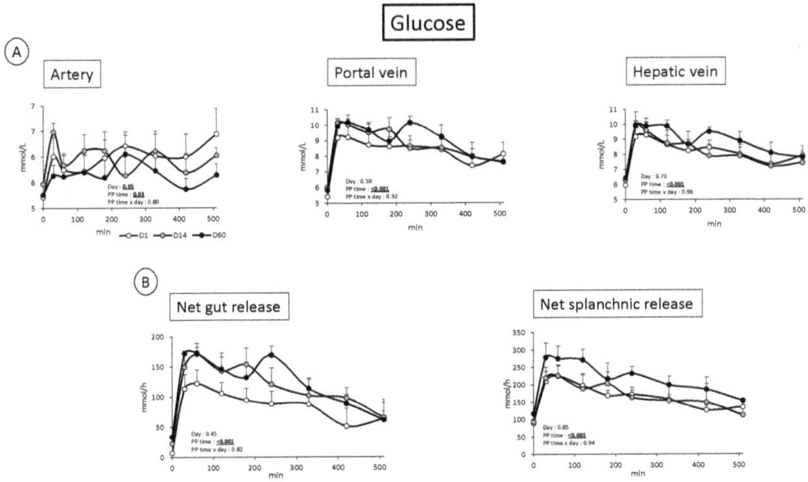

Figure 1. (**A**) Glucose concentration (mmol/L) after meal intake (post prandial state, PP: 0 to 510 min) in artery, portal vein, and hepatic vein. (**B**) Net glucose gut and splanchnic (gut + liver) fluxes (mmol/h) after meal intake (Post prandial state, PP: 0 to 510 min) before (D1) and after 14 days (D14) and 60 days (D60) of adaptation to HFHS diet. Two-way RM ANOVA, * D14 significantly different from D1, † D60 significantly different from D1, ‡ D60 significantly different from D14; Values are means ± SEM. For calculations and statistical treatment of the data and for details, see Materials and Methods.

On the contrary, and consistently to what observed in the fasted state, arterial plasma as well as portal and hepatic veins lactate concentration were more elevated 60 min after feeding at D60 compared to D1 (PP time × day effect: $p < 0.001$ in the three vessels, Figure 3). This was accompanied with a significant (PP time × Day: $p < 0.002$) increased net gut and splanchnic release of lactate. Lastly, insulin was increased post-prandially (particularly 30 min after meal ingestion) to a similar extend at D1 and D60 whereas this increase was of a greater importance at D14 (day effect: 0.03) (Figure 2).

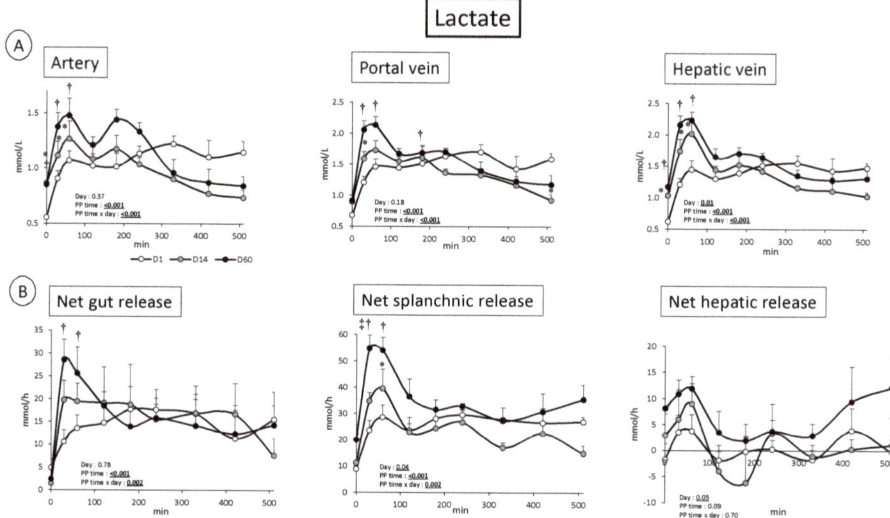

Figure 2. (**A**) Lactate concentration (mmol/L) after meal intake (Post prandial state, PP: 0 to 510 min) in artery, portal vein and hepatic vein; and (**B**) net lactate gut, splanchnic (gut + liver) and hepatic fluxes (mmol/h) after meal intake (post prandial state, PP: 0 to 510 min) before (D1) and after 14 days (D14) and 60 days (D60) of adaptation to HFHS diet. Two way RM ANOVA, * D14 significantly different from D1, † D60 significantly different from D1, ‡ D60 significantly different from D14; Values are means ± SEM. For calculations and statistical treatment of the data and for details, see Materials and Methods.

3.2.2. Nitrogenous Nutrients

Similarly to what observed in the fasted state, urea levels were lower in the fed state in all blood vessels at D14 ($p < 0.05$), and to a lesser extent at D60 (Figure 3). Again, the small differences in concentrations between vessels did not allow us to determine significant variations in net gut or hepatic uptake/release of urea throughout the experimental period.

In the fed state, and as could be expected, a significant (PP time effect: $p < 0.01$) increased TAA, IAA and NIAA concentration in the 3 vessels was observed with a peak 1 and 3 h post meal ingestion (IAA data shown in Figure 3). In artery, only glycine and proline levels were not increased or did not tend to be altered by meal ingestion (Figure 5, data not shown for Proline). Similarly to many other AA, glutamine was increased in the fed state in all vessels (PP time effect: $p < 0.001$, $p < 0.001$, and $p = 0.079$ in artery, portal vein, and hepatic vein, respectively) (Figure 4).

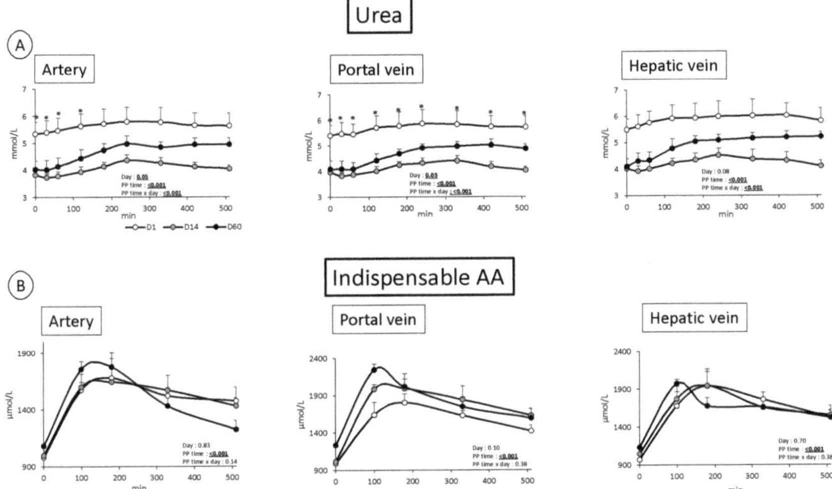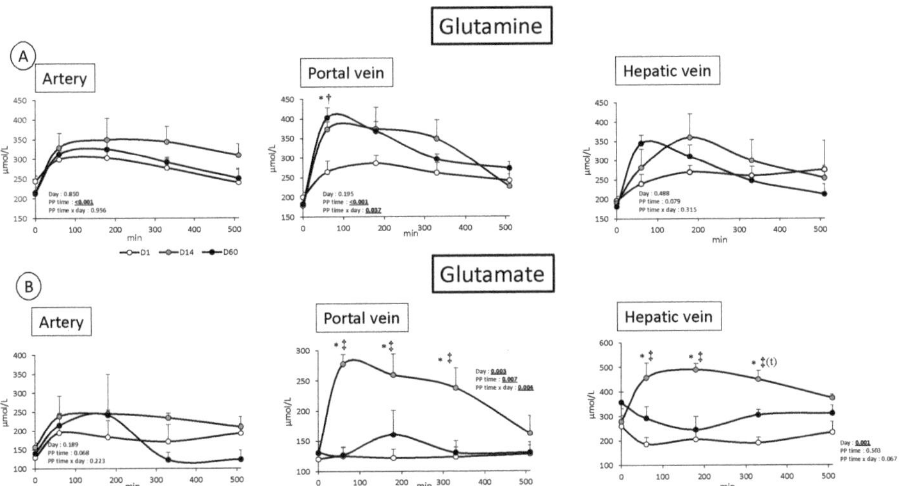

Figure 3. (**A**) Urea concentration (mmol/L) after meal intake (Post prandial state, PP: 0 to 510 min) in artery, portal vein and hepatic vein; (**B**) indispensable amino acids (AA) concentration (μmol/L) after meal intake (Post prandial state, PP: 0 to 510 min) in artery, portal vein and hepatic vein; before (D1) and after 14 days (D14) and 60 days (D60) of adaptation to HFHS diet. Two way RM ANOVA, * D14 significantly different from D1, † D60 significantly different from D1, ‡ D60 significantly different from D14; Values are means ± SEM. For calculations and statistical treatment of the data and for details, see Materials and Methods.

Figure 4. (**A**) Glutamine concentration (μmol/L) after meal intake (post prandial state, PP: 0 to 510 min) in artery, portal vein and hepatic vein; and (**B**) glutamate concentration (μmol/L) after meal intake (post prandial state, PP: 0 to 510 min) in artery, portal vein and hepatic vein; before (D1) and after 14 days (D14) and 60 days (D60) of adaptation to HFHS diet. Two way RM ANOVA, * D14 significantly different from D1, † D60 significantly different from D1, ‡ D60 significantly different from D14; Values are means ± SEM. For calculations and statistical treatment of the data and for details, see Materials and Methods.

However, an important increased glutamine level in the portal vein was observed at D14 and D60 relatively to D1 (PP time × day effect = 0.037) whereas PP effect is lower ($p = 0.08$) in hepatic vein (Figure 4), suggesting an increased release by the gut associated with an increased utilization by the liver. Glutamate followed the same pattern of change in the portal vein as glutamine, but only at D14 (day effect: $p = 0.003$, PP time × day effect: 0.004) (D1 and D60 similar) (Figure 5). However, the increased glutamate concentration observed in the portal vein was also present in the hepatic vein (day: $p = 0.001$), suggesting a different role of the liver towards glutamate utilization relatively to glutamine. Taurine concentration is, similarly to glutamate, increased (or tended to) in two vessels in the fed state (PP time effect: $p = 0.068$ and $p = 0.007$ in artery and portal vein, respectively, not significant in the hepatic vein) (Figure 5). In the portal vein, a more important post prandial increased level of taurine is observed at D14 relatively to D1 (Day: $p = 0.02$) (Figure 5). No such profile existed for taurine both in artery and hepatic vein. Lastly, glycine presented a very specific profile (not seen for all other AA): an increased fasted level in the three vessels was observed both at D14 and D60 ($p < 0.01$ for all vessels) (Figure 5). Associated with this, in the fed state, glycine concentrations were decreased at D14 and D60 whereas glycine levels at D1 increased in all vessels (PP time effect: $p < 0.001$, $p = 0.08$ and $p < 0.001$ in artery, portal vein, and hepatic vein, respectively) (Figure 5).

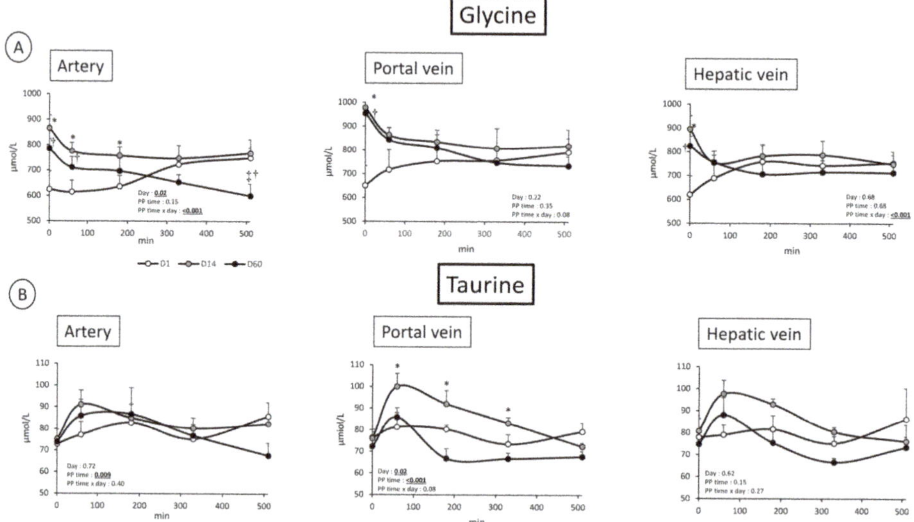

Figure 5. (**A**) Glycine concentration (μmol/L) after meal intake (Post prandial state, PP: 0 to 510 min) in artery, portal vein and hepatic vein; and (**B**) taurine concentration (μmol/L) after meal intake (post prandial state, PP: 0 to 510 min) in artery, portal vein and hepatic vein; before (D1) and after 14 days (D14) and 60 days (D60) of adaptation to HFHS diet. Two way RM ANOVA, * D14 significantly different from D1, † D60 significantly different from D1, ‡ D60 significantly different from D14; Values are means ± SEM. For calculations and statistical treatment of the data and for details, see Materials and Methods.

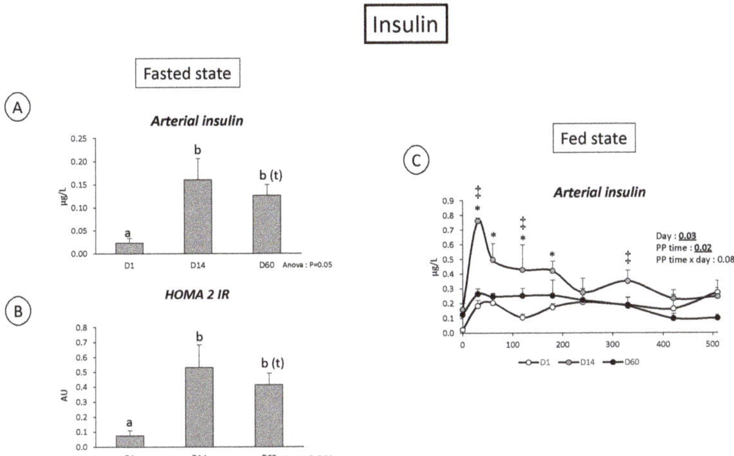

Figure 6. (**A**) Arterial insulin concentration (µg/L) in the fasted before (D1) and after 14 days (D14) and 60 days (D60) of adaptation to HFHS diet. (**B**) HOMA 2 IR (UA) in the fasted state before (D1) and after 14 days (D14) and 60 days (D60) of adaptation to HFHS diet. (**C**) Insulin concentration (µg/L) after meal intake (Post prandial state, PP: 0 to 510 min) in artery before (D1) and after 14 days (D14) and 60 days (D60) of adaptation to HFHS diet. Fasted state: One way RM ANOVA: with different letters: significantly different; (t): tendency. Fed state: Two way RM ANOVA, * D14 significantly different from D1, † D60 significantly different from D1, ‡ D60 significantly different from D14; Values are means ± SEM. For calculations and statistical treatment of the data and for details, see Materials and Methods.

4. Discussion

In the present study we assessed the effect of two-month overfeeding on nutrients uptake/release by the splanchnic area in the adult mini pig in the fasted state for a wide range of molecules and in the fed state for selected metabolites/molecules (glucose, lactate, urea, insulin, amino acids). We mention that the data presented in the fasted state at D1 represent the metabolic status of an animal adapted to a diet providing energy and proteins capable to maintain body weight stable and the post prandial state at D1 is the first day they receive the HFHS diet.

Although the animals did not present diabetes or fasting hyperglycemia at the end of the HFHS overfeeding period (and as already shown in a previous work on other animals but a rather similar diet [18]), their utilization of energy and nitrogenous nutrients at the splanchnic level was strongly altered to handle the unusual nutrients overflow both in the fed and the fasted states due to overfeeding. Some short term (after 14 days of overfeeding) and longer-term adaptive mechanisms are discussed in the present study.

4.1. Metabolic Adaptations to Overfeeding in Pigs: Impact on Nitrogenous and Energy Nutrients at the Gut, Hepatic, and Whole Body Levels

As the primary tissue in contact with the diet changes, the gut leaves large quantities of glucose to reach the portal vein when animals are fed the HFHS diet. As could be expected following the meal intake, the net plasma gut release is increased post-prandially as well as at the portal, hepatic and to a lesser extend arterial level from D1 to D60. These kinetic patterns are not differentially altered between D1 and D60, as shown by an absence of difference of significant net plasma gut release of glucose between D1 and D60. This is not the case in the fasted state where glucose concentrations are not altered in all vessels, but net plasma gut release increased, suggesting an intense utilization of glucose by other tissues and possibly by peripheral tissues, as generally observed in the fasted state (e.g., muscle and adipose tissues). The fact that the net glucose release by the liver and splanchnic area (Table 4) remains stable in the fasted state throughout time shows that contrarily to what generally occurs in well installed IR, no increased hepatic glucose release/production occurs in our model [26–29]. Residual amounts of glucose released by digestion of carbohydrates from the previous meal can be one of the explanations to the significant increased glucose net plasma gut release observed in the fasted state at D60 compared to D1. As HFHS meal supplies important amounts of nutrients, and particularly lipids, carbohydrates hydrolysis and digestion rate may be delayed between D1 (animals were adapted to the Control diet) and D14/D60 where the gut had been adapted to HFHS diet. Another possibility is an increased gluconeogenesis by the gut due to the increased supply of gluconeogenic precursors, as previously suggested [16], such as luminal amino acids (alanine for instance), plasma lactate or even small amounts of short chain fatty acids produced by microbiota from dietary fibers (tendency for an increased net gut release of propionate, suggesting an increased synthesis of SCFA by microbiota in the gut, Table 4). It should be noted that the increased overall food intake between D1 and D60 is associated with and increased supply of dietary fibers, potentially capable to stimulate microbiota activity. The increased SCFA synthesis by microbiota observed in the present study (as shown by net portal release of propionate) has already been reported in the literature on obesity (and related diseases) and illustrates the complex role of SFCA in the relationship between microbiota and host (SCFA increased both in fiber-supplemented [30] but also in obese individuals [31]). Indeed, as suggested by [31], according to their signaling pathway or metabolic mechanism they are involved in, they can either promote (via stimulation of triglycerides accumulation) or prevent (Histone deacetylases inhibition and/or GPR 41 and 43 activation) hepatic steatosis. In the present work, our hypothesis is that overfeeding (more than the relative oversupply of lipids) associated with an increased overall supply of dietary fibers may have maintained the activity of carbohydrates degrading microbes, leading to increased portal release of some SCFA.

To handle the important supply of nutrients, particularly glucose, and notably avoid even more massive glucose release, the gut adapts by oxidizing it into lactate, which is then released into the portal vein (more increased in the fed states (Figure 2) at D14 and D60 relative to D1). These increased lactate levels are visible in all vessels including artery. Such an increased fasting lactate level is known to occur progressively in obesity [32] and during IR installation [33], as observed in our model with D14 values intermediate between D1 and D60. As glucose supply from the diet is important due to overfeeding in the fed state, a conversion of glucose into pyruvate followed by its reduction into L-lactate (anaerobic glycolysis [34]) is a mechanism that can limit hyperglycemia [35]. Aside from a production in the gut, and as the portal vein also drains visceral adipose tissue, an increased lactate production from adipose tissues located at the gut level cannot be excluded [14]. In the fasted state, other mechanisms can explain the increased lactate levels. The sites of metabolic adaptation are also different (located at the muscle/adipose tissues and liver levels). The lactate that reaches the liver may not entirely be metabolized into glucose [35], and lactate net release from the liver has been shown to increase between D1 and D60, suggesting a limited capacity for lactate uptake (notably for gluconeogenesis).

The oversupply of food does not only disturb the energy metabolism but also amino acid utilization for energy, protein metabolism, and urea synthesis/elimination. First, levels of urea in all vessels are decreased both in the fasted state and the fed state. Such a decreased plasma urea content has already been discussed in a previous paper [19], in rodents fed a cafeteria diet [36] and corresponds to a nitrogen-sparing effect due to overfeeding [37]. The increased availability of all nutrients (lipids, glucose, AA, other N-based products, like nucleotides) tend to limit AA oxidation and, consequently, urea production. Due to this (and also the increased supply of proteins in HFHS-fed animals), a relatively small, but significant, increased AA plasma level (TAA, and among them particularly some NIAA (glycine) and BCAA) is observed. The reduced activity of the urea cycle observed is supported by the increased level of fasting plasma ornithine which accumulates instead of being metabolized into citrulline via carbamoyl-P synthetase and NH_3 incorporation into the urea cycle. For technical reasons, we could not measure plasma ammonia levels but data from pigs fed a similar diet (unpublished results) did not present significant alterations of ammonia levels throughout the obesity development. This is also in line with the increased concentrations we observe for glycine and serine known to be direct precursors for ammonia synthesis [38], and which accumulate instead of being catabolized into ammonia (and later into urea). Hence, in absence of necessity to detoxify ammonia, urea cycle remains at a low level. Our present model of oversupply of both energy and AA ends up with a mechanism of decreased urea and probably ammonia synthesis and relative increased level of certain AA. This lies a question un-answered: where and under what form the nitrogen excess can be utilized? A stimulation of muscle growth and AA utilization for protein synthesis could occur in our pigs (even if adults) associated with an increase in animal size (as previously shown in a similar study [18]) and may help to avoid massive hyper-aminoacidemia. This could be mediated by insulin that is highly responsive to meal intake at D14, suggesting that growth stimulation might be maximum at D14 (very low urea/high AA levels at D14) and intermediate between D14 and D60 (insulin less stimulated by meal and urea levels intermediate Figures 2 and 4). Other fates for nitrogen could be: N_2 in expired air or urinary nitrates, nitrites, uric acid, peptides or amino acids. An increased N excretion as AA in urine is possible as the presence of AA in urine has been spotted in a recent work on the same model [18].

Looking at specific amino acids, a significant alteration of post prandial profile of the glutamate/glutamine couple was observed in our study. Indeed, glutamine portal concentration was increased both at D14 and D60 vs. D1 (Figure 4). This suggests that this major fuel for small intestine (along with glutamate) [39] is replaced by other energy nutrients, a mechanism already demonstrated elsewhere [40]. In parallel, glutamate post prandial concentration in the portal vein was also significantly increased, but only at D14, whereas it was not altered at D1 and D60. One explanation is that glutamine, used by the gut as a fuel at D1, could have been replaced by glucose as a fuel at D14 and glutamate and glucose at D60. At D1, as it was the first time that the gut was challenged by the diet, catabolism of all energy nutrients was stimulated. Due to the long term overfeeding, at D14, a prioritization of glucose gut uptake and utilization may have occurred via a stimulation by insulin (increased in PP state at D14 (Figure 6)), as glucose uptake may be at least partially regulated by insulin receptors present in intestine epithelium, [41]. Lastly, the glutamate excess present in the enterocytes, due to glutamate dietary supply and potentially glutamine excess, could ultimately lead to an increased utilization of glutamate by the gut to limit overall hyperaminoacidemia. This increased utilization of glutamate by the gut and activation of Krebs cycle [42] could lead to an increased production of lactate, in absence of total catabolism, and concur to explain the increased lactate portal concentration and net plasma gut release observed at D60. This could also explain why urea levels are less decreased at D60 compared to D14 (D60 plasma concentrations were intermediate between D1 and D60).

Our dietary intervention was also characterized by an increased supply of lipids (palm oil) leading to an increased arterial cholesterol level (Table 1). Interestingly, glycine and taurine, two amino acids involved in recycling of biliary acids in human and pigs [43] presented an increased level in all vessels in the fasted state for glycine and in the portal vein in the fed state at D14 for taurine (Figure 5). As they combine to bile acids within the liver to be further recycled into the bile, their utilization in the liver is increased when dietary supply of lipids is increased. However, the fact that the increased availability is observed in the portal vein in the fasted state for glycine and in the fed state for taurine with a probable utilization within the liver for both AA, the metabolic fate of these two AA in relation to lipids intake should require further investigation. Interestingly, glycine has also been demonstrated, when administrated in the diet of rats, to limit non esterified fatty acids and lipids accumulation in adipocytes, via a stimulation of mitochondrial activity [44]. Aside from the already observed "buffering effect" of adipose tissues to limit hyperlipidemia [18]. Could the increased plasma glycine observed in the fasted state in our animals be another endogenous mechanism capable to counteract the increased plasma lipids concentration (in our study, triglycerides are not altered)?

4.2. Analysis of the Kinetics of Evolution of the Arterial Metabolites in Our Model of Overfed Mini Pigs

As already discussed in the present paper but also in previous published works [18], the metabolic shifts observed in our model represents the early stages of development of obesity and diet-induced metabolic disturbances ultimately leading to insulin resistance (IR). Increased arterial fasted cholesterol (total, LDL and HDL), lactate, insulin and HOMA2 IR are observed between D1 and D60. In parallel, short chain fatty acids (acetate, isobutyrate, or isovalerate) and several amino acids (AA): leucine, isoleucine, valine, tryptophane, glycine, ornithine, and proline increased, whereas urea and methionine decreased between D1 and D60.

However, the pattern of change of these parameters can differ. Some parameters are significantly modified only at D14 (arterial isobutyrate, isovalerate, and LDL cholesterol) and represent a short lasting diet-dependent adaptations. A second group of parameters are modified early (D14) and remained stable at D60 (AA: methionine, proline, isoleucine, glycine, phenylalanine, lactate, urea, fructosamine, insulin, HOMA, total and HDL cholesterol), suggesting that these alterations can be very closely linked to the rapid shift from a "maintenance" diet to a "high fat high sucrose" diet but could also be considered as very early markers of a metabolic shift and/or development of insulin resistance. Looking at the potential regulatory role of glycine on lipids and cholesterol handling (see above), this AA should be studied in detail. Lastly, some parameters (acetate, leucine, valine, tryptophane, and serine associated with the progressive pigs' weight) are also particularly interesting as they are significantly (or tend to be) increased more progressively and more lately than the other parameters, suggesting a progressive shift of their utilization/metabolism at the whole body level (less driven by the direct impact of the change of diet and associated nutrients). Leucine and valine are already considered as predictors of insulin resistance [45] and their slow increase over the entire experimental period confirms their relevance in the development of insulin resistance/obesity phenotype [19]. Acetate is less studied but its concentration is increased in the caecum and plasma as well as its overall turnover in insulin resistant or high fat fed animals/humans [46–48]. The role of the microbiota in this increased acetate production of obese/IR states, particularly in the fasted state may be significant [46] and may lead to altered glucose-stimulated insulin production, fat storage and could ultimately lead to insulin resistance. Unfortunately, due to high variability between animals, no increased gut or splanchnic net acetate production could be observed in the present study (Table 4) whereas it is the case for propionate and butyrate. A reduction of acetate utilization by muscles for oxidation is a possible explanation for the increased arterial acetate levels, as a significant part of acetate utilization takes place in muscle [49]. The establishment of a resistance for acetate oxidation in muscle can be hypothesized as acetate oxidation has been shown insulin-sensitive in rodents' hindquarters [50]. Lastly, tryptophan has also been shown (among aromatic amino acids), to be increased and tightly correlated to obesity, adiponectin and intrahepatic fat content [51–53].

In conclusion, we have shown that two months of overfeeding in pigs led to important metabolic shifts which may have led to the maintenance of a relatively stable glycemia. This was achieved thanks to important shifts in lactate and amino acids/urea metabolism. A progressive increased release of lactate by the gut and the liver in the fasted and the fed state observed in the present study could trigger other metabolic perturbations leading to nutritional-related pathologies and should be examined in detail. The analysis of the profile of arterial metabolites throughout the experimental period showed, as could be anticipated, a progressive increased BCAA levels (as detailed elsewhere [19]) but also acetate, tryptophan and, potentially, glycine, which are less studied in the context of obesity development, should require further investigation as potential markers of metabolic shift towards insulin resistance.

Author Contributions: Conceptualization: I.S.-A., S.P., and D.R.; methodology: I.S.-A., A.-B.M., B.C., J.D., N.H., C.M., E.P.-G., and S.P.; software: I.S.-A. and S.P.; validation: I.S.-A., S.P., and C.M.; formal analysis: I.S.-A., S.P., M.A.B., and D.R.; investigation: I.S.-A., S.P., B.C., J.D., and D.R.; resources: I.S.-A., D.D., D.R., and S.P.; data curation: I.S.-A. and S.P.; writing—original draft preparation: I.S.-A., A.-B.M., D.R., and S.P.; writing—review and editing: I.S.-A., D.D., S.P., and D.R.; visualization: I.S.-A. and S.P.; supervision, I.S.-A. and S.P.; project administration: I.S.-A., D.R., and S.P.; funding acquisition: I.S.-A., D.D., D.R., and S.P.

Funding: Institut de la Recherche Agronomique (INRA).

Acknowledgments: The authors acknowledge P. Lhoste, D. Durand, F. Bechereau, and the personnel of Animal Facility for technical assistance.

Conflicts of Interest: The authors declare no conflict of interest.

References

1. The GBD 2015 Obesity Collaborators. Health Effects of Overweight and Obesity in 195 Countries over 25 Years. *N. Engl. J. Med.* **2017**, *377*, 13–27. [CrossRef] [PubMed]
2. Kim, M.S.; Lee, M.S.; Kown, D.Y. Inflammation-mediated obesity and insulin resistance as targets for nutraceuticals. *Ann. N. Y. Acad. Sci.* **2011**, *1229*, 140–146. [CrossRef] [PubMed]
3. Newsholme, P.; Cruzat, V.; Arfuso, F.; Keane, K. Nutrient regulation of insulin secretion and action. *J. Endocrinol.* **2014**, *221*, R105–R120. [CrossRef] [PubMed]
4. Morgan, K.; Uyuni, A.; Nandgiri, G.; Mao, L.; Castaneda, L.; Kathirvel, E.; French, S.W.; Morgan, T.R. Altered expression of transcription factors and genes regulating lipogenesis in liver and adipose tissue of mice with high fat diet-induced obesity and nonalcoholic fatty liver disease. *Eur. J. Gastroenterol. Hepatol.* **2008**, *20*, 843–854. [CrossRef] [PubMed]
5. Trayhurn, P. Adipocyte biology. *Obes. Rev. Off. J. Int. Assoc. Study Obes.* **2007**, *8* (Suppl. 1), 41–44. [CrossRef] [PubMed]
6. Turner, N.; Cooney, G.J.; Kraegen, E.W.; Bruce, C.R. Fatty acid metabolism, energy expenditure and insulin resistance in muscle. *J. Endocrinol.* **2014**, *220*, T61–T79. [CrossRef] [PubMed]
7. McNelis, J.C.; Olefsky, J.M. Macrophages, immunity, and metabolic disease. *Immunity* **2014**, *41*, 36–48. [CrossRef] [PubMed]
8. Ley, R.E.; Turnbaugh, P.J.; Klein, S.; Gordon, J.I. Microbial ecology: human gut microbes associated with obesity. *Nature* **2006**, *444*, 1022–1023. [CrossRef] [PubMed]
9. Le Chatelier, E.; Nielsen, T.; Qin, J.; Prifti, E.; Hildebrand, F.; Falony, G.; Almeida, M.; Arumugam, M.; Batto, J.M.; Kennedy, S.; et al. Richness of human gut microbiome correlates with metabolic markers. *Nature* **2013**, *500*, 541–546. [CrossRef] [PubMed]
10. Konrad, D.; Wueest, S. The gut-adipose-liver axis in the metabolic syndrome. *Physiology (Bethesda)* **2014**, *29*, 304–313. [CrossRef] [PubMed]
11. Youn, J.H.; Kim, J.K.; Buchanan, T.A. Time courses of changes in hepatic and skeletal muscle insulin action and GLUT4 protein in skeletal muscle after STZ injection. *Diabetes* **1994**, *43*, 564–571. [CrossRef] [PubMed]
12. Picarel-Blanchot, F.; Berthelier, C.; Bailbe, D.; Portha, B. Impaired insulin secretion and excessive hepatic glucose production are both early events in the diabetic GK rat. *Am. J. Physiol.* **1996**, *271*, E755–E762. [CrossRef] [PubMed]
13. Mithieux, G. Metabolic effects of portal vein sensing. *Diabetes Obes. Metab.* **2014**, *16* (Suppl. 1), 56–60. [CrossRef] [PubMed]
14. Item, F.; Konrad, D. Visceral fat and metabolic inflammation: The portal theory revisited. *Obes. Rev.* **2012**, *13* (Suppl. 2), 30–39. [CrossRef] [PubMed]
15. Coate, K.C.; Kraft, G.; Shiota, M.; Smith, M.S.; Farmer, B.; Neal, D.W.; Williams, P.; Cherrington, A.D.; Moore, M.C. Chronic overeating impairs hepatic glucose uptake and disposition. *Am. J. Physiol. Endocrinol. Metab.* **2015**, *308*, E860–E867. [CrossRef] [PubMed]
16. De Vadder, F.; Kovatcheva-Datchary, P.; Zitoun, C.; Duchampt, A.; Backhed, F.; Mithieux, G. Microbiota-Produced Succinate Improves Glucose Homeostasis via Intestinal Gluconeogenesis. *Cell Metab.* **2016**, *24*, 151–157. [CrossRef] [PubMed]
17. Shen, J.; Obin, M.S.; Zhao, L. The gut microbiota, obesity and insulin resistance. *Mol. Asp. Med.* **2013**, *34*, 39–58. [CrossRef] [PubMed]
18. Polakof, S.; Remond, D.; Bernalier-Donadille, A.; Rambeau, M.; Pujos-Guillot, E.; Comte, B.; Dardevet, D.; Savary-Auzeloux, I. Metabolic adaptations to HFHS overfeeding: How whole body and tissues postprandial metabolic flexibility adapt in Yucatan mini-pigs. *Eur. J. Nutr.* **2018**, *57*, 119–135. [CrossRef] [PubMed]
19. Polakof, S.; Remond, D.; David, J.; Dardevet, D.; Savary-Auzeloux, I. Time-course changes in circulating branched-chain amino acid levels and metabolism in obese Yucatan minipig. *Nutrition* **2017**, *50*, 66–73. [CrossRef] [PubMed]

20. Mohamed, A.B.; Rémond, D.; Chambon, C.; Sayd, T.; Hebraud, M.; Capel, F.; Cohade, B.; Hafnaoui, N.; Béchet, D.; Coudy-Gandilhon, C.; et al. A mix of dietary fermentable fibers improves lipids handling by the liver of overfed minipigs. *J. Nutr. Biochem.* **2018**, *65*, 72–82. [CrossRef] [PubMed]

21. Katz, M.L.; Bergman, E.N. Simultaneous measurements of hepatic and portal venous blood flow in the sheep and dog. *Am. J. Physiol.* **1969**, *216*, 946–952. [CrossRef] [PubMed]

22. Barbe, F.; Menard, O.; Le Gouar, Y.; Buffiere, C.; Famelart, M.H.; Laroche, B.; Le Feunteun, S.; Dupont, D.; Remond, D. The heat treatment and the gelation are strong determinants of the kinetics of milk proteins digestion and of the peripheral availability of amino acids. *Food Chem.* **2013**, *136*, 1203–1212. [CrossRef] [PubMed]

23. Huntington, G.B. Portal blood flow and net absorption of ammonia-nitrogen, urea-nitrogen, and glucose in nonlactating Holstein cows. *J. Dairy Sci.* **1982**, *65*, 1155–1162. [CrossRef]

24. Rodriguez-Lopez, J.M.; Cantalapiedra-Hijar, G.; Durand, D.; Isserty-Thomas, A.; Ortigues-Marty, I. Influence of the para-aminohippuric acid analysis method on the net hepatic flux of nutrients in lactating cows. *J. Anim. Sci.* **2014**, *92*, 1074–1082. [CrossRef] [PubMed]

25. Pouteau, E.; Meirim, I.; Metairon, S.; Fay, L.B. Acetate, propionate and butyrate in plasma: Determination of the concentration and isotopic enrichment by gas chromatography/mass spectrometry with positive chemical ionization. *J. Mass Spectrom.* **2001**, *36*, 798–805. [CrossRef] [PubMed]

26. Gallagher, E.J.; Leroith, D.; Karnieli, E. Insulin resistance in obesity as the underlying cause for the metabolic syndrome. *Mount Sinai J. Med.* **2010**, *77*, 511–523. [CrossRef] [PubMed]

27. Wahren, J.; Ekberg, K. Splanchnic regulation of glucose production. *Annu. Rev. Nutr.* **2007**, *27*, 329–345. [CrossRef] [PubMed]

28. Gastaldelli, A.; Baldi, S.; Pettiti, M.; Toschi, E.; Camastra, S.; Natali, A.; Landau, B.R.; Ferrannini, E. Influence of obesity and type 2 diabetes on gluconeogenesis and glucose output in humans: A quantitative study. *Diabetes* **2000**, *49*, 1367–1373. [CrossRef] [PubMed]

29. Basu, R.; Chandramouli, V.; Dicke, B.; Landau, B.; Rizza, R. Obesity and type 2 diabetes impair insulin-induced suppression of glycogenolysis as well as gluconeogenesis. *Diabetes* **2005**, *54*, 1942–1948. [CrossRef] [PubMed]

30. Canfora, E.E.; Meex, R.C.R.; Venema, K.; Blaak, E.E. Gut microbial metabolites in obesity, NAFLD and T2DM. *Nat. Rev. Endocrinol.* **2019**. [CrossRef] [PubMed]

31. Chu, H.; Duan, Y.; Yang, L.; Schnabl, B. Small metabolites, possible big changes: A microbiota-centered view of non-alcoholic fatty liver disease. *Gut* **2019**, *68*, 359–370. [CrossRef] [PubMed]

32. Crawford, S.O.; Ambrose, M.S.; Hoogeveen, R.C.; Brancati, F.L.; Ballantyne, C.M.; Young, J.H. Association of lactate with blood pressure before and after rapid weight loss. *Am. J. Hypertens.* **2008**, *21*, 1337–1342. [CrossRef] [PubMed]

33. Lovejoy, J.; Newby, F.D.; Gebhart, S.S.; DiGirolamo, M. Insulin resistance in obesity is associated with elevated basal lactate levels and diminished lactate appearance following intravenous glucose and insulin. *Metabolism* **1992**, *41*, 22–27. [CrossRef]

34. Clara, R.; Schumacher, M.; Ramachandran, D.; Fedele, S.; Krieger, J.P.; Langhans, W.; Mansouri, A. Metabolic Adaptation of the Small Intestine to Short- and Medium-Term High-Fat Diet Exposure. *J. Cell. Physiol.* **2017**, *232*, 167–175. [CrossRef] [PubMed]

35. Adeva-Andany, M.; Lopez-Ojen, M.; Funcasta-Calderon, R.; Ameneiros-Rodriguez, E.; Donapetry-Garcia, C.; Vila-Altesor, M.; Rodriguez-Seijas, J. Comprehensive review on lactate metabolism in human health. *Mitochondrion* **2014**, *17*, 76–100. [CrossRef] [PubMed]

36. Sabater, D.; Agnelli, S.; Arriaran, S.; Fernandez-Lopez, J.A.; Romero Mdel, M.; Alemany, M.; Remesar, X. Altered nitrogen balance and decreased urea excretion in male rats fed cafeteria diet are related to arginine availability. *BioMed Res. Int.* **2014**, *2014*, 959420. [CrossRef] [PubMed]

37. Herrero, M.C.; Angles, N.; Remesar, X.; Arola, L.; Blade, C. Splanchnic ammonia management in genetic and dietary obesity in the rat. *Int. J. Obes. Relat. Metab. Disord.* **1994**, *18*, 255–261. [PubMed]

38. Alemany, M. The problem of nitrogen disposal in the obese. *Nutr. Res. Rev.* **2012**, *25*, 18–28. [CrossRef] [PubMed]

39. Bertolo, R.F.; Burrin, D.G. Comparative aspects of tissue glutamine and proline metabolism. *J. Nutr.* **2008**, *138*, 2032S–2039S. [CrossRef] [PubMed]

40. Van Der Schoor, S.R.; Reeds, P.J.; Stoll, B.; Henry, J.F.; Rosenberger, J.R.; Burrin, D.G.; Van Goudoever, J.B. The high metabolic cost of a functional gut. *Gastroenterology* **2002**, *123*, 1931–1940. [CrossRef] [PubMed]

41. Ussar, S.; Haering, M.F.; Fujisaka, S.; Lutter, D.; Lee, K.Y.; Li, N.; Gerber, G.K.; Bry, L.; Kahn, C.R. Regulation of Glucose Uptake and Enteroendocrine Function by the Intestinal Epithelial Insulin Receptor. *Diabetes* **2017**, *66*, 886–896. [CrossRef] [PubMed]

42. Burrin, D.G.; Stoll, B. Metabolic fate and function of dietary glutamate in the gut. *Am. J. Clin. Nutr.* **2009**, *90*, 850S–856S. [CrossRef] [PubMed]

43. Solaas, K.; Ulvestad, A.; Soreide, O.; Kase, B.F. Subcellular organization of bile acid amidation in human liver: A key issue in regulating the biosynthesis of bile salts. *J. Lipid Res.* **2000**, *41*, 1154–1162. [PubMed]

44. El Hafidi, M.; Perez, I.; Zamora, J.; Soto, V.; Carvajal-Sandoval, G.; Banos, G. Glycine intake decreases plasma free fatty acids, adipose cell size, and blood pressure in sucrose-fed rats. *Am. J. Physiol.* **2004**, *287*, R1387–R1393. [CrossRef] [PubMed]

45. McCormack, S.E.; Shaham, O.; McCarthy, M.A.; Deik, A.A.; Wang, T.J.; Gerszten, R.E.; Clish, C.B.; Mootha, V.K.; Grinspoon, S.K.; Fleischman, A. Circulating branched-chain amino acid concentrations are associated with obesity and future insulin resistance in children and adolescents. *Pediatr. Obes.* **2013**, *8*, 52–61. [CrossRef] [PubMed]

46. Perry, R.J.; Peng, L.; Barry, N.A.; Cline, G.W.; Zhang, D.; Cardone, R.L.; Petersen, K.F.; Kibbey, R.G.; Goodman, A.L.; Shulman, G.I. Acetate mediates a microbiome-brain-beta-cell axis to promote metabolic syndrome. *Nature* **2016**, *534*, 213–217. [CrossRef] [PubMed]

47. Fernandes, J.; Vogt, J.; Wolever, T.M. Kinetic model of acetate metabolism in healthy and hyperinsulinaemic humans. *Eur. J. Clin. Nutr.* **2014**, *68*, 1067–1071. [CrossRef] [PubMed]

48. Piloquet, H.; Ferchaud-Roucher, V.; Duengler, F.; Zair, Y.; Maugere, P.; Krempf, M. Insulin effects on acetate metabolism. *Am. J. Physiol. Endocrinol. Metab.* **2003**, *285*, E561–E565. [CrossRef] [PubMed]

49. Mittendorfer, B.; Sidossis, L.S.; Walser, E.; Chinkes, D.L.; Wolfe, R.R. Regional acetate kinetics and oxidation in human volunteers. *Am. J. Physiol.* **1998**, *274*, E978–E983. [CrossRef] [PubMed]

50. Karlsson, N.; Fellenius, E.; Kiessling, K.H. Influence of acetate on glucose metabolism in the perfused hind-quarter of the rat. *Acta Physiol. Scand.* **1976**, *98*, 347–355. [CrossRef] [PubMed]

51. Nakamura, H.; Jinzu, H.; Nagao, K.; Noguchi, Y.; Shimba, N.; Miyano, H.; Watanabe, T.; Iseki, K. Plasma amino acid profiles are associated with insulin, C-peptide and adiponectin levels in type 2 diabetic patients. *Nutr. Diabetes* **2014**, *4*, e133. [CrossRef] [PubMed]

52. Haufe, S.; Witt, H.; Engeli, S.; Kaminski, J.; Utz, W.; Fuhrmann, J.C.; Rein, D.; Schulz-Menger, J.; Luft, F.C.; Boschmann, M.; et al. Branched-chain and aromatic amino acids, insulin resistance and liver specific ectopic fat storage in overweight to obese subjects. *Nutr. Metab. Cardiovasc. Dis.* **2016**, *26*, 637–642. [CrossRef] [PubMed]

53. Wijekoon, E.P.; Skinner, C.; Brosnan, M.E.; Brosnan, J.T. Amino acid metabolism in the Zucker diabetic fatty rat: Effects of insulin resistance and of type 2 diabetes. *Can. J. Physiol. Pharmacol.* **2004**, *82*, 506–514. [CrossRef] [PubMed]

 *nutrients*

Article

Post Meal Energy Boluses Do Not Increase the Duration of Muscle Protein Synthesis Stimulation in Two Anabolic Resistant Situations

Laurent Mosoni, Marianne Jarzaguet, Jérémie David, Sergio Polakof, Isabelle Savary-Auzeloux, Didier Rémond and Dominique Dardevet *

Unité de Nutrition Humaine, INRA, Université Clermont Auvergne, UMR1019, F-63000 Clermont-Ferrand, France; laurent.mosoni@inra.fr (L.M.); marianne.jarzaguet@inra.fr (M.J.); jeremie.david@inra.fr (J.D.); sergio.polakof@inra.fr (S.P.); isabelle.savary-auzeloux@inra.fr (I.S.-A.); didier.remond@inra.fr (D.R.)
* Correspondence: dominique.dardevet@inra.fr; Tel.: +33-0-473-624-505; Fax: +33-0-473-624-638

Received: 14 February 2019; Accepted: 27 March 2019; Published: 29 March 2019

Abstract: Background: When given in the long term, whey proteins alone do not appear to be an optimal nutritional strategy to prevent or slow down muscle wasting during aging or catabolic states. It has been hypothesized that the digestion of whey may be too rapid during a catabolic situation to sustain the anabolic postprandial amino acid requirement necessary to elicit an optimal anabolic response. Interestingly, it has been shown recently that the duration of the postprandial stimulation of muscle protein synthesis in healthy conditions can be prolonged by the supplementary ingestion of a desynchronized carbohydrate load after food intake. We verified this hypothesis in the present study in two different cases of muscle wasting associated with anabolic resistance, i.e., glucocorticoid treatment and aging. Methods: Multi-catheterized minipigs were treated or not with glucocorticoids for 8 days. Muscle protein synthesis was measured sequentially over time after the infusion of a ^{13}C phenylalanine tracer using the arterio-venous method before and after whey protein meal ingestion. The energy bolus was given 150 min after the meal. For the aging study, aged rats were fed the whey meal and muscle protein synthesis was measured sequentially over time with the flooding dose method using ^{13}C Valine. The energy bolus was given 210 min after the meal. Results: Glucocorticoid treatment resulted in a decrease in the duration of the stimulation of muscle protein synthesis. The energy bolus given after food intake was unable to prolong this stimulation despite a simultaneous increase of insulin and glucose following its absorption. In old rats, a similar observation was made with no effect of the energy bolus on the duration of the muscle anabolic response following whey protein meal intake. Conclusions. Despite very promising observations in healthy situations, the strategy aimed at increasing muscle protein synthesis stimulation by giving an energy bolus during the postprandial period remained inefficient in our two anabolic resistance models.

Keywords: aging; catabolic state; anabolic resistance; protein synthesis; energy bolus

---

1. Introduction

Skeletal muscle size and function are directly related to the amount of muscle protein content and quality. Muscle protein homeostasis is ensured by continuous turn-over, including the degradation of non-functional muscle proteins which are replaced by the synthesis of new ones. This equilibrium between muscle protein synthesis and breakdown varies throughout the day. During the post-absorptive period, in the absence of food intake, muscle proteolysis exceeds synthesis to provide essential amino acids to other organs, whereas an anabolic response is initiated when muscle protein synthesis is higher than proteolysis after food intake [1]. It is now well-established that the postprandial anabolic response in skeletal muscle occurs through the stimulation of protein synthesis initiated by increased essential amino acid bioavailability, whereas the decrease in proteolysis mainly occurs through the stimulation of insulin secretion in response to feeding [2,3]. Although decreased physical activity has been identified as a cause, muscle atrophy has also been explained by the presence of anabolic resistance following food intake in several physio-pathological catabolic states as diverse as aging, cancer, and disuse [4–10]. In each case, increasing dietary protein intake above the current recommended dietary allowances (RDA at 0.8 g/kg/day) [11] has been proposed to overcome the anabolic resistance observed. Among dietary proteins, rapidly digested and leucine-rich proteins (i.e., whey) have been shown to be the most efficient for the acute stimulation of muscle protein synthesis during aging, disuse or glucocorticoid treatment when compared to casein, a protein with lower leucine content and whose digestion rate is slower [9,12–14]. However, when given in the long-term, whey proteins alone do not appear to be an optimal nutritional strategy to prevent or slow down muscle wasting during aging [15,16] or catabolic states [10,17–19]. This could be explained by the nature and intensity of the catabolic state but also by the fact that the digestion of whey may be too rapid during a catabolic situation to sustain the anabolic postprandial amino acid requirement necessary to elicit an optimal anabolic response [20]. Indeed, the stimulation of muscle protein synthesis has a defined and limited duration during the postprandial period [21,22]. Atherton et al. showed that whey ingestion is only able to stimulate muscle protein synthesis for 2 h [21]. The leucine content of a complete meal drives peak activation but not the duration of skeletal muscle protein synthesis [23]. Interestingly, it has been shown recently that the duration of the postprandial stimulation of muscle protein synthesis in healthy conditions could nevertheless be prolonged by maintaining the cellular energy status with the supplementary ingestion of a desynchronized carbohydrate load after food intake (±150 min) [24]. Such a prolongation would be very useful to slow down the loss of muscle mass during aging, or in other catabolic situations. To our knowledge, the use of such an energy bolus at a precise time after a meal has never been tested in catabolic situations. Thus, the present study aimed to determine whether an energy bolus given 30 min after peak postprandial muscle protein synthesis stimulation could also increase muscle protein synthesis during a catabolic situation. We tested two catabolic situations: glucocorticoid treatment in young mini pigs, and chronic loss of muscle mass during aging in the rat. Unfortunately, in both situations, the energy bolus was unable to re-stimulate muscle protein synthesis.

2. Materials and Methods

In the present manuscript, we studied two different muscle wasting conditions for which we have clearly described an anabolic resistance in skeletal muscle following food intake. In these two studies, we showed that whey, a rapidly digested protein rich in leucine, was more efficient than casein at the same protein content for (re)initiating an anabolic response, mainly by increasing muscle protein synthesis postprandially [9,10]. Thus, some of these results which represent the "control" groups in these manuscripts have been already published. In the present publication, the original results are the results in the same conditions, but associated with the desynchronized energy bolus.

2.1. Glucocorticoid Treatment Study

2.1.1. Animal Housing, Surgery and Ethics Statement

The present study was approved by the Animal Care and Use Committee of Auvergne (CEMEA Auvergne; Permit Number: CE 68-12) and the Ministère de l'Enseignement Supérieur et de la Recherche (no. 02125.02) and described in full by Revel et al. [10]. For the study, 18 adult male Yucatan mini pigs (averaging 20 kg) were housed individually in subject pens (1 × 1.5 m) in a ventilated room with controlled temperature (21 °C). They were fed twice daily with 220 g/d of a concentrated feed containing 16% protein, 1% fat, 4% cellulose, and 5% ash (Porcyprima; Sanders Centre Auvergne, Aigueperse, France) and had free access to water. Three weeks before the experiment, the minipigs went into surgery and were fitted with catheters in the inferior cava vein and the aorta. For the cava vein, the catheter was inserted just downstream of the junction with the iliac veins. A transit time ultrasonic blood flow probe (6 mm probe, R-series; Transonic Systems, Inc., Ithaca, NY, USA) was implanted around the distal aorta 1–2 cm before it splits into the iliac arteries. Catheters and probe cables were exteriorized through the skin of the right flank of the animal. A minimum of 2 weeks was allowed for recovery from surgery before initiating the experiment. Surgical procedures, as well as post-surgical care, were described previously in detail by Rémond et al. [25].

2.1.2. Tracer Infusion Procedures

On the day of the experiment, after an overnight fasting period, the animals were separated into 3 groups (n = 6 per group): control group (WHEY CONTROL), glucocorticoid treated group (WHEY DEXA), and the glucocorticoid group with a desynchronized energy bolus during the postprandial period (WHEY DEXA BOLUS). For all groups, after basal blood samples were withdrawn from the artery and iliac vein, a priming dose of labeled [ring U-^{13}C] L-Phenylalanine (4.2 µmol·kg^{-1}) was injected and then [ring U-^{13}C] L-Phenylalanine was infused (4.2 µmol.kg^{-1}·h^{-1}) for 550 min through the hepatic vein. During the first 150 min, the animals remained deprived of food and samples of arterial and iliac venous blood were simultaneously withdrawn at 90, 120, and 150 min, and represented the post-absorptive period (PA). At t = 150 min, the animals consumed a test meal (250 g) containing whey as a protein source (13%), as well as lipids (6%), and carbohydrates (57%). The meal was entirely consumed within 10–15 min, arterial and iliac venous blood was sampled every 30 min during the remaining 400 min (postprandial period, PP). The ultrasonic blood flow probe was used to record blood flow continuously throughout the whole experiment. For the glucocorticoid-treated group (WHEY DEXA), 8 days before the experiment, the animals received a dose of 0.4 mg/kg/day of dexamethasone added to the feed every morning (0.15 mg/kg) and evening (0.25 mg/kg). Food intake was not modified by the treatment. The animals were subjected to the same tracer infusion protocol described above except that dexamethasone was consumed with the test meal. The WHEY DEXA BOLUS group followed the same protocol except that 150 min after the test meal ingestion (30 min after the expected peak of protein synthesis observed in our previous experiment [10]), the animals were allowed to consume 100 g of glucose/ saccharose (50/50) to initiate an energy bolus similar to 70% of the carbohydrate content of the test meal, as described by Wilson et al. [24]. In the experiment performed by Wilson et al., the bolus was similar to 100% of the carbohydrate content of the test meal and was given 45 min after peak muscle protein synthesis.

2.1.3. Samples Analysis

The plasma ^{13}C enrichment of phenylalanine was measured by gas chromatography–mass spectrometry (GC–MS, model HP5975C/7890A, Agilent, Santa Clara, CA, USA) with the use of tertiary-buthyldimethylsilyl derivatives and by ion monitoring with m/Z 336 and 342, as previously described [10,26].

Plasma concentrations of amino acids were determined by ion exchange chromatography on deproteinized samples [10]. Plasma glucose was assayed using an enzymatic method on an autoanalyzer (Pentra 400, Horiba, Montpellier, France) and the insulin concentration was assayed by ELISA (Mercodia, Uppsala, Sweden).

2.1.4. Calculations

Muscle protein synthesis was calculated using the arterio-venous differences method according to the equations previously described by Bruins et al. [27] and Paddon-jones et al. [28]. Phenylalanine was selected to represent amino acid kinetics because it is neither produced nor metabolized in skeletal muscle [29]. In this tissue, the disposal and production of phenylalanine reflect protein synthesis and protein breakdown. The complete and detailed equations used to calculate muscle protein synthesis were described by Revel et al. [10].

2.1.5. Statistics and Analysis of Results

Data are presented as means $\pm$ SE. Repeated time variance analyses were performed to test the effect of time, status (WHEY CONTROL, WHEY DEXA, WHEY DEXA BOLUS) and interaction time $\times$ status. LSD post-hoc tests were used to compare mean values at each time (Statview, SAS Institute, Cary, NC, USA).

2.2. Aging Study

2.2.1. Animals and Diets

The present study was approved by the Animal Care and Use Committee of Auvergne (CEMEA Auvergne; Permit Number: C2EA-02) and the Ministry of Higher Education and Research (no. 2016101911586999). Twenty-month-old male Wistar rats (Charles River, L'Arbresle, France) were housed individually and kept in a controlled environment (temperature maintained at 22 °C; 12:12 light: dark cycle). The average weight of the rats was close to 600 g. After an adaptation period during which the animals were fed regular chow (Safe A04, Augy, France), they were divided into 2 groups, fasted overnight and fed the next morning with a meal (6 g). The majority of this amount (>60%) was consumed within the first 30 min and withdrawn after 1 h if any food remained uneaten. One group received the meal made with whey (13%) as a protein source (+ 6% lipids and 71% carbohydrates) and the second group received the same meal but also 4.5 g of glucose/sucrose (50/50), which was similar to 104% of the amount of carbohydrate in the test meal 210 min after the beginning of the meal. Once again, our aim was to mimic the study of Wilson et al. [24] but in heavier and older rats. Our bolus was given 30 min after peak muscle protein synthesis. The animals were then sacrificed before (0) and 90, 125, 180, 240 and 270 min (only for the energy bolus group) after the beginning of the meal under 4% isoflurane anesthesia (n = 10/time point). An abdominal incision was made and blood was withdrawn from the abdominal aorta with syringes containing EDTA. The gastrocnemius muscles were rapidly removed, weighed and freeze-clamped in liquid nitrogen, and stored at -80 °C. Regarding the muscle protein synthesis measurement, 40 min before the sacrifice, animals were injected intravenously with L-valine (150 µmol per 100 g body weight) containing 100% L-[1-^{13}C] valine (Euriso-Top) (see [9] for the detailed protocol).

2.2.2. Analytical Procedures

Glucose concentrations were measured enzymatically using commercial kits (Horiba, Montpellier, France). Plasma insulin levels were also assessed using a commercial ELISA kit (Mercodia, Uppsala, Sweden). To measure protein synthesis, the muscles were powdered in liquid nitrogen in a ball mill (Dangoumeau, Prolabo, Paris, France). A 200 mg aliquot of frozen muscle powder was homogenized in 2 mL of 10% trichloroacetic acid (TCA). Proteins were hydrolyzed in 6 N HCl at 110 °C for 48 h. HCl was removed by evaporation and amino acids purified by cation exchange chromatography. The enrichment of [1-^{13}C] valine in muscle proteins was measured on the basis of its N-acetyl-propyl derivatives by gas chromatography–combustion-isotope ratio mass spectrometry (GC–C-IRMS). The plasma ^{13}C enrichment of valine was measured by gas chromatography–mass spectrometry (GC–MS, model HP5975C/7890A, Agilent, Santa Clara, CA, USA) with the use of tertiary-buthyldimethylsilyl derivatives (see [9] for the detailed protocol). The absolute synthesis rate (ASR: total muscle protein synthesized) was calculated from the product of the protein fractional synthesis rate (FSR) and the protein content of the tissue, and expressed in mg/d. FSR (in %/day) was calculated from the formula: FSR = Sb $\times$ 100/Sa $\times$ t, where Sb is muscle protein-bound [1-^{13}C] valine enrichment (minus natural basal enrichment of protein), Sa is the mean enrichment of plasma valine during tracer incorporation, and t is the incorporation time in days calculated between the time of tracer injection and the time of muscle sampling, i.e., "PA" between -40 min and 0 min; "90 min" between 50 and 90 min; "125 min" between 85 and 125 min; "180 min" between 140 and 180 min, "240 min" between 200 and 240 min and "270 min" between 230 and 270 min. The mean Sa enrichment was the Sa (t1/2) value calculated from the linear regression obtained in the tissue between the time of injection and time t.

2.2.3. Statistics and Analysis of Results

Data are presented as means $\pm$ SE. Separate groups of animals were used each time; thus a repeated time variance analysis could not be performed. We used a two-way variance analysis to discriminate between the effect of time of measurement, and the effect of nutritional status (WHEY or WHEY BOLUS). The interaction was not testable since, by construction, the bolus was given at time 210 min and affected only time 240 and time 270. LSD post-hoc tests were performed to compare mean values at each time (Statview, SAS Institute, Cary, NC, USA).

3. Results

3.1. Minipigs Treated with Glucocorticoids

3.1.1. Plasma Glucose, Insulin, and Arterial Amino Acid Concentrations

After the whey-based meal, an increase in both glucose and insulin was observed (Figure 1A,B). Plasma glucose increased rapidly (1.2-fold between 180–240 min) and gradually decreased to the basal values throughout the remaining postprandial period (300–550 min). Insulin was secreted rapidly with maximum stimulation at 210 min (7–8-fold), and then insulin decreased back to the basal values throughout the rest of the postprandial period.

Plasma insulin and glucose increased in WHEY DEXA minipigs during the early postprandial period (150–300 min), when compared to the control situation. During the late postprandial period (300–550 min), no significant differences were observed for either plasma insulin or glucose between the WHEY CONTROL and the WHEY DEXA minipigs (Figure 1A,B). When the glucocorticoid-treated mini pigs received the energy bolus at 300 min (WHEY DEXA BOLUS), a significant increase in both plasma glucose and insulin was recorded in the late postprandial period (300–550 min) when compared to the two other groups: WHEY CONTROL and WHEY DEXA (Figure 1A,B).

Plasma leucine also increased rapidly (approximately 2-fold) after the WHEY CONTROL meal intake (Figure 1C). This difference was maintained during the first two-thirds of the postprandial period, then finally decreased rapidly during the late postprandial period (>400 min) (Figure 1C). The same pattern was recorded for the essential amino acids (EAA) (Figure 1D).

Post-absorptive leucine concentrations were significantly increased after the glucocorticoid treatment (time 0, 90 and 120 min: +35–40%) (Figure 1C). By contrast, the glucocorticoid treatment did not significantly alter plasma leucine and EAA kinetics following whey meal intake (Figure 1D). When the energy bolus was given at 300 min, no significant changes were recorded in leucine or EAA kinetics in the late postprandial period (300–550 min) in the WHEY DEXA BOLUS when compared to the WHEY CONTROL and WHEY DEXA groups (Figure 1D).

3.1.2. Muscle Protein Synthesis

When fed the whey meal, muscle protein synthesis increased to a peak at 240 min after which it decreased slowly until the end of the postprandial period studied (Figure 1E). In the WHEY DEXA group, muscle protein synthesis was significantly lower in the post-absorptive state and was only transiently significantly stimulated by the whey protein during the early part of the post-prandial period (150–270 min) (Figure 1E). In the late post-prandial period, muscle protein synthesis remained significantly lower in the WHEY DEXA group when compared to the WHEY CONTROL group. When the energy bolus was given at 300 min, no modification of muscle protein synthesis was recorded when compared to the WHEY DEXA group and it stayed significantly different to the values recorded in the WHEY CONTROL group (Figure 1E).

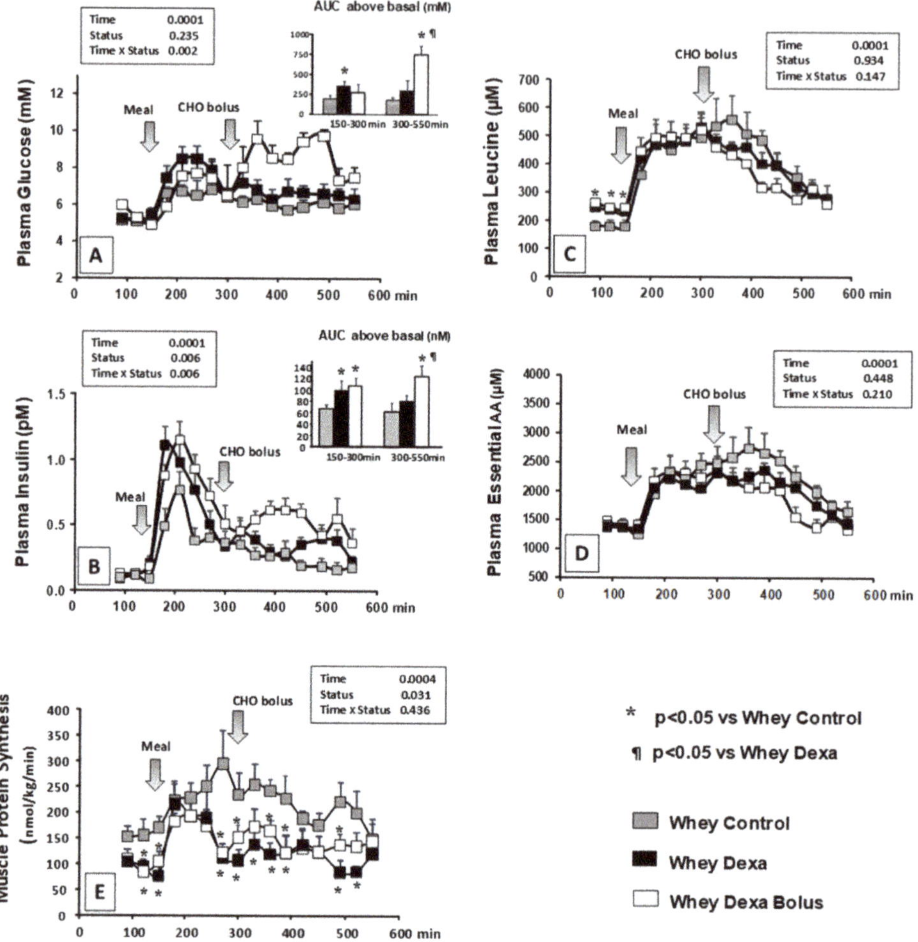

Figure 1. Plasma glucose (**A**), insulin (**B**), leucine (**C**) and essential amino acids (**D**) in control (WHEY CONTROL), in glucocorticoid-treated (WHEY DEXA) and in glucocorticoid-treated with the carbohydrates (CHO) bolus (WHEY DEXA BOLUS) minipigs. The areas under the curve (AUC) for glucose and insulin are presented in inserts. Postprandial AUCs were calculated by subtracting the post-absorptive values (t = 90–150 min) to each postprandial values. (**E**): Muscle protein synthesis kinetics for the post-absorptive period (before t = 150 min) and after the meal (after 150 min) in control (WHEY CONTROL), in glucocorticoid-treated (WHEY DEXA) and glucocorticoid-treated + carbohydrate (CHO) bolus (WHEY DEXA BOLUS) minipigs. The CHO bolus was given at t = 300 min. Data are presented as means ± SEM. (*n* = 6).

3.2. Aging Study

Plasma Glucose and Insulin

Plasma glucose increased significantly after the ingestion of the whey protein meal and glycemia peaked 180 min after food intake (Figure 2A). Insulin was also significantly increased after food intake to reach maximal values at 125 and 180 min (Figure 2B). When the energy bolus was given at 210 min, plasma glucose remained elevated and was significantly increased when compared to the WHEY CONTROL Group (Figure 2A). At 240 and 270 min, plasma glucose was 31 and 21% significantly higher, respectively, than at 240 min without the sugar bolus (Figure 2B). By contrast, insulinemia was not significantly increased after the intake of the energy bolus and even tended to be lower (−31%) than the value recorded at 240 min in the WHEY CONTROL (Figure 2B).

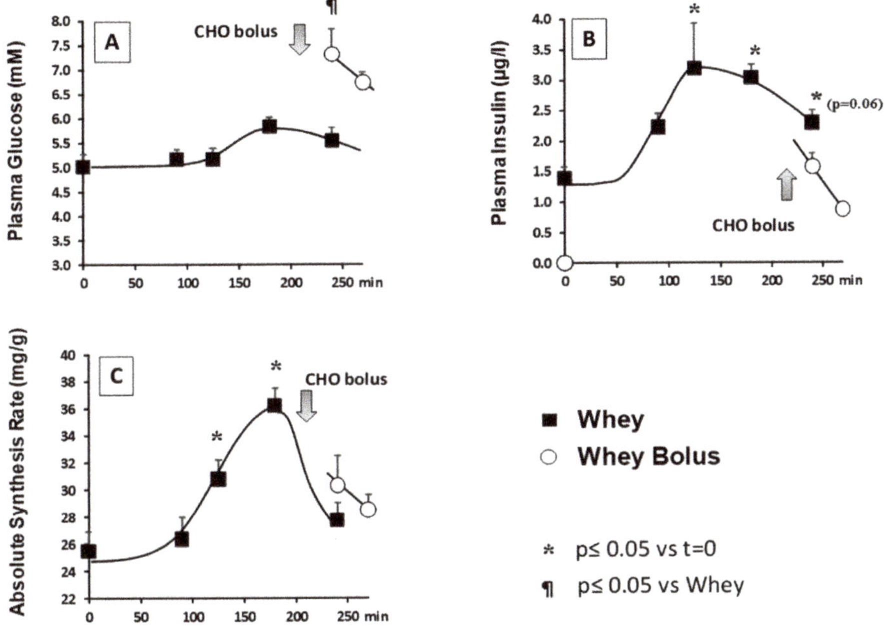

Figure 2. Plasma glucose (**A**), insulin (**B**) and muscle protein synthesis (**C**) in aged rats fed with whey or with whey followed by an energy bolus (CHO) 210 min after food intake. Data are presented as means ± SEM. (*n* = 10).

3.3. Muscle Protein Synthesis

Muscle protein synthesis was significantly stimulated after the ingestion of the whey protein meal and maximum stimulation was recorded at 180 min (Figure 2C). After 180 min, muscle protein synthesis decreased to become similar to the post-absorptive value at 240 min. When the energy bolus was given at 210 min, it did not prevent a decrease in muscle protein synthesis in the Whey Bolus group and its value was not significantly different from the values recorded in the Whey group (Figure 2C).

4. Discussion

We previously showed that casein was unable to initiate the stimulation of muscle nitrogen balance or muscle protein synthesis when given at the amount fulfilling healthy adult requirements in the two anabolic resistant states presented, i.e., glucocorticoids treatment and aging [9,10]. The reason for this anabolic resistance was explained by the lower sensitivity of the main signaling pathway (i.e., mTOR signaling pathway), which leads to the stimulation of muscle protein synthesis in response to food intake and particularly to dietary leucine intake and availability [30–33]. These observations, which are now well accepted, have led to recommending in such situations the preferential intake of rapidly digested and leucine-rich proteins such as whey proteins. Indeed, when the availability of essential amino acids and/or leucine is increased, the defect in the stimulation of muscle protein synthesis is attenuated or reversed during both glucocorticoid treatment [33] and aging [34]. Long-term supplementation of free leucine or leucine-rich proteins was therefore tested in both elderly humans and aged rodents. Verhoeven et al. [16] tested a 3-month leucine supplementation (7.5 g/day) and showed that it did not augment skeletal muscle mass or strength in healthy elderly men. Animal studies also showed no beneficial effect of leucine or whey protein supplementation on muscle mass in aged rodents unless it was given in high amounts within a high protein diet [15]. In a recent review, Woo et al. [35] concluded that to date, evidence suggests that nutritional intervention including high-quality protein or leucine does have benefits in the elderly, but mainly if combined with exercise. Similar conclusions could be drawn from glucocorticoid-induced anabolic resistance in which leucine supplementation remained without effect against the deleterious effect of dexamethasone on muscle fiber atrophy and strength loss [18]. We previously showed that whey proteins could initiate an anabolic response in the skeletal muscle in glucocorticoid-treated minipigs, but that the response remained transient and was only visible during the early post-prandial period (i.e., first 2 h), in contrast to the 7 h-stimulated muscle protein accretion obtained in healthy conditions [10]. In other catabolic states, with muscle wasting and anabolic resistance, such as bed rest and immobilization, leucine supplementation and whey proteins supplementation have also been shown as not being optimal for preserving muscle mass and strength [19,36–39].

Recently, it has been shown that muscle protein synthesis stimulation after feeding is of finite duration even if amino acid availability is still high and mTOR signaling pathways remain activated [21,22]. It has been postulated that muscles can sense they are "full," and this phenomenon has been named the "muscle full" effect [40,41]. Few studies have been carried out to elucidate the mechanisms involved in this "muscle full" effect and whether it can be prevented in order to prolong the anabolic response after food intake. They found that it was the decrease in muscle energy status associated with the increased activity of the intracellular energy sensor AMPK, which leads to blocking the stimulation of muscle protein synthesis [24,42]. However, when post meal supplements of carbohydrates (increased available energy) were given, the activation of AMPK was prevented, and the post-prandial muscle anabolic response was significantly extended in young growing rodents [24,42]. This could constitute a new strategy for slowing down muscle loss during the catabolic state.

Therefore, for the first time, we tested this nutritional strategy in catabolic models: glucocorticoid-treated young pigs and aging rats (a milder anabolic resistant state). In glucocorticoid treated pigs, the carbohydrate bolus further increased plasma glucose and insulin levels but remained without effect on the duration of muscle protein synthesis stimulation. Similarly, in aging rats, no modification in the duration of this stimulation was recorded. We hypothesize that glucocorticoids, which generate insulin resistance at the skeletal muscle level, had altered glucose uptake and metabolism and then rendered the energy bolus inefficient for maintaining or increasing ATP during the remaining postprandial period. In the aged rodents, it is noteworthy that our energy bolus dramatically increased plasma glucose, as expected, but did not increase plasma insulin. Since this increase of insulin was recorded in younger animals in which the energy bolus prolonged muscle protein synthesis, it is possible that during aging the lack of insulin elevation after the carbohydrate bolus did not allow simultaneous glucose uptake and metabolism in skeletal muscles and was unable to correct the energy deficit and AMPK activation. This is consistent because it has been shown that elderly subjects have a lower pulse amplitude and less responsive insulin secretion regarding oscillations in glucose (see [43] for a review). Furthermore, as mentioned for the glucocorticoid treatment, aging is also associated with glucose intolerance/insulin resistance, which may have further prevented glucose uptake during the energy bolus intake.

In addition, specific defects in AMPK pathways are possible. Indeed, it was shown that dexamethasone treatment caused intracellular ATP deprivation and robust AMPK activation [44]. This could explain the lack of effect of the energy bolus, which is assumed to increase ATP availability and reduce AMPK activation. Similarly, aging can disrupt AMPK signaling. For instance, mitochondrial biogenesis is reduced during aging in muscle, in particular in response to alteration in initial signaling through AMPK [45].

Other methodological differences exist between our study and the studies examining the effects of leucine or carbohydrate supplements for regulating protein synthesis duration [23,24]. The young mature rats were trained to consume three meals per day and were slightly food restricted (−20%). Since AMPK is a major sensor of energy availability in tissues, this food restriction could stimulate AMPK activity. Our older rats were not food-restricted and had much higher stores of energy than the young adult rats, which could also explain the lack of effect of our energy bolus on protein synthesis.

Limiting Points of the Study

In the case of the glucocorticoid-treated minipigs, we may also hypothesize that, contrary to rodents, the CHO bolus was unable to prolong the duration of the anabolic response to food intake in healthy control animals fed a whey protein meal. We did not perform trials with this group in the present study, so we cannot exclude that the lack of effect of the CHO bolus was not related to the glucocorticoid treatment *per se* but was possibly also related to species-specific metabolic response to this nutritional strategy. Another possibility is that since glucocorticoid treatment reduced the overall response of muscle protein synthesis to feeding, and in particular led to an earlier peak of protein synthesis, our energy bolus could have been given too late to achieve the stimulation of muscle protein synthesis in this group. Finally, for technical reasons, we were unable to measure AMP/ATP and/or P-AMPK/AMPK ratios, which would have been a beneficial addition to our study. However, we had already shown that in both the models tested, resistance to the anabolic effect of food intake was indeed correlated with a defect in the activation of the Akt/mTOR signaling pathway, and thus the energy bolus was a pertinent strategy for trying to prevent this resistance and prolong its duration [9,10]. Further studies are necessary to confirm that the "energy bolus strategy" is not possible in catabolic situations.

5. Conclusions

The use of a simple energy bolus to prolong the anabolic response to a meal would have been a very interesting tool for preventing muscle loss during catabolic states. The carbohydrate bolus was indeed efficient in young rats fed whey protein meals to sustain muscle protein synthesis [24]. However, our attempt to prolong the already weak anabolic effect of whey in our catabolic model and to increase the anabolic muscle response of whey during aging using a desynchronized carbohydrate bolus during the postprandial period did not succeed. We emphasize that nutritional strategies aimed at preventing the adverse effect of catabolic states on skeletal muscle requires additional studies and would probably need a combination of several levers, including physical activity.

Author Contributions: Conceptualization, D.D., L.M., I.S.-A. and S.P.; methodology, M.J., J.D., D.R., and D.D.; validation, D.D., L.M., I.S.-A., D.R. and S.P.; formal analysis, D.D. and L.M.; investigation, M.J., J.D., D.R., L.M. and D.D.; resources, D.D. and D.R.; data curation, D.D.; writing—original draft preparation, D.D. and L.M.; writing—review and editing, D.D., L.M., I.S.-A., D.R. and S.P.; funding acquisition, D.D.

Funding: This research received funding from the Agence Nationale de la Recherche (ARN) through the Carnot Institute "Qualiment" and from the "Olga Triballat" Award.

Acknowledgments: The authors are grateful to the personnel of the Animal Facility (C. de L'Homme, P. Denis, A. Cissoire, B. Coahde, P. Lhoste) for their technical assistance.

Conflicts of Interest: The authors declare no conflict of interest.

References

1. Dardevet, D.; Remond, D.; Peyron, M.A.; Papet, I.; Savary-Auzeloux, I.; Mosoni, L. Muscle wasting and resistance of muscle anabolism: The "anabolic threshold concept" for adapted nutritional strategies during sarcopenia. *Sci. World J.* **2012**, *2012*, 269531. [CrossRef]
2. Atherton, P.J.; Wilkinson, D.J.; Smith, K. Chapter 9—Feeding Modulation of Amino Acid Utilization: Role of Insulin and Amino Acids in Skeletal Muscle. In *The Molecular Nutrition of Amino Acids and Proteins*; Dardevet, D., Ed.; Academic Press: Boston, MA, USA, 2016; pp. 109–124. [CrossRef]

3. Prod'homme, M.; Rieu, I.; Balage, M.; Dardevet, D.; Grizard, J. Insulin and amino acids both strongly participate to the regulation of protein metabolism. *Curr. Opin. Clin. Nutr. Metab. Care* **2004**, *7*, 71–77. [CrossRef] [PubMed]

4. Mosoni, L.V.M.; Serrurier, B.; Prugnaud, J.; Obled, C.; Guezennec, C.Y.; Mirand, P.P. Altered response of protein synthesis to nutritional state and endurance training in old rats. *Am. J. Physiol.* **1995**, *268*, E328–E335. [CrossRef] [PubMed]

5. Cuthbertson, D.; Smith, K.; Babraj, J.; Leese, G.; Waddell, T.; Atherton, P.; Wackerhage, H.; Taylor, P.M.; Rennie, M.J. Anabolic signaling deficits underlie amino acid resistance of wasting, aging muscle. *FASEB J.* **2004**, *19*, 422–424. [CrossRef]

6. Horstman, A.; Olde Damink, S.; Schols, A.; van Loon, L. Is Cancer Cachexia Attributed to Impairments in Basal or Postprandial Muscle Protein Metabolism? *Nutrients* **2016**, *8*, 499. [CrossRef] [PubMed]

7. Rudrappa, S.S.; Wilkinson, D.J.; Greenhaff, P.L.; Smith, K.; Idris, I.; Atherton, P.J. Human Skeletal Muscle Disuse Atrophy: Effects on Muscle Protein Synthesis, Breakdown, and Insulin Resistance—A Qualitative Review. *Front. Physiol.* **2016**, *7*. [CrossRef] [PubMed]

8. Magne, H.; Savary-Auzeloux, I.; Migné, C.; Peyron, M.A.; Combaret, L.; Rémond, D.; Dardevet, D. Contrarily to whey and high protein diets, dietary free leucine supplementation cannot reverse the lack of recovery of muscle mass after prolonged immobilization during ageing. *J. Physiol.* **2012**, *590*, 2035–2049. [CrossRef]

9. Jarzaguet, M.; Polakof, S.; David, J.; Migné, C.; Joubrel, G.; Efstathiou, T.; Rémond, D.; Mosoni, L.; Dardevet, D. A meal with mixed soy/whey proteins is as efficient as a whey meal in counteracting the age-related muscle anabolic resistance only if the protein content and leucine levels are increased. *Food Funct.* **2018**, *9*, 6526–6534. [CrossRef]

10. Revel, A.; Jarzaguet, M.; Peyron, M.A.; Papet, I.; Hafnaoui, N.; Migné, C.; Mosoni, L.; Polakof, S.; Savary-Auzeloux, I.; Rémond, D.; et al. At same leucine intake, a whey/plant protein blend is not as effective as whey to initiate a transient postprandial muscle anabolic response during a catabolic state in mini pigs. *PLoS ONE* **2017**, *12*, e0186204. [CrossRef]

11. Rand, W.M.; Pellett, P.L.; Young, V.R. Meta-analysis of nitrogen balance studies for estimating protein requirements in healthy adults. *Am. J. Clin. Nutr.* **2003**, *77*, 109–127. [CrossRef] [PubMed]

12. Dangin, M.; Guillet, C.; Garcia-Rodenas, C.; Gachon, P.; Bouteloup-Demange, C.; Reiffers-Magnani, K.; Fauquant, J.; Ballèvre, O.; Beaufrère, B. The Rate of Protein Digestion affects Protein Gain Differently during Aging in Humans. *J. Physiol.* **2003**, *549*, 635–644. [CrossRef] [PubMed]

13. Magne, H.; Savary-Auzeloux, I.; Rémond, D.; Dardevet, D. Nutritional strategies to counteract muscle atrophy caused by disuse and to improve recovery. *Nutr. Res. Rev.* **2013**, *26*, 149–165. [CrossRef]

14. Ferrando, A.A.; Paddon-Jones, D.; Hays, N.P.; Kortebein, P.; Ronsen, O.; Williams, R.H.; McComb, A.; Symons, T.B.; Wolfe, R.R.; Evans, W. EAA supplementation to increase nitrogen intake improves muscle function during bed rest in the elderly. *Clin. Nutr.* **2010**, *29*, 18–23. [CrossRef] [PubMed]

15. Mosoni, L.; Gatineau, E.; Gatellier, P.; Migné, C.; Savary-Auzeloux, I.; Rémond, D.; Rocher, E.; Dardevet, D. High Whey Protein Intake Delayed the Loss of Lean Body Mass in Healthy Old Rats, whereas Protein Type and Polyphenol/Antioxidant Supplementation Had No Effects. *PLoS ONE* **2014**, *9*, e109098. [CrossRef]

16. Verhoeven, S.; Vanschoonbeek, K.; Verdijk, L.B.; Koopman, R.; Wodzig, W.K.W.H.; Dendale, P.; van Loon, L.J.C. Long-term leucine supplementation does not increase muscle mass or strength in healthy elderly men. *Am. J. Clin. Nutr.* **2009**, *89*, 1468–1475. [CrossRef]

17. Choudry, H.A.; Pan, M.; Karinch, A.M.; Souba, W.W. Branched-Chain Amino Acid-Enriched Nutritional Support in Surgical and Cancer Patients. *J. Nutr.* **2006**, *136*, S314–S318. [CrossRef]

18. Nicastro, H.; Zanchi, N.E.; da Luz, C.R.; de Moraes, W.M.A.M.; Ramona, P.; de Siqueira Filho, M.A.; Chaves, D.F.S.; Medeiros, A.; Brum, P.C.; Dardevet, D.; et al. Effects of leucine supplementation and resistance exercise on dexamethasone-induced muscle atrophy and insulin resistance in rats. *Nutrition* **2012**, *28*, 465–471. [CrossRef] [PubMed]

19. Nicastro, H.; Artioli, G.G.; dos Santos Costa, A.; Solis, M.Y.; da Luz, C.R.; Blachier, F.; Lancha, A.H. An overview of the therapeutic effects of leucine supplementation on skeletal muscle under atrophic conditions. *Amino Acids* **2011**, *40*, 287–300. [CrossRef] [PubMed]

20. Lacroix, M.; Bos, C.; Léonil, J.; Airinei, G.; Luengo, C.; Daré, S.; Benamouzig, R.; Fouillet, H.; Fauquant, J.; Tomé, D.; et al. Compared with casein or total milk protein, digestion of milk soluble proteins is too rapid to sustain the anabolic postprandial amino acid requirement. *Am. J. Clin. Nutr.* **2006**, *84*, 1070–1079. [CrossRef]

21. Atherton, P.J.; Etheridge, T.; Watt, P.W.; Wilkinson, D.; Selby, A.; Rankin, D. Muscle full effect after oral protein: Time-dependent concordance and discordance between human muscle protein synthesis and mTORC1 signaling. *Am. J. Clin. Nutr.* **2010**, *92*, 1080–1088. [CrossRef]

22. Bohé, J.; Low, J.F.A.; Wolfe, R.R.; Rennie, M.J. Latency and duration of stimulation of human muscle protein synthesis during continuous infusion of amino acids. *J. Physiol.* **2001**, *532*, 575–579. [CrossRef]

23. Norton, L.E.; Layman, D.K.; Bunpo, P.; Anthony, T.G.; Brana, D.V.; Garlick, P.J. The Leucine Content of a Complete Meal Directs Peak Activation but Not Duration of Skeletal Muscle Protein Synthesis and Mammalian Target of Rapamycin Signaling in Rats. *J. Nutr.* **2009**, *139*, 1103–1109. [CrossRef]

24. Wilson, G.J.; Layman, D.K.; Moulton, C.J.; Norton, L.E.; Anthony, T.G.; Proud, C.G.; Rupassara, S.I.; Garlick, P.J. Leucine or carbohydrate supplementation reduces AMPK and eEF2 phosphorylation and extends postprandial muscle protein synthesis in rats. *Am. J. Physiol. Endocrinol. Metab.* **2011**, *301*, E1236–E1242. [CrossRef] [PubMed]

25. Rémond, D.; Buffière, C.; Godin, J.P.; Mirand, P.P.; Obled, C.; Papet, I.; Dardevet, D.; Williamson, G.; Breuillé, D.; Faure, M. Intestinal Inflammation Increases Gastrointestinal Threonine Uptake and Mucin Synthesis in Enterally Fed Minipigs. *J. Nutr.* **2009**, *139*, 720–726. [CrossRef] [PubMed]

26. Rieu, I.; Magne, H.; Savary-Auzeloux, I.; Averous, J.; Bos, C.; Peyron, M.A.; Combaret, L.; Dardevet, D. Reduction of low grade inflammation restores blunting of postprandial muscle anabolism and limits sarcopenia in old rats. *J. Physiol.* **2009**, *587*, 5483–5492. [CrossRef] [PubMed]

27. Bruins, M.J.; Soeters, P.B.; Lamers, W.H.; Deutz, N.E. l-Arginine supplementation in pigs decreases liver protein turnover and increases hindquarter protein turnover both during and after endotoxemia. *Am. J. Clin. Nutr.* **2002**, *75*, 1031–1044. [CrossRef]

28. Paddon-Jones, D.; Sheffield-Moore, M.; Zhang, X.J.; Volpi, E.; Wolf, S.E.; Aarsland, A.; Ferrando, A.A.; Wolfe, R.R. Amino acid ingestion improves muscle protein synthesis in the young and elderly. *Am. J. Physiol. Endocrinol. Metab.* **2004**, *286*, E321–E328. [CrossRef]

29. Tourian, A.; Goddard, J.; Puck, T.T. Phenylalanine hydroxylase activity in mammalian cells. *J. Cell. Physiol.* **1969**, *73*, 159–170. [CrossRef] [PubMed]

30. Dardevet, D.; Sornet, C.; Balage, M.; Grizard, J. Stimulation of in vitro rat muscle protein synthesis by leucine decreases with age. *J. Nutr.* **2000**, *130*, 2630–2635. [CrossRef]

31. Kimball, S.R.; Shantz, L.M.; Horetsky, R.L.; Jefferson, L.S. Leucine regulates translation of specific mRNAs in L6 myoblasts through mTOR-mediated changes in availability of eIF4E and phosphorylation of ribosomal protein S6. *J. Biol. Chem.* **1999**, *274*, 11647–11652. [CrossRef]

32. Rieu, I.; Sornet, C.; Grizard, J.; Dardevet, D. Glucocorticoid excess induces a prolonged leucine resistance on muscle protein synthesis in old rats. *Exp. Gerontol.* **2004**, *39*, 1315–1321. [CrossRef]

33. Paddon-Jones, D.; Sheffield-Moore, M.; Creson, D.L.; Sanford, A.P.; Wolf, S.E.; Wolfe, R.R.; Ferrando, A.A. Hypercortisolemia alters muscle protein anabolism following ingestion of essential amino acids. *Am. J. Physiol. Endocrinol. Metab.* **2003**, *284*, E946–E953. [CrossRef]

34. Rieu, I.; Balage, M.; Sornet, C.; Giraudet, C.; Pujos, E.; Grizard, J.; Mosoni, L.; Dardevet, D. Leucine supplementation improves muscle protein synthesis in the elderly men independently of hyperaminoacidemia. *J. Physiol.* **2006**, *575*, 305–315. [CrossRef] [PubMed]

35. Woo, J. Nutritional interventions in sarcopenia: Where do we stand? *Curr. Opin. Clin. Nutr. Metab. Care* **2018**, *21*, 19–23. [CrossRef] [PubMed]

36. Trappe, S.; Creer, A.; Minchev, K.; Slivka, D.; Louis, E.; Luden, N.; Trappe, T. Human soleus single muscle fiber function with exercise or nutrition countermeasures during 60 days of bed rest. *Am. J. Physiol. Regul. Integr. Comp. Physiol.* **2008**, *294*, R939–R947. [CrossRef] [PubMed]

37. Trappe, S.; Creer, A.; Slivka, D.; Minchev, K.; Trappe, T. Single muscle fiber function with concurrent exercise or nutrition countermeasures during 60 days of bed rest in women. *J. Appl. Physiol.* **2007**, *103*, 1242–1250. [CrossRef] [PubMed]

38. Magne, H.; Savary-Auzeloux, I.; Peyron, M.A.; Migne, C.; Combaret, L.; Rémond, D.; Dardevet, D. Which Nutritional Protein Strategy to Recover Muscle Mass after Prolonged Immobilization during Aging? In Proceedings of the 34th Congress of the European Society of Clinical Nutrition and Metabolism (ESPEN), Barcelone, Spain, 8–11 September 2012; pp. 6–7.

39. Verney, J.; Martin, V.; Ratel, S.; Chavanelle, V.; Bargetto, M.; Etienne, M.; Chaplais, E.; Le Ruyet, P.; Bonhomme, C.; Combaret, L.; et al. Soluble milk proteins improve muscle mass recovery after immobilization-induced muscle atrophy in old rats but do not improve muscle functional property restoration. *J. Nutr. Health Aging* **2016**, *21*, 1133–1141. [CrossRef] [PubMed]

40. Phillips, B.E.; Hill, D.S.; Atherton, P.J. Regulation of muscle protein synthesis in humans. *Curr. Opin. Clin. Nutr. Metab. Care* **2012**, *15*, 58–63. [CrossRef] [PubMed]

41. Dideriksen, K.; Reitelseder, S.; Holm, L. Influence of Amino Acids, Dietary Protein, and Physical Activity on Muscle Mass Development in Humans. *Nutrients* **2013**, *5*, 852–876. [CrossRef]

42. Wilson, G.; Moulton, C.; Garlick, P.; Anthony, T.; Layman, D. Post-Meal Responses of Elongation Factor 2 (eEF2) and Adenosine Monophosphate-Activated Protein Kinase (AMPK) to Leucine and Carbohydrate Supplements for Regulating Protein Synthesis Duration and Energy Homeostasis in Rat Skeletal Muscle. *Nutrients* **2012**, *4*, 1723–1739. [CrossRef]

43. Chang, A.M.; Halter, J.B. Aging and insulin secretion. *Am. J. Physiol. Endocrinol. Metab.* **2003**, *284*, E7–E12. [CrossRef] [PubMed]

44. Liu, J.; Peng, Y.; Wang, X.; Fan, Y.; Qin, C.; Shi, L.; Tang, Y.; Cao, K.; Li, H.; Long, J.; et al. Mitochondrial Dysfunction Launches Dexamethasone-Induced Skeletal Muscle Atrophy via AMPK/FOXO3 Signaling. *Mol. Pharm.* **2016**, *13*, 73–84. [CrossRef] [PubMed]

45. Kim, Y.; Triolo, M.; Hood, D.A. Impact of Aging and Exercise on Mitochondrial Quality Control in Skeletal Muscle. *Oxid. Med. Cell. Longev.* **2017**, *2017*. [CrossRef] [PubMed]

Article

Wheat Albumin Increases the Ratio of Fat to Carbohydrate Oxidation during the Night in Healthy Participants: A Randomized Controlled Trial

Shinichiro Saito [1,*], Toshitaka Sakuda [1], Aiko Shudo [2], Yoko Sugiura [2] and Noriko Osaki [1]

[1] Biological Research Laboratories, Kao Corporation, 2-1-3 Bunka Sumida-ku, Tokyo 131-8501, Japan; sakuda.toshitaka@kao.com (T.S.); osaki.noriko@kao.com (N.O.)

[2] Health Care Food Research Laboratories, Kao Corporation, 2-1-3 Bunka Sumida-ku, Tokyo 131-8501, Japan; shudou.aiko@kao.com (A.S.); sugiura.yoko@kao.com (Y.S.)

* Correspondence: saito.shinichiro@kao.com; Tel.: +81-3-5630-7456

Received: 25 December 2018; Accepted: 17 January 2019; Published: 18 January 2019

Abstract: Not only are energy expenditure (EE) and the respiratory quotient (RQ) parameters of the energy nutrient utilization and energy balance, they are also related to the development of obesity. In this study, post-meal night-time energy metabolism was investigated following the oral ingestion of wheat albumin (WA) with a late evening meal. A randomly assigned, double-blind, placebo-controlled crossover trial for a single oral ingestion in healthy participants was completed. The participants ingested the placebo (PL) or WA (1.5 g) containing tablets 3 minutes before the late evening meal at 22:00 hour, and energy metabolism was measured using a whole-room indirect calorie meter until wake-up. The participants were in bed from 00:00 hour until 06:30 hour. Twenty healthy participants completed the trial and were included in the analyses. Night-time RQ and carbohydrate oxidation were significantly lower following the WA treatment as compared with the PL treatment. Although the total EE was not significantly different between treatments, postprandial fat oxidation was significantly higher following the WA treatment as compared with the PL treatment. In conclusion, WA has the potential to shift the energy balance to a higher ratio of fat to carbohydrate oxidation during the night.

Keywords: energy expenditure; fat oxidation; human; respiratory quotient; wheat albumin

---

1. Introduction

Obesity is well known to result from an imbalance between energy consumption and expenditure. Therefore, increases in physical activity or proper diet therapy are recommended to maintain body weight. In addition, recent studies have reported that a lower respiratory quotient (RQ) or ratio of fat to carbohydrate oxidation was associated with weight gain over several years in non-diabetic Pima Indians [1] and non-obese Italian women [2], even after the adjustment for energy expenditure (EE), suggesting that the RQ is an independent predictor against the development of obesity. Moreover, in individuals with obesity, carbohydrate oxidation did not change while fat oxidation was attenuated by aging [3]. Thus, in addition to the imbalance between energy consumption, a lower ratio of fat to carbohydrate oxidation is a risk factor for weight gain.

A previous study showed that wheat albumin (WA), which has a long history of human consumption as a natural food constituent, is a potential agent against postprandial hyperglycemia [4]. As carbohydrate overloading stimulates carbohydrate oxidation together with fat storage and reduced fat oxidation [5], we hypothesized that carbohydrate loading during the night would induce a lower ratio of fat and carbohydrate oxidation, even in healthy individuals, presumably due to the lower physical activity during the night. Based on our unpublished pilot study, we predicted that this phenomenon would be improved by WA via its suppressive effect on the glucose response after meals.

Thus, we investigated energy metabolism during the night with or without a single oral ingestion of WA with a late evening meal.

2. Materials and Methods

2.1. Ethics Approval and Consent to Participate

This study was performed in accordance with the tenets of the Declaration of Helsinki (2013) and was approved by the Ethical Committee of the Kao Corporation (Tokyo, Japan). After receiving a full explanation of the study, all participants provided written informed consent. The study was registered with the University Hospital Medical Information Network (UMIN) clinical registry prior to the enrolment of the first participant as UMIN000020151; registered 18 December 2015 at http://www.umin.ac.jp/ctr/index.htm.

2.2. Study Design

This study was a randomized, double-blind, placebo-controlled, crossover trial performed under supervision by the physician in charge with a 5-day washout period. The study protocol is shown in Figure 1. Between the screening and the second visit, the participants were instructed to maintain and record their dietary, alcohol, and smoking habits, and level of physical activity. The participants were prohibited from drinking alcohol and undergoing heavy exercise the day before the visits and from smoking cigarettes on the days of the visits based on the report showing that the energy expenditure was increased by exercise or smoking [6]. The participants consumed a designated meal (3310 kJ, protein (P):fat (F):carbohydrate (C) = 14:25:61 as the energy value) for dinner in the evening until 21:00 hour one day before the trial, and had a designated breakfast (2249 kJ, P:F:C = 11:51:38 as the energy value) at 08:00 hour and designated lunch (2887 kJ, P:F:C = 17:18:65 as the energy value) at 12:00 hour on the day of the trial. The participants were not allowed any energy intake other than the designated meals from after dinner on the day before the visit until after completion of the visit. After bathing, the anthropometric parameters of the participants, such as body weight and blood pressure, were measured, as shown in Table 1, and then they entered the whole-room indirect calorie meter (chamber room) at 19:00 hour and underwent low-energy activities such as watching TV, using the PC, or reading books in a sitting position until 21:00 hour to habituate to the conditions of the chamber room (25 °C, 40% humidity). At 21:00 hour, the participants' baseline energy metabolism levels were measured in a sitting position for 30 minutes. Blood samples were obtained at 21:30 hour via the specific window on the chamber door to collect samples with no air exchange between the inside and outside of the room. The participants ingested PL or WA (1.5 g) containing tablets 3 min before the designated late evening meal (Japanese style meal, such as rice, grilled fish, grilled chicken, and boiled vegetables; 3423 kJ, P:F:C = 16:19:65 as the energy value) at 22:00 hour and their energy metabolism was measured until wake-up. The participants slept from 00:30 hour until 06:30 hour and their sleep quality was measured using an Actigraph (ActiGraph, Pensacola, FL, USA). Water consumption was controlled during the measurement period. The treatment allocation was concealed throughout the study (from screening to finalizing the dataset) from the people involved, including the participants, the caregivers, the physicians, the manufacturers of the test tablets, the person in charge of the allocation, and the outcome assessors.

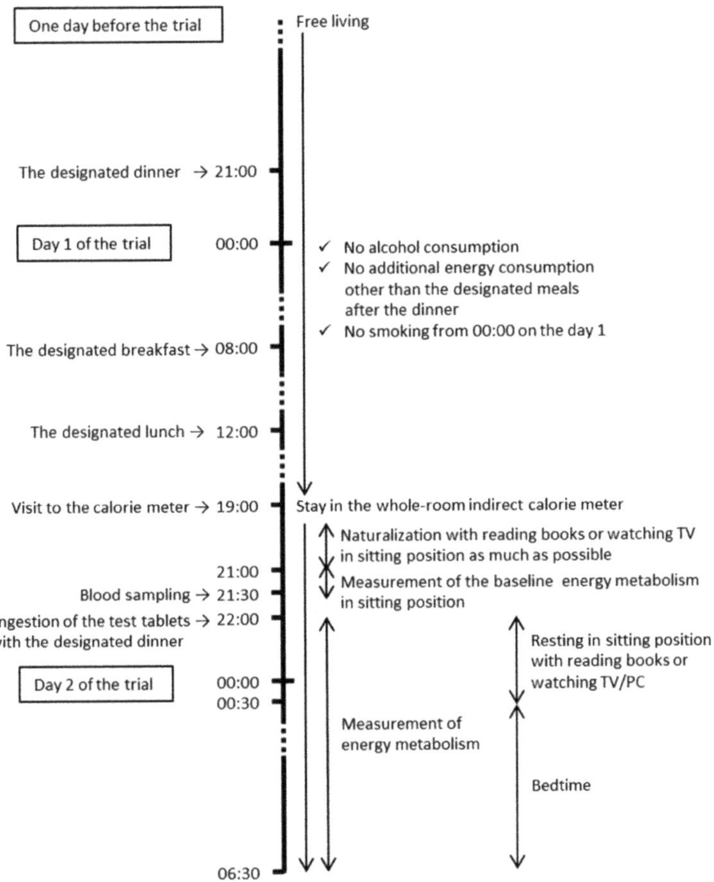

Figure 1. The study protocol.

Table 1. Characteristics of the participants.

Parameter	Value
Number of participants (male/female)	20 (14/6)
Age, years	38 ± 10
Body weight, kg	62.4 ± 9.8 (M, 67.7 ± 7.7; F, 51.9 ± 4.0)
Body mass index, kg/m^2	21.8 ± 1.9
Fat mass, kg	12.5 ± 2.9
Fat free mass, kg	49.8 ± 8.2
Systolic blood pressure, mmHg	124 ± 18
Diastolic blood pressure, mmHg	73 ± 14
Glucose, mmol/L	5.36 ± 0.31
Insulin, pmol/L	19.2 ± 7.2
Triglyceride, mg/dL	0.894 ± 0.485

Data are means ± standard deviations. M, male; F, female.

2.3. Participants

A sufficient sample size for the primary outcome of postprandial fat oxidation was estimated to be 20 participants based on the outcomes of our unpublished pilot study outcome (power 0.8 and type I error 0.05). In this study, potential participants were screened from men and women aged from 24 to 59 years. Participants were excluded if they met the following exclusion criteria: (1) Presence of liver, kidney, or heart disease; respiratory, endocrine, metabolism, nervous system, or consciousness dysfunction; diabetes; or other diseases; (2) previous experience of seizures based on circulatory-diseases or under treatment for a condition of this type; (3) presence of arrhythmia; (4) taking medications for hyperglycemia, lipidemia, or hypertension; (5) experience of seizures based on a neural disease; (6) surgery within two months before the trial; (7) previous gastrectomy or enterectomy; (8) allergies to any constituents in the test meal or tablets; (9) unpleasant feeling during blood drawing; (10) claustrophobia; (11) insomnia; (12) chronic headache; (13) donation of 200 mL or more of blood within one month before the trial or 400 mL or more of blood within three months before the trial; (14) a habitual bed-time of after 1:00 a.m. on weekdays; (15) heavy smoker (>20 cigarettes); (16) weight change of more than 2.0 kg one month prior to informed consent; (17) shift worker or engaged in night work and shift operations three months prior to informed consent; (18) pregnant or expecting pregnancy; (19) taking supplements or food for a specific use of health authorized by the government; (20) were not checked by a doctor regarding their health condition or their health examination results for the past two years prior to the trial could not be accessed; (21) did not disclose their age and their latest health examination result; (22) did not reply to the questionnaire about living situation and condition; and (23) could not indicate their menstruation situation. The participants were randomly assigned to each sequence (ingestion order) with stratified randomization for glucose, triglyceride, age, and sex using computer-generated random numbers under blind conditions.

2.4. Test Tablets

The test diet was three tablets containing a total of 1.5 g WA for a single oral administration. The PL tablets did not contain any WA. These were prepared using identical ingredients including flavors and preservatives except for the WA, with a weight of 1.1 g for each tablet. The energy value of a single dose was 14.7 kJ for the WA tablet and 10.9 kJ for the PL tablet. The tablets could not be distinguished by appearance, taste, or odor and were provided to the participants after concealment.

2.5. Whole-Room Indirect Calorie Meter

EE and substrate utilization for each participant were measured in the respiratory chamber. Whole-room indirect calorimeter measurements were obtained by the previously described methods [7]. In brief, the room temperature, humidity, and fresh airflow were set to 25 °C, 40%, and 70 L/minute, respectively. Oxygen consumption (VO_2) and carbon dioxide production (VCO_2) were calculated using the method reported by Henning et al. [8]. VO_2 and VCO_2 were calculated across a 60-minute period to obtain the values of EE, RQ, fat oxidation, and carbohydrate oxidation for the transient response analysis [9,10]. Protein oxidation was estimated based on urinary nitrogen excretion. All urine samples were collected and weighed while participants were in the whole-room indirect calorie meter and measured in triplicate using a chemiluminescent nitrogen analyzer (TN-100, Mitsubishi Chemical, Kanagawa, Japan).

2.6. Blood Samples

Collected blood samples for the measurements as shown in Table 1 were centrifuged at $1000 \times g$ for 15 minutes at 4 °C to isolate the serum or plasma and were measured by SRL, Inc. (Tokyo, Japan) or LSI Medience Co. (Tokyo, Japan).

2.7. Statistics

The primary outcome of this study was the difference in fat oxidation during the night between PL and WA treatments. To determine the effect of WA on the primary outcome, the mean value during the night from 22:00 hour to 06:30 hour was estimated and assessed using a mixed model adjusted by order of the treatment (No significant effect of order of the treatment was observed). A two-sided

p-value $\leq$ 0.05 was considered to be statistically significant. All statistical analyses were performed using the IBM SPSS Statistics version 19 (IBM Co., Armonk, NY, USA).

3. Results

3.1. Characteristics of the Participants

Thirty-five individuals were screened, and 21 were recruited. Of those recruited, one participant dropped out of the study right after the consumption of the PL tablet on the first visit due to an unfavorable feeling due to the texture and flavor of the PL tablet. Twenty participants completed the study and were included in the analyses. From screening until the second visit, no considerable habitual changes were recorded. The characteristics of the participants are presented in Table 1. The dietary records from the three days before the measurements are indicated in Table 2. There were no significant differences in dietary status before the visits or percentage of sleep during the trial between treatments (PL and WA, 92.9 $\pm$ 4.3% and 91.9 $\pm$ 3.5%, respectively, mean $\pm$ standard deviation).

Table 2. Dietary records of the three days before the treatment.

Parameter	PL	WA
Energy, kJ/day	8299 $\pm$ 1616	8464 $\pm$ 1610
Protein, g/day	72.1 $\pm$ 18.0	69.5 $\pm$ 17.6
Fat, g/day	64.2 $\pm$ 19.4	68.7 $\pm$ 22.7
Carbohydrate, g/day	249.2 $\pm$ 54.5	250.5 $\pm$ 46.1

Data are means $\pm$ standard deviations. There were no significant differences between treatments. PL: placebo, WA: wheat albumin.

3.2. Substrate Utilization

In the study, the test tablets were ingested 3 minutes before the designated meals, and the substrate oxidation was compared between treatments. The mean fat oxidation during the night increased significantly following WA ingestion as compared with the PL, whereas the mean carbohydrate oxidation significantly decreased. There was no significant difference in the mean protein oxidation (Figure 2). Fat mass is reported to be associated with fat oxidation [11]. In this study, the fat mass before the WA treatment was not significantly different compared with that before the PL treatments (PL and WA, 12.5 $\pm$ 2.9 kg and 12.5 $\pm$ 3.2, respectively, mean $\pm$ standard deviation). The fat oxidation on the fat mass during the WA treatment was significantly increased compared with that during the PL treatment.

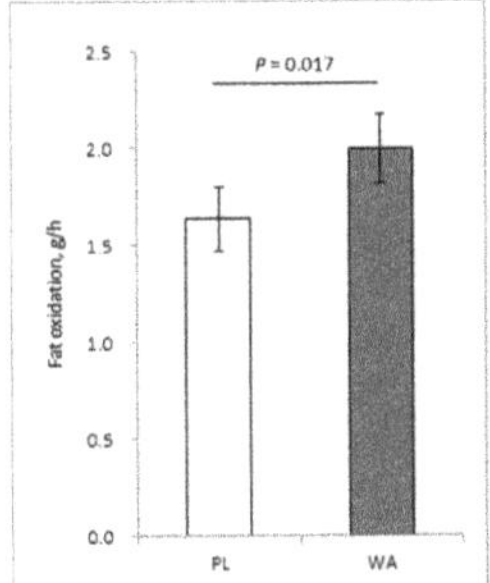
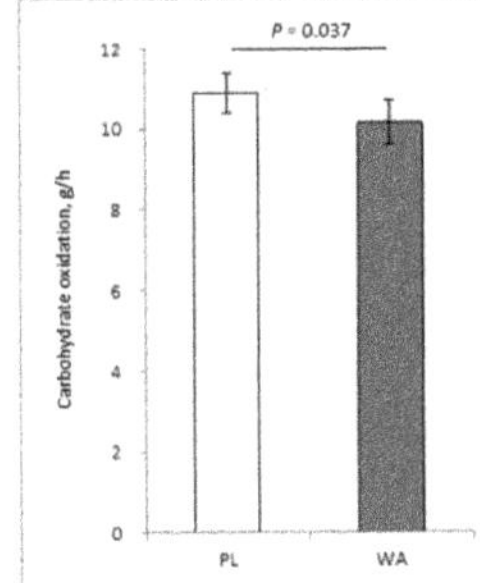
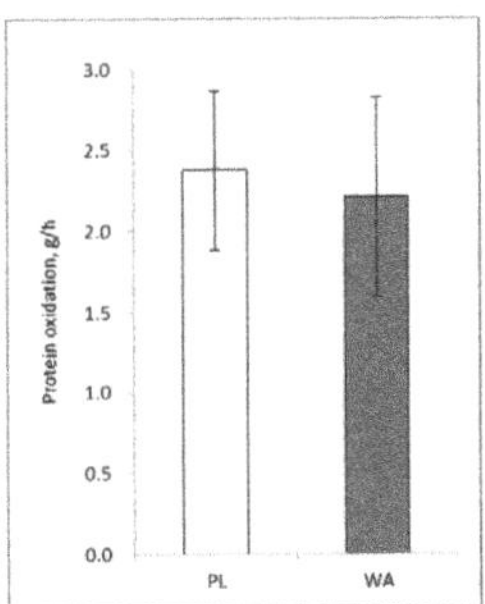

Figure 2. (**Left**) Fat oxidation. (**Middle**) Carbohydrate oxidation. (**Right**) Protein oxidation. Data are means $\pm$ standard errors during the night from 22:00 hour to 06:30 hour. Significant differences between the treatments were assessed with a mixed model adjusted by order of treatment.

3.3. EE and RQ

There were no significant differences in the baseline EE (PL and WA, 1189 ± 182 kJ and 1150 ± 156, respectively, mean ± standard deviation) and the baseline RQ (PL and WA, 0.816 ± 0.020 L/L/h and 0.805 ± 0.026, respectively, mean ± standard deviation). There was no significant difference in the mean EE during the night between treatments; however, the mean RQ was significantly lower after WA ingestion than after PL ingestion (Table 3), suggesting a trend in fuel utilization of greater fat oxidation during the night following the WA treatment. Fat free mass change is reported to be associated with EE [12]. In this study, the fat free mass before the WA treatment was not significantly different compared with that before the PL treatment (PL and WA, 49.8 ± 8.2 kg and 49.8 ± 8.0, respectively, mean ± standard deviation). The EE on the fat free mass was also not significantly different between the treatments.

Table 3. Energy expenditure (EE) and respiratory quotient (RQ).

Parameter	PL	WA
EE, kJ/h	1148 ± 139	1142 ± 122
RQ, L/L/h	0.904 ± 0.027	0.890 ± 0.032 **

Data are means ± standard deviations during night from 22:00 hour to 06:30 hour. The asterisk denotes significant differences between the treatments assessed with a mixed model adjusted by order of the treatment; ** $p < 0.01$.

4. Discussion

In this study, the effect of the WA treatment on substrate oxidation during the night was investigated in healthy humans. Our previous unpublished study showed a moderate suppressive effect of WA on the glucose response during the night. The underlying mechanism of the effect is thought to be its inhibitory action on alpha-amylase activity [4], which induces a lower carbohydrate absorption and may also accompany the lower glucose-dependent insulinotropic polypeptide (GIP) response. Indeed, the plasma concentration of GIP was lowered significantly by the WA treatment when compared to the PL treatment in our previous unpublished study. A higher postprandial concentration of GIP, an incretin that stimulates insulin secretion from the pancreatic beta-cells, has been directly associated with a lower metabolic rate [13] and stimulates fat accumulation in adipose tissue as an exopancreatic function [14]. Additionally, higher blood GIP levels induced by chronic GIP treatment were shown to reduce fat utilization in high-fat diet-fed mice [15]. Thus, the potential mechanism underlying enhanced fat oxidation by WA treatment may be associated with the GIP lowering effect.

In the present study, fat oxidation was enhanced and RQ was lowered following WA treatment, but EE was not changed. An imbalanced energy balance between energy consumption and the expenditure is directly associated with obesity [16]. Therefore, the anti-obesity effect of tWA via acute increased EE was not expected. However, a low ratio of fat to carbohydrate oxidation was reported to be a predictor of future weight gain in several studies, such as in Pima Indians [1] and healthy women [2]. An assumable underlying mechanism was reported to be enhanced lipogenesis induced by the overflow of carbohydrate metabolism [5]. Therefore, a lowering effect on the ratio of fat to carbohydrate oxidation may reduce the overflow. Although we observed an acute effect on RQ, repeated nightly treatment of WA might have an impact on body fat accumulation overnight. Thus, an intervention study with a focus on weight gain in such patients with night eating syndrome [17] is of interest.

The difference of mean fat oxidation between the treatments was 0.35 g/h (PL and WA, 1.64 ± 0.73 g/h and 1.99 ± 0.79, respectively, mean ± standard deviation). Based on this, the estimated cumulative fat oxidation during the night from 22:00 hour to 06:30 hour (8.5 hours) was 3.0 g. The impact of this amount is estimated to be 1.08 kg as fat under simple calculation if the repeated WA treatment was performed for one year. However, given that there was no difference in energy expenditure, it remains unknown if this would result in any change in body weight or composition.

The limitations and potential biases in this study were the imbalanced gender of the participants (male:female = 14:6) and the use of a single race (Japanese). The included authors are employees of the manufacturer of the test diet.

5. Conclusions

WA has the potential to shift the energy balance to a higher ratio of fat to carbohydrate oxidation during the night.

Author Contributions: S.S. was responsible for the conception and design of the study and drafting the manuscript. T.S. made substantial contributions to the conception and design. A.S. and Y.S. made substantial contributions to the collection of data. N.O. gave final approval of the version to be published as a general supervisor of the study treatment.

Funding: This study was conducted under financial support from the Kao Corporation.

Acknowledgments: We thank Shigeru Kobayashi, Tokyo Rinkai Hospital, who was the physician in charge.

Conflicts of Interest: The authors are employees of the Kao Corporation. The test tablets were prepared by the Kao Corporation.

References

1. Zurlo, F.; Lillioja, S.; Esposito-Del Puente, A.; Nyomba, B.L.; Raz, I.; Saad, M.F.; Swinburn, B.A.; Knowler, W.C.; Bogardus, C.; Ravussin, E. Low ratio of fat to carbohydrate oxidation as predictor of weight gain: Study of 24-h RQ. *Am. J. Physiol.* **1990**, *259*, E650–E657. [CrossRef] [PubMed]

2. Marra, M.; Scalfi, L.; Contaldo, F.; Pasanisi, F. Fasting respiratory quotient as a predictor of long-term weight changes in non-obese women. *Ann. Nutr. Metab.* **2004**, *48*, 189–192. [CrossRef] [PubMed]

3. Solomon, T.P.; Marchetti, C.M.; Krishnan, R.K.; Gonzalez, F.; Kirwan, J.P. Effects of aging on basal fat oxidation in obese humans. *Metabolism* **2008**, *57*, 1141–1147. [CrossRef] [PubMed]

4. Kodama, T.; Miyazaki, T.; Kitamura, I.; Suzuki, Y.; Namba, Y.; Sakurai, J.; Torikai, Y.; Inoue, S. Effects of single and long-term administration of wheat albumin on blood glucose control: Randomized controlled clinical trials. *Eur. J. Clin. Nutr.* **2005**, *59*, 384–392. [CrossRef] [PubMed]

5. Schutz, Y. Concept of fat balance in human obesity revisited with particular reference to de novo lipogenesis. *Int. J. Obes. Relat. Metab. Disord.* **2004**, *28*, S3–S11. [CrossRef] [PubMed]

6. Walker, J.; Collins, L.C.; Nannini, L.; Stamford, B.A. Potentiating effects of cigarette smoking and moderate exercise on the thermic effect of a meal. *Int. J. Obes. Relat. Metab. Disord.* **1992**, *16*, 341–347. [PubMed]

7. Hibi, M.; Masumoto, A.; Naito, Y.; Kiuchi, K.; Yoshimoto, Y.; Matsumoto, M.; Katashima, M.; Oka, J.; Ikemoto, S. Nighttime snacking reduces whole body fat oxidation and increases LDL cholesterol in healthy young women. *Am. J. Physiol. Regul. Integr. Comp. Physiol.* **2013**, *304*, R94–R101. [CrossRef] [PubMed]

8. Henning, B.; Lofgren, R.; Sjostrom, L. Chamber for indirect calorimetry with improved transient response. *Med. Biol. Eng. Comput.* **1996**, *34*, 207–212. [CrossRef] [PubMed]

9. Brouwer, E. On simple formulae for calculating the heat expenditure and the quantities of carbohydrate and fat oxidized in metabolism of men and animals, from gaseous exchange (Oxygen intake and carbonic acid output) and urine-N. *Acta Physiol. Pharmacol. Neerl.* **1957**, *6*, 795–802. [PubMed]

10. Weir, J.B. New methods for calculating metabolic rate with special reference to protein metabolism. *J. Physiol.* **1949**, *109*, 1–9. [CrossRef] [PubMed]

11. Frisancho, A.R. Reduced rate of fat oxidation: A metabolic pathway to obesity in the developing nations. *Am. J. Hum. Biol.* **2003**, *15*, 522–532. [CrossRef] [PubMed]

12. Piaggi, P.; Thearle, M.S.; Bogardus, C.; Krakoff, J. Lower energy expenditure predicts long-term increases in weight and fat mass. *J. Clin. Endocrinol. Metab.* **2013**, *98*, E703–E707. [CrossRef] [PubMed]

13. Zhou, H.; Yamada, Y.; Tsukiyama, K.; Miyawaki, K.; Hosokawa, M.; Nagashima, K.; Toyoda, K.; Naitoh, R.; Mizunoya, W.; Fushiki, T.; et al. Gastric inhibitory polypeptide modulates adiposity and fat oxidation under diminished insulin action. *Biochem. Biophys. Res. Commun.* **2005**, *335*, 937–942. [CrossRef] [PubMed]

14. Seino, Y.; Fukushima, M.; Yabe, D. GIP and GLP-1, the two incretin hormones: Similarities and differences. *J. Diabetes Investig.* **2010**, *1*, 8–23. [CrossRef] [PubMed]

15. Shimotoyodome, A.; Suzuki, J.; Fukuoka, D.; Tokimitsu, I.; Hase, T. RS4-type resistant starch prevents high-fat diet-induced obesity via increased hepatic fatty acid oxidation and decreased postprandial GIP in C57BL/6J mice. *Am. J. Physiol. Endocrinol. Metab.* **2010**, *298*, E652–E662. [CrossRef] [PubMed]
16. Flatt, J.P.; Ravussin, E.; Acheson, K.J.; Jequier, E. Effects of dietary fat on postprandial substrate oxidation and on carbohydrate and fat balances. *J. Clin. Investig.* **1985**, *76*, 1019–1024. [CrossRef] [PubMed]
17. McCuen-Wurst, C.; Ruggieri, M.; Allison, K.C. Disordered eating and obesity: Associations between binge-eating disorder, night-eating syndrome, and weight-related comorbidities. *Ann. N. Y. Acad. Sci.* **2018**, *1411*, 96–105. [CrossRef] [PubMed]

Correction

Correction: Sartorius et al. "Postprandial Effects of a Proprietary Milk Protein Hydrolysate Containing Bioactive Peptides in Prediabetic Subjects" *Nutrients* 2019, *11*, 1700

Tina Sartorius [1], Andrea Weidner [1], Tanita Dharsono [1], Audrey Boulier [2], Manfred Wilhelm [3] and Christiane Schön [1,*]

[1]   BioTeSys GmbH, Schelztorstr. 54–56, 73728 Esslingen, Germany; t.sartorius@biotesys.de (T.S.); a.weidner@biotesys.de (A.W.); t.dharsono@biotesys.de (T.D.)

[2]   Ingredia S.A., 51 Avenue F. Lobbedez CS 60946, 62033 Arras CEDEX, France; a.boulier@ingredia.com

[3]   Department of Mathematics, Natural and Economic Sciences, Ulm University of Applied Sciences, Albert-Einstein-Allee 55, 89081 Ulm, Germany; manfred.wilhelm@thu.de

*   Correspondence: c.schoen@biotesys.de; Tel.: +49-711-3105-7145

Received: 24 April 2020; Accepted: 28 April 2020; Published: 29 April 2020

Abstract: Milk proteins have been hypothesized to protect against type 2 diabetes (T2DM) by beneficially modulating glycemic response, predominantly in the postprandial status. This potential is, amongst others, attributed to the high content of whey proteins, which are commonly a product of cheese production. However, native whey has received substantial attention due to its higher leucine content, and its postprandial glycemic effect has not been assessed thus far in prediabetes. In the present study, the impact of a milk protein hydrolysate of native whey origin with alpha-glucosidase inhibiting properties was determined in prediabetics in a randomized, cross-over trial. Subjects received a single dose of placebo or low- or high-dosed milk protein hydrolysate prior to a challenge meal high in carbohydrates. Concentration–time curves of glucose and insulin were assessed. Incremental areas under the curve (iAUC) of glucose as the primary outcome were significantly reduced by low-dosed milk peptides compared to placebo (p = 0.0472), and a minor insulinotropic effect was seen. A longer intervention period with the low-dosed product did not strengthen glucose response but significantly reduced HbA1c values (p = 0.0244). In conclusion, the current milk protein hydrolysate of native whey origin has the potential to modulate postprandial hyperglycemia and hence may contribute in reducing the future risk of developing T2DM.

Keywords: alpha-glucosidase inhibitor; biopeptides; blood glucose; glycemic control; hyperglycemia; milk peptides; postprandial; prediabetes; pre-meal; type 2 diabetes

---

The authors wish to make a correction to the published version of their paper [1].

Unfortunately, we noticed a mistake regarding the description of the bioactive dipeptide in our paper. Instead of "arginine–proline (AP)" the correct amino acid dipeptide is "alanine–proline (AP)". In the published version of our paper [1], this affects two positions, namely:

1.   Page 2 of 16: "We used Pep2Dia® as an investigational product containing a bioactive alanine–proline (AP) dipeptide with alpha-glucosidase inhibiting properties".

2.   Page 4 of 16: "2.3 Intervention/ The investigational product (Pep2Dia®) was a milk protein hydrolysate from native whey protein containing a bioactive alanine–proline (AP) dipeptide (between 0.15% and 0.4%) with alpha-glucosidase inhibiting properties."

The authors apologize to the readers for any inconvenience caused by the change. This change does not impact the content of the paper, the overall results or scientific conclusions. The original manuscript will remain online on the article webpage, with a reference to this correction.

Author Contributions: Project administration and resources: C.S. and T.D. designed the study; Investigation: the study was undertaken at the study site of BioTeSys GmbH under supervision of A.W., T.D., and C.S. Formal analysis and visualization: C.S. and M.W. planned and performed statistical analysis and created the figures. Writing: T.S. and C.S. drafted the manuscript and made the final approval of the published version. A.B. contributed to discussion and manuscript revision. All authors significantly contributed to results interpretation, critical manuscript revision, and approval of the final manuscript. All authors have read and agreed to the published version of the manuscript.

Funding: This research was funded by Ingredia S.A., France.

Acknowledgments: The authors would like to thank all subjects who took part in this clinical trial.

Conflicts of Interest: C.S., A.W., T.D., T.S.: employees of contracted research organization. M.W. is an independent statistician supporting with data analysis. The study was financially supported by Ingredia S.A., France. The sponsors contributed to the discussion about the study design and selection of outcome measures prior to the study start. Planning and organization of the study and its realization, data analysis and report generating were independently undertaken solely by BioTeSys GmbH and M.W. The authors from BioTeSys GmbH and M.W. declare no conflict of interest regarding the publication of this paper.

References

1. Sartorius, T.; Weidner, A.; Dharsono, T.; Boulier, A.; Wilhelm, M.; Schön, C. Postprandial Effects of a Proprietary Milk Protein Hydrolysate Containing Bioactive Peptides in Prediabetic Subjects. *Nutrients* **2019**, *11*, 1700. [CrossRef] [PubMed]

Article

Postprandial Effects of a Proprietary Milk Protein Hydrolysate Containing Bioactive Peptides in Prediabetic Subjects

Tina Sartorius [1], Andrea Weidner [1], Tanita Dharsono [1], Audrey Boulier [2], Manfred Wilhelm [3] and Christiane Schön [1,*]

[1] BioTeSys GmbH, Schelztorstr. 54–56, 73728 Esslingen, Germany
[2] Ingredia S.A., 51 Avenue F. Lobbedez CS 60946, 62033 Arras CEDEX, France
[3] Department of Mathematics, Natural and Economic Sciences, Ulm University of Applied Sciences, Albert-Einstein-Allee 55, 89081 Ulm, Germany
* Correspondence: c.schoen@biotesys.de; Tel.: +49-711-3105-7145

Received: 17 June 2019; Accepted: 19 July 2019; Published: 23 July 2019

Abstract: Milk proteins have been hypothesized to protect against type 2 diabetes (T2DM) by beneficially modulating glycemic response, predominantly in the postprandial status. This potential is, amongst others, attributed to the high content of whey proteins, which are commonly a product of cheese production. However, native whey has received substantial attention due to its higher leucine content, and its postprandial glycemic effect has not been assessed thus far in prediabetes. In the present study, the impact of a milk protein hydrolysate of native whey origin with alpha-glucosidase inhibiting properties was determined in prediabetics in a randomized, cross-over trial. Subjects received a single dose of placebo or low- or high-dosed milk protein hydrolysate prior to a challenge meal high in carbohydrates. Concentration–time curves of glucose and insulin were assessed. Incremental areas under the curve (iAUC) of glucose as the primary outcome were significantly reduced by low-dosed milk peptides compared to placebo ($p = 0.0472$), and a minor insulinotropic effect was seen. A longer intervention period with the low-dosed product did not strengthen glucose response but significantly reduced HbA$_{1c}$ values ($p = 0.0244$). In conclusion, the current milk protein hydrolysate of native whey origin has the potential to modulate postprandial hyperglycemia and hence may contribute in reducing the future risk of developing T2DM.

Keywords: alpha-glucosidase inhibitor; biopeptides; blood glucose; glycemic control; hyperglycemia; milk peptides; postprandial; prediabetes; pre-meal; type 2 diabetes

---

1. Introduction

Insulin resistance, a condition established by genetic and environmental factors, leads to impaired glucose tolerance due to an imbalance between insulin sensitivity and insulin secretion. This so-called prediabetic status plays an important pathophysiological role in the development of type 2 diabetes mellitus (T2DM) and is a hallmark of obesity, dyslipidemias, and other major risk factors contributing to the metabolic syndrome [1]. Over the years, our understanding of insulin resistance has improved tremendously, while T2DM is expected to have increasing detrimental effects on the health of populations and healthcare systems. Aside from preventive activities, to combat sedentary lifestyles and unbalanced diets in particular, reducing postprandial glycemia is as important as lowering fasting blood glucose levels to limit (or at least delay) the appearance of T2DM in at-risk individuals, and even modest postprandial hyperglycemia may lead to β-cell dysfunction [2,3]. As such, numerous studies have consistently demonstrated that pathophysiological abnormalities associated with an increased postprandial hyperglycemia ≥155 mg/dL (value of 1 h postload glucose concentration) including impaired insulin sensitivity, β-cell dysfunction, and increased glucose intestinal absorption, which are linked to an increased risk for future T2DM [2]. Most anti-diabetic agents that are currently available reduce fasting blood glucose levels but have little impact on postprandial glycemic excursions and thus do not normalize postprandial hyperglycemia [4]. In this context, simple dietary modifications, proper nutrition, and exercise may modify postprandial derailments; also, there is growing interest in food components that may beneficially modulate glycemic response, predominantly in the postprandial status.

There are many scientific reports highlighting the role of biologically active peptides derived from food proteins (e.g., milk, eggs, plant proteins), and clinical studies revealed that protein-rich dairy products are beneficial for reducing the risk of developing T2DM due to their glycemic and insulinotropic effect to improve glycemic status [5–8]. Thereby, the possible protective mechanism has been ascribed to the protein fraction [9–11]. Such biologically active protein fragments are released from parent proteins after enzymatic action (e.g., protein hydrolysis in the digestive tract) and positively influence various functions of the human body by interacting with enzymes or receptors [12,13]. Moreover, it is assumed that milk proteins have more effects on metabolic response in subjects with disturbed glucose metabolism [14]. Milk comprises two protein fractions, the slowly digestible casein and the fast digestible whey fraction [15]. Whey from native origin is produced by direct filtration of pasteurized skimmed milk and is therefore a more native protein compared to whey protein from cheese production with more denatured protein character and less leucine content than native whey [16]. A further degradation of proteins is achieved by hydrolysis in mainly di- and tri-peptides via proteolytic enzymes in the digestive tract, resulting in a complex mixture of peptides of different length and free amino acids [17]. The hydrolysis process may change the kinetic pattern of the specific protein fractions, as shown for casein with more rapid digestion characteristics [15,18,19] and for whey with an improved absorption in a perfused human jejunum model [20]. Milk-derived bioactive peptides can be encrypted in both casein (α-, β-, and γ-casein) and whey proteins (β-lactoglobulin, α-lactalbumin, serum albumin, immunoglobulins, lactoferrin, and protease-peptone fractions) [21]. Also, milk protein hydrolysates were previously analyzed for their antidiabetic properties due to inhibition of alpha-glucosidase, a carbohydrate degrading digestive enzyme [22,23]. Counteracting alpha-glucosidase action through inhibition delays carbohydrate hydrolysis and consequently extends its digestion time (gastric emptying), which results in reduced glucose absorption from the gastrointestinal tract. Thus, postprandial levels of blood glucose and insulin are reduced [23,24]. The alpha-glucosidase inhibitor acarbose was shown to be an effective and valuable option in delaying or preventing the progression to T2DM [25]. However, this oral antidiabetic drug is known to cause gastrointestinal side effects when used as long-term therapy. Thus, in the last few years, progress has been attained in the search for new peptide alpha-glucosidase inhibitors, either synthetic or of natural origin.

The purpose of our study was to evaluate whether a proprietary milk protein hydrolysate of native whey origin containing alpha-glucosidase inhibiting bioactive peptides might improve postprandial glucose profile after single dosage or after a six week intervention period in prediabetic subjects. Therefore, both glucose response and insulin secretion were assessed from individual concentration–time curves. We used Pep2Dia® as an investigational product containing a bioactive arginine-proline (AP) dipeptide with alpha-glucosidase inhibiting properties. Based on literature and on proprietary in vitro studies according to Kang et al. [26] (European Patent EP 3,107,556), there is evidence that the current milk protein hydrolysate acts on the inhibition of alpha-glucosidase with an IC_{50} value of 0.0025 mg/mL and thereby reduces glucose absorption from the gastrointestinal tract [27].

2. Materials and Methods

2.1. Study Subjects

From September 2018 to January 2019, a total of 21 subjects were included in the monocentric study at BioTeSys GmbH (Esslingen am Neckar, Germany). Overall, 84 non-smoking female and male people aged 30–70 years with a body mass index (BMI) of 19–35 kg/m^2 were pre-screened for eligibility, from which 30 subjects were screened to ascertain their eligibility. The main inclusion criteria were prediabetic HbA$_{1c}$ values between 5.7% to 6.4% and/or fasting glucose ≥5.6 mmol/L (≥100 mg/dL) and <7.0 mmol/L (<125 mg/dL), confirmed twice on two separate days if HbA$_{1c}$ value was <5.7%. The subjects had to be in good physical and mental health represented by the medical history, physical examination, electrocardiogram, vital signs, and results of biochemistry and hematology. Finally, 21 subjects (8 men, 13 women) were included and all completed the study successfully, as shown in Figure 1.

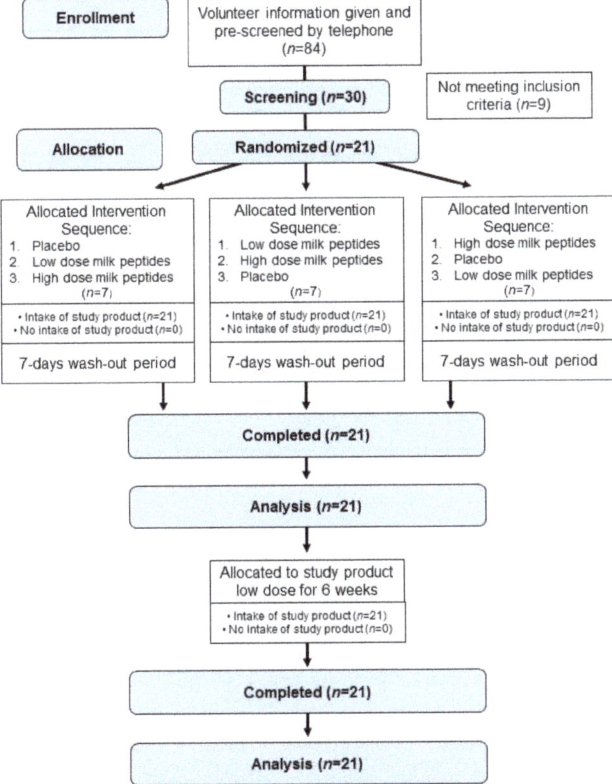

Figure 1. Subject recruitment flow chart.

The main exclusion criteria were a relevant history or presence of any medical disorder potentially interfering with this study (e.g., malabsorption, chronic gastro-intestinal diseases, severe depression, cardiovascular disease occurrence within the last 3 months, etc.), regular intake of medications or supplements known to affect glucose tolerance, diagnosed type 2 diabetics with medical treatment, and drug, alcohol, and medication abuses. Medications for treatment of chronic diseases that do not affect the metabolism of the study product were permitted and were judged individually regarding interference with the study by an investigator. Any concomitant chronic disease medication and medication used for the treatment of adverse events (AEs) was documented.

With regard to the 84 subjects given information and pre-screened by phone interview, the inaptitude was due to fasting glucose levels ≤100 mg/dL or >125 mg/dL in previous laboratory reports from subjects' physician, BMI >35 kg/m^2, metformin medication, lack of interest/response, and time collision.

This study was conducted in orientation towards the guidelines of the Declaration of Helsinki and Good Clinical Practice. The protocol and all documents were approved by the Institutional Review Board (IRB) of Landesärztekammer Baden-Württemberg with the reference number F-2018-062. A written informed Consent Form was obtained from all participants prior to screening evaluations. The present study was registered with ClinicalTrials.gov (ID: NCT03,932,695).

2.2. Study Design

The study was performed as a randomized, double-blind, placebo-controlled, monocentric, 3-way-cross-over study with 21 eligible subjects under fasted conditions at the study site of BioTeSys GmbH, Esslingen, Germany. A CONSORT 2010 checklist of information is included in the Supplementary Materials (Table S1). There was a wash-out period of 7 days between the study days to assess postprandial glucose response after a challenge meal. Within the cross-over study design, subjects received all interventions randomly allocated to 3 sequence groups.

Following an overnight fasting period of at least 10 h, a permanent venous catheter was inserted, and baseline blood (time points were 10 min and 5 min prior to challenge meal) was examined at the three visits within the cross-over study (kinetic days). Subjects received a single dose of placebo or 1400 mg (low dose) or 2800 mg (high dose) bioactive peptides from milk protein hydrolysate 15 min prior to a challenge meal high in carbohydrates (consisting of white bread, jam, and butter standardized to 75 g carbohydrates), and blood was further sampled at 15, 30, 45, 60, 90, 120, 150, and 180 min after the intake of the challenge meal. All participants received the two dosages of the study product and the placebo, and effects were compared to the placebo.

Additionally, an open-label single arm phase was performed with a daily intake of the low dose milk peptide concentration for 6 weeks to estimate effects over a longer period. After the 6 week intervention period, the postprandial assessment after the intake of 1400 mg bioactive peptides from milk protein hydrolysate 15 min prior to a challenge meal was repeated comparable to the cross-over phase. Subjects were encouraged not to change their food habits and physical activity during the study. Therefore, nutrition habit questionnaires were filled in during screening after the single dose cross-over study (= before 6 week intervention) and after 6 week intervention within the open-label single arm design. Thereby, subjects were asked about their food habits using a semi-quantitative short questionnaire assessing different food categories (fruit, vegetables, sausage, meat, intake of dairy products, sweets including beverages).

Blood analysis comprised determination of glucose and insulin plasma concentration over time at defined intervals besides blood routine parameters such as hematogram or total cholesterol. Subjects were asked to avoid alcohol 24 h before each study visit and to consume standardized meals 24 h prior to each visit to control for external confounding factors. In detail, breakfast was individually standardized, and for lunch, tortellini with pesto was served. Furthermore, a standardized snack (apple and cookie) was provided, and bread with cream cheese "Frischkäse" and cucumber had to be consumed as dinner. Additionally, subjects were not allowed to consume food or drink anything other than water for at least 10 h before testing, and no strenuous physical activity or endurance sports were allowed within 24 h before the study visits. In the morning of the study visits, subjects were instructed to drink a minimum of 200 mL water after waking up before they came to the study site.

2.3. Intervention

The investigational product (Pep2Dia®) was a milk protein hydrolysate from native whey protein containing a bioactive arginine-proline (AP) dipeptide (between 0.15% and 0.4%) with alpha-glucosidase inhibiting properties. The proprietary compound was prepared by Ingredia S.A. (Arras CEDEX, France) and was produced from native whey extracted by filtration according to Boutrou et al. [28]. Furthermore, a protease was used to perform the respective procedure. The protein is composed of 100% soluble protein with mainly β-lactoglobulin and α-lactalbumin. The profile of the peptides in the investigational product was as follows: 94.5% with a molecular weight (MW) <5000 Da, 0.5% with 5000–10,000 Da MW, and 5% with a MW > 10,000 Da. The products were provided in capsules (vegetable fiber) with 350 mg of milk protein hydrolysate per capsule (which includes, on average, 0.96 mg AP peptide). Maltodextrin with dextrose equivalent of 9 (DE9) was used as placebo with 350 mg per capsule. For single dose intake, 4 verum capsules and 4 placebo capsules (low dose) or, for high dose, 8 verum capsules were taken 15 min prior to a challenge meal. In the open-label single arm phase, subjects consumed the investigational product (4 verum capsules) daily 15 min prior to lunch, and at the study visit, subjects ingested 4 verum capsules after an overnight fasting period 15 min prior to a challenge meal. Manufacturing and encapsulation were carried out in compliance with Good Manufacturing Practice conditions, and all excipients as well as capsule shells met the current European food regulations. Size, shape, color, odor, and secondary packaging were identical between verum and placebo capsules to ensure double-blind conditions. Capsules were provided by Ingredia S.A. (Arras CEDEX, France), and subjects received either placebo or low dose (1400 mg) or high dose (2800 mg) milk protein hydrolysate.

2.4. Sample Collection and Processing

Venous blood samples were taken at screening visits to assess safety parameters (differentiated hematogram and clinical laboratory). At the same day, analyses with standard methods were performed at an accredited laboratory (Synlab Medizinisches Versorgungszentrum Leinfelden-Echterdingen, Germany). Blood samples were centrifuged at 3000× g for 10 min at 4 °C, and aliquots for spare samples for the determination of glucose and insulin were taken. Plasma glucose was analyzed using the Atellica® CH analyzer (Siemens Healthcare GmbH, Germany; assay: Atellica CH Glucose Hexokinase_3, Ref. 11,097,592) with enzymatic UV detection based on the glucose hexokinase method. Briefly, glucose-6-phosphate formed from glucose and ATP by hexokinase was oxidized by NAD^+ in a reaction catalyzed by glucose-6-phosphate dehydrogenase to give NADH, which was quantitated spectrophotometrically at 340/410 nm. Serum insulin was analyzed using the Atellica® IM analyzer (Siemens Healthcare GmbH, Germany; assay: Atellica IM IRI, Ref. 10,995,628) with insulin detection based on a sandwich-type of electrochemiluminescence immunoassay using two monoclonal antibodies against insulin. Thereby, insulin quantification was linked to the number of relative light units (RLUs). Fasting blood glucose was controlled in finger prick samples using the HemoCue Glucose 201+ Analyzer (HITADO GmbH, Möhnesee, Germany) on the morning of each study day.

2.5. Methods for Safety (Adverse Events, Concomitant Medication, and Tolerability)

During the study intervention, the subjects documented any adverse events and concomitant medication. The tolerability was assessed at the end of the study days. The subjects rated overall tolerability to three categories from "well tolerated", "slightly unpleasant", or "very unpleasant".

2.6. Data Analysis and Statistics

Based on previous data [29] reporting a reduction of postprandial glucose levels after a challenge meal with different milk proteins with up to 18% reduction, a conservative assumption with a reduction of 11% was applied for the prior sample size calculation, resulting in an effect size of d = 0.74. Based on the following input details—alpha error problem of $\alpha = 0.05$, actual power of 80%, correlation between groups of 0.5—a sample size of $n = 17$ subjects was estimated, which was applied for the 3-way cross-over design in phase I. Considering a drop-out rate of 15% and equally sized sequence groups for the 3-way cross-over design in phase I, the study was performed with $n = 21$ subjects. The part II open-label phase was planned to be exploratory as a first proof of concept study to estimate long-term effects and to gain first experiences for further clinical studies. Pharmacokinetic parameters were individually calculated with the blood concentration–time curves. As the primary efficacy endpoint, the incremental area under the observed concentration–time curve above the baseline (iAUC), more precisely $iAUC_{0-180\,min}$, was calculated by applying the trapezoidal rule with the y-axis, defined by glucose plasma concentration, and the x-axis defined via sampling time points. Secondary efficacy target variables were $iAUC_{0-180\,min}$ of insulin, total $AUC_{0-180\,min}$, and ΔC_{max} of glucose and insulin. Primary and secondary endpoints were analyzed using a linear mixed model of iAUC with treatment (3 levels), period (3 levels), sequence (3 levels), and baseline blood glucose level within study periods as fixed effects and subject as random effect. Due to the 7 days wash-out period, examination of possible carry-over effects was not foreseen. The residuals of this model were checked for normality using the Shapiro–Wilk test with an alpha level of 0.05. If applicable, data were log transformed prior to analysis. Multiple pairwise comparisons of least squares means of primary and secondary endpoints were adjusted by the method of Dunnett–Hsu in order to assess differences between the two active treatments and placebo. Data of the cross-over design are presented as least square means with 95% confidence interval (CI).

Moreover, in the open-label study period, besides HbA_{1c} values, the homeostasis model assessment (HOMA) index and the Matsuda index were used to evaluate the impact of the study product intake during a longer period on insulin sensitivity, and comparisons were performed between the baseline and the end of intervention. Additionally, during the open-label study period of 6 weeks, the pharmacokinetic endpoints after the challenge test were compared with placebo during the study phase I. Data were evaluated using a paired t-test. In case of non-normal distribution of data, a Wilcoxon signed-rank test was applied. Data of the open label phase are presented as means with 95% CI. All 21 subjects were included in the analysis. Statistical tests were performed two-sided, and p values < 0.05 were statistically significant. Statistical evaluation, summary tables, and graphs were generated using GraphPad Prism software (La Jolla, CA, USA) and SAS V9.4 statistical software (SAS Institute, Cary, North Carolina).

3. Results

3.1. Subject Characteristics

The investigated study population was a non-smoking prediabetic study group, on average 62.4 years (95% CI: 60.0–64.9) old with a BMI of 28.1 kg/m^2 (95% CI: 26.3–30.0). A total of 21 subjects ($n = 13$ women, $n = 8$ men) completed the study.

Table 1 presents the participants' demographic data and screening data. Vital signs and blood routine parameters were within normal range. None of the subjects were vegetarian or vegan, and 52% of the participants practiced sports on a regular basis.

Table 1. Demographic and screening data.

Variable	Prediabetics ($n = 21$)	
	Mean	95% CI
Age (years)	62.4	(60.0–64.9)
BMI (kg/m^2)	28.1	(26.3–30.0)
Systolic BP (mmHg)	134.7	(127.3–142.1)
Diastolic BP (mmHg)	83.9	(79.6–88.2)
HbA$_{1c}$ (%)	5.83	(5.69–5.97)
Fasting plasma glucose (mg/dL)	109.6	(105.3–113.9)

BMI: body mass index; BP: blood pressure; HbA$_{1c}$: glycated haemoglobin.

Subjects were advised not to change their eating habits, which were controlled by a semi-quantitative nutrition habit questionnaire comprising 24 food categories. There was no significant change over intervention period ($p = 0.2148$). Fasting baseline values of glucose and insulin did not differ among the single dose treatment days ($p > 0.05$). Regarding the homeostasis model assessment of insulin resistance (HOMA-IR), a parameter that estimates insulin sensitivity considering the relation between fasting insulin and fasting glucose, was—on average—clearly above the cut-off level of two [30], indicating insulin resistance and not different among the testing days (placebo: 2.95; low dose: 2.80; high dose: 2.96; six week intervention: 2.84; Figure 2).

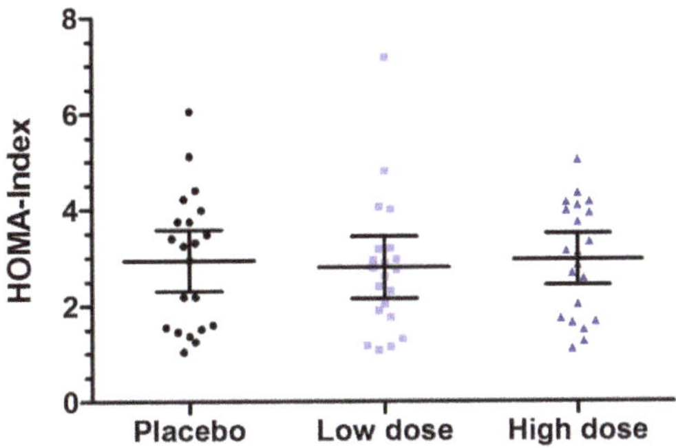

Figure 2. Distribution of homeostasis model assessment (HOMA) index at baseline at the testing days before intake of placebo or milk protein hydrolysate containing bioactive milk peptides in low and high doses. Data represent mean ±95% CI. Data represent mean ±95% CI. No statistical difference between baseline conditions.

3.2. Milk Peptides and Their Postprandial Effect on Glucose Response after Single Dose Intake

There was a significant increase of plasma glucose concentration over time after the challenge meal ($p < 0.0001$ after all study interventions). The concentration–time curves indicate that milk peptides have an impact on postprandial blood glucose profile in prediabetic subjects (Figure 3). No dose linearity between low dose (1400 mg) and high dose (2800 mg) milk peptides could be revealed, and the effects were even slightly more distinct after single dose intake of low dose milk peptides in comparison to the high dose.

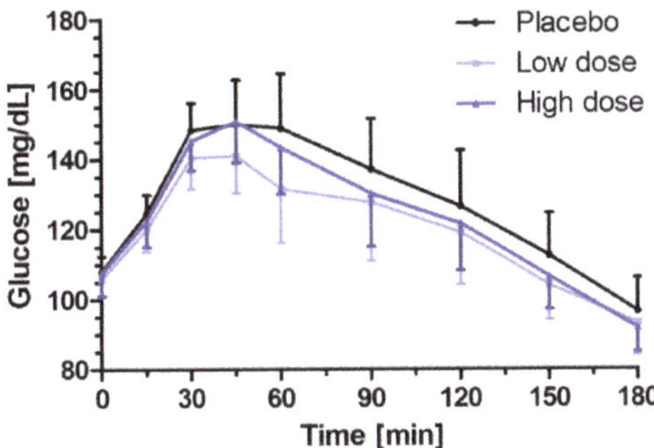

Figure 3. Glucose responses to a single intake of placebo or low- or high-dosed milk protein hydrolysate containing bioactive peptides consumed 15 min prior to a challenge meal. Data represent mean ±95% CI.

In terms of $iAUC_{0-180\,min}$ glucose, single dose intake of low dose milk peptides resulted in significantly reduced values compared to the placebo (3441.1 vs. 4312.0 mg/dL × min, $p = 0.0472$), whereas the high dose milk peptides were not statistically different to the placebo ($p = 0.1749$) (Table 2). The secondary endpoint ΔC_{max}, the maximum increase of glucose above baseline, confirmed the significant postprandial glucose lowering effect of the low dose milk peptides with a mean increase in plasma glucose of 44.8 mg/dL (95% CI: 35.9–53.8) vs. 52.8 mg/dL (95% CI: 43.9–61.8) for placebo ($p = 0.0237$) vs. 49.1 mg/dL (95% CI: 40.1–58.0) for high dose milk peptides (Table 2). In addition, analyses of total $AUC_{0-180\ min}$ and C_{max} revealed statistical significance for the low dose milk peptides in comparison to the placebo (low dose vs. placebo: $AUC_{0-180\ min}$: 21,931 vs. 23,073 mg/dL × min, $p = 0.0313$; C_{max}: 152.1 mg/dL (95% CI: 143.1–161.0) vs. 160.1 mg/dL (95% CI: 151.1–169.0), $p = 0.0237$).

Table 2. Postprandial glucose incremental areas under the curve (iAUC) and ΔC_{max} of glucose after single dose intake of placebo or low-or high-dosed milk protein hydrolysate. Group means referred to least squares (LS) means.

Glucose	Placebo	Low-Dosed Milk Protein Hydrolysate			High-Dosed Milk Protein Hydrolysate		
Variable	LS Mean 95% CI	LS Mean 95% CI	Treatment Difference 95% CI	p	LS Mean 95% CI	Treatment Difference 95% CI	p
$\text{iAUC}_{0-180 \text{ min}}$ (mg/dL $\times$ min)	4312.0 (2938.4–5685.6)	3441.1 (2066.9–4815.3)	−870.9 (−1732.4−−9.5)	0.0472	3685.8 (2312.6–5059.0)	−626.2 (−1480.5–228.1)	0.1749
ΔC_{max} (mg/dL)	52.8 (43.9–61.8)	44.8 (35.9–53.8)	−8.00 (−15.02−−0.98)	0.0237	49.1 (40.1–58.0)	−3.78 (−10.75–3.19)	0.3627

3.3. Milk Peptides and Their Postprandial Effect on Insulin Response after Single Dose Intake

The impact of milk peptides on insulin release as a response to the challenge meal was a minor evident (Figure 4). There was a slight reduction of iAUC of insulin after an intake of low dose milk peptides in comparison to the placebo [low dose: 6339.8 µU/mL × min (95% CI: 4997.5–8042.5); placebo: 6844.5 µU/mL × min (95% CI: 5396.5–8681.9)]; however, the difference was not significant ($p = 0.4296$). No difference to the placebo was seen in the high dose milk peptides (7212.0 µU/mL × min (95% CI: 5396.5–8681.9) vs. 6844.5 µU/mL × min (95% CI: 5396.5–8681.9), $p = 0.6606$). The maximum increase in plasma insulin (ΔC_{max}) after the challenge meal was lower for low dose milk peptides compared to the placebo and the high dose milk peptides [low dose: 66.2 µU/mL (95% CI: 54.8–80.1); high dose: 74.1 µU/mL (95% CI: 61.3–89.5); placebo: 71.4 µU/mL (95% CI: 59.0–86.3)]. ΔC_{max} of both low and high dose milk peptides were not different to the placebo (low dose: $p = 0.5536$; high dose: $p = 0.8573$). These results indicate a negligible insulinotropic effect of the current milk protein hydrolysate.

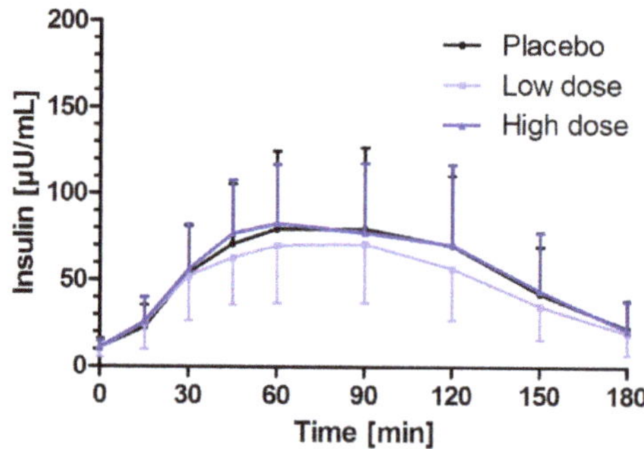

Figure 4. Insulin responses to low and high dose milk protein hydrolysate containing bioactive peptides consumed 15 min prior to a challenge meal. Data represent mean ±95% CI.

3.4. Six Week Intervention with Low Dose Milk Peptides

Plasma concentrations of fasting blood glucose and fasting insulin after six weeks of low dose milk peptides intervention were comparable with the fasting conditions prior to the challenge meal with single dose placebo intervention [baseline vs. six week intervention: 108.0 mg/dL (95% CI: 103.6–112.4) vs. 106.8 mg/dL (95% CI: 102.4–111.1) for glucose ($p = 0.5165$); 11.01 mg/dL (95% CI: 8.68–13.34) vs. 10.67 µU/mL (95% CI: 8.22–13.12) for insulin ($p = 0.3352$)]. Approximation of whole-body insulin sensitivity, which combines both hepatic and peripheral tissue insulin sensitivity, was performed by assessment of the Matsuda index; 61.9% of subjects were in the pathological range with values <4, 9.5% in the borderline range with values between 4 and 6, and 28.6% were in the normal (healthy) range with values of 6–12 at baseline, defined as the condition prior to the six week intervention period. Daily intake of low dose milk peptides for six weeks did not result in a change of HOMA-IR (baseline vs. six week intervention: 2.87 (95% CI: 2.28–3.45) vs. 2.84 (95% CI 2.14–3.53); $p = 0.5202$)), but resulted in a slight increase of the Matsuda index by trend [baseline vs. six week intervention: 4.32 (95% CI: 3.21–5.43) vs. 4.59 (95% CI: 3.48–5.71); $p = 0.0952$)] (Figure 5a). There was a significant reduction of HbA$_{1c}$ levels after a six week intervention treatment with low dose milk peptides resulting in HbA$_{1c}$ values of 5.69% (95% CI: 5.58–5.79) compared to baseline values of 5.78% (95% CI: 5.67–5.89) with $p = 0.0244$ (Figure 5b). Notably, 11 out of 21 subjects (52.4%) completed with HbA$_{1c}$ levels <5.7% after the six week intervention period.

In accordance with the single dose treatment, the concentration–time curve of postprandial plasma glucose concentration in response to the challenge meal after six week intervention with low dose milk peptides was below the placebo intervention at all time points (0–180 min) (Figure 6). However, the six week intervention period did not strengthen the acute postprandial glucose response in comparison with the single dose intake of low dose milk peptides, as iAUC values of glucose were similar [single dose vs. six week intervention: 3423 mg/dL × min (95% CI: 2181–4664) vs. 3577 mg/dL × min (95% CI: 2305–4849); $p = 0.6766$)] but statistically different to placebo ($p = 0.037$).

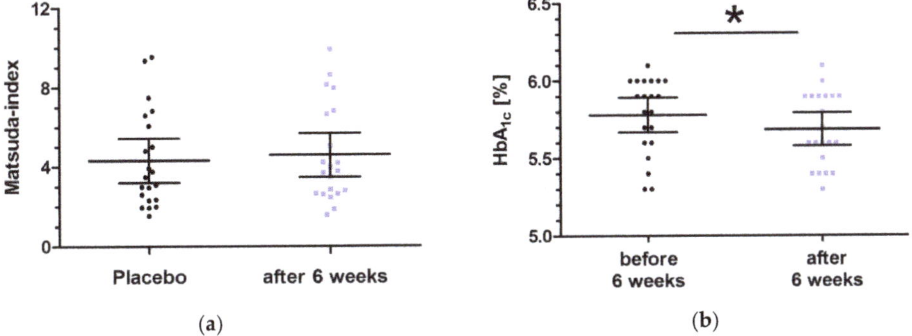

(**a**) (**b**)

Figure 5. Distribution of Matsuda index (**a**) and of HbA$_{1c}$ [%] values (**b**) after the six week intervention period with low dose milk protein hydrolysate containing bioactive milk peptides. Data represent mean ±95% CI. Statistical difference as indicated: * $p < 0.05$.

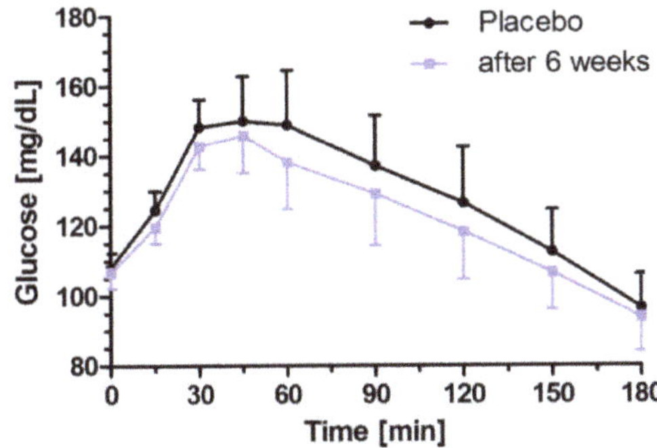

Figure 6. Glucose responses after six week intervention with low dose milk protein hydrolysate containing bioactive milk peptides and after challenge meal. Data represent mean ±95% CI.

Moreover, glucose response analyses in terms of ΔC_{max} and total $AUC_{0-180\,min}$ supported the abovementioned primary endpoints and confirmed the significant postprandial glucose lowering effects after low dose milk peptides intervention over a longer period of six weeks (placebo vs. six week intervention: 53.1 mg/dL (95% CI: 44.1–62.1) vs. 47.5 mg/dL (95% CI: 39.2–55.9), $p = 0.0399$ for ΔC_{max}, and 23,211 mg/dL × min (95% CI: 21,280–25,143) vs. 22,099 mg/dL × min (95% CI: 20,393–23,804); $p = 0.0408$ for $AUC_{0-180\,min}$)).

Low dose milk peptides intervention over a period of six weeks had no impact on the insulin response compared to the single dose intake regarding iAUC and ΔC_{max}. Again, although descriptively, there was (on average) a slight reduction by trend of iAUC of insulin in comparison to the placebo with $p = 0.0952$ for iAUC insulin (after six weeks: 7434 μU/mL × min (95% CI: 5770–9097); placebo: 8163 μU/mL × min (95% CI: 5962–10,363)). This was confirmed by ΔC_{max} values of insulin (after 6 weeks: 76.7 μU/mL (95% CI: 61.4–92.1); placebo: 80.4 μU/mL (95% CI: 61.7–99.1), $p = 0.3048$)).

3.5. Safety Assessment

All subjects (100%) rated the tolerability of the study products as "well tolerated" during the kinetic days of single dose intake and after the six week intervention period with low dose milk peptides. During the assessment of postprandial glucose response after a challenge meal after single dose intake, no adverse events (AEs) were reported. In terms of the six week intervention period with low dose milk peptides, a total of 14 adverse events were assessed by 10 subjects (predominantly headaches (6 x) and common cold (5 x)). Of those AEs, one serious adverse event (SAE) was reported on one surgery accompanied with hospitalization. None of the AEs were related to the study product.

4. Discussion

In the present study, we investigated the impact of a proprietary milk protein hydrolysate from native whey origin containing a bioactive AP dipeptide on postprandial glucose and insulin responses after a challenge meal in prediabetic subjects after a single dosage regimen or over a longer period of six weeks. Based on literature and on proprietary in vitro studies according to Kang et al. [26] (European Patent EP 3107,556), there is evidence that the current milk protein hydrolysate containing bioactive AP dipeptides acts on the inhibition of alpha-glucosidase with an IC_{50} value of 0.0025 mg/mL and thereby reduces glucose absorption from the gastrointestinal tract [27]. The amount of the bioactive AP dipeptide per capsule is, on average, 0.96 mg and thus in line with the content of already published bioactive peptides of whey protein origin [31], irrespective of the metabolic effects. The concentration–time curves indicated that the study product has the potential to counteract postprandial hyperglycemia in prediabetic subjects. After single dose application, effects on glucose response were slightly more distinct by intake of low-dosed milk peptides (1400 mg) 15 min prior to a challenge meal in comparison to the high dose (2800 mg) in terms of reduced iAUC glucose. Compared to placebo, a significant difference was seen for the low dosage ($p = 0.0472$) but not for the high dose. Of note, no linear dose–response relationship could be revealed. This might have been due to the multi-peptide characteristics, and interactions of single components in different concentrations might have been responsible for the limited dose–response. However, this needs further exploration in future studies. In addition, the secondary endpoints ΔC_{max}, $AUC_{0–180\,min}$, and C_{max} supported the findings for the primary endpoint iAUC and confirmed the significant postprandial glucose lowering effects after single dose intake of the low-dosed milk peptide.

Milk protein hydrolysates were previously analyzed for their antidiabetic properties with an alpha-glucosidase inhibiting effect [22,23]. Thereby, potent peptide fractions of a whey protein concentrate were identified with high biological activities of peptide fractions with a molecular weight lower than 33 kDa [32]. In what way the biological activity due to molecular weight might be causative for the postprandial glucose response exceeds the objective of the current study. Compared to already published literature in which the pre-meal effect of milk proteins (whey proteins) were analyzed in subjects with and without T2DM, an absent glucose response was demonstrated in both groups [14] owing to the insulinotropic rather than the glycemic effect of whey protein, which has higher amounts of lysine, threonine, tryptophan, leucine, and isoleucine [33]. Of note, recent in vitro data using preadipocytes revealed that the tripeptides IPP (Ile-Pro-Pro) and VPP (Val-Pro-Pro), which are derived from milk casein, enhance insulin sensitivity and contribute toward the prevention of insulin resistance in the presence of tumor necrosis factor [34]. VPP-mediated improved insulin sensitivity was also confirmed in diet-induced obese mice by decreasing pro-inflammatory cytokines in adipose tissue [35]. Further, it is known that whey and casein proteins differentially affect postprandial glucose and insulin response. It was shown that insulin secretion was greater with whey protein than with casein, whereas incretin responses in terms of GLP-1 tended to be lower with casein than with whey protein [36].

Analysis of iAUC of insulin release of the individual concentration–time curves revealed that the study product's impact on insulin release was minor, evident from the response to the challenge meal. However, after single dose application there was, although descriptively, on average a slight reduction of iAUC of insulin after intake of the low-dosed milk peptide in comparison to the placebo but without reaching statistical significance.

One has to take into account that differential patterns in insulin response after milk protein intake were reported between studies, which may be the result of a number of fundamental differences in study design, such as preload design and the type of milk proteins, protein amount, or altered milk peptides/bioactive peptide sequences. Results of our study contrast with previous literature demonstrating a significantly reduced glucose response with a concomitant increase in insulin AUC by intake of 18 g milk protein (whey) to 25 g glucose [37]. This effect was ascribed to amino acid availability, which may potentiate the increased insulin response since plasma amino acids also increased in a dose-dependent manner. Similarly, a combination of whey and free amino acids induced a rapid insulinotropic effect, which influenced early glycemia [38]. Further literature demonstrated an insulinotropic effect of milk proteins or whey protein in terms of higher insulin release [14,39,40]. It is discussed that the insulinotropic properties appear to originate from a specific postprandial plasma amino acid pattern with predominantly isoleucine, leucine, lysine, threonine, and valine, the main amino acids of whey protein [38]. Whether the difference in the respective glycemic and/or the insulinotropic responses of the current milk protein hydrolysate with bioactive peptides might be related to a different incretin pattern or to changes in plasma amino acid concentration was not clarified in the present study. However, one has to mention that the current milk protein hydrolysate is of native whey origin, the cleanest and the least processed whey protein available, whereas most of clinical trials used regular whey protein from cheese whey (e.g., [14,33,37,38,41]). Due to the process of creating native whey, namely filtration of pasteurized skimmed milk, more proteins remain intact and thus there is a higher leucine content than the more common whey protein concentrate from cheese production [42]. Of note, it has been shown that intake of native whey protein induces greater leucine blood concentrations than other whey protein supplements [16]. Whether the higher leucine content of the native whey protein might be causative for the more glycemic than insulinotropic response after the challenge meal is speculative but might be an explanation to already published data from other groups using regular whey protein from cheese whey.

In addition, it may be considered that a glycemic response does not necessarily impact insulin release. In this context, the inconsistency between glycemic and insulinotropic responses to fresh milk and two fermented milk products in healthy subjects was previously addressed [39]. However, it is known that whey protein in particular tends to be less glycemic and more insulinotropic [40], and casein, another bioactive milk component, was reported to reduce the postprandial rise in blood glucose by an increased insulin response and blood glucose disposal in T2DM subjects when coingested with carbohydrates [43–45]. Interestingly, one study assessed the glycemic response following consumption of liquid protein preloads of whey (55 g) and casein (55 g) in comparison with lactose (56 g) and glucose (56 g) controls in overweight, prediabetic subjects [41]. Although a significant reduction in glucose response was shown, insulin concentrations were not affected. Furthermore, no impact on post-meal insulinaemia in accordance with a 16% reduction in post-meal glycemia over 360 min in overweight subjects further supports observations of the current study product [46]. Thus, one might assume that the current milk protein hydrolysate containing bioactive peptides may influence plasma glucose via insulin-independent mechanisms. This is supported by in vitro experiments demonstrating the alpha-glucosidase acting mode of action for the study product (unpublished data). Therefore, one might speculate that the slight reduction in insulin release might be a secondary response due to lower postprandial increase of glucose.

We further assessed the effect of the milk protein hydrolysate with bioactive peptides for a longer period of six weeks with a daily intake of 1400 mg of the study product. Notably, this intervention resulted in a slight improvement of whole-body insulin sensitivity (hepatic and peripheral tissue insulin sensitivity) as assessed by the Matsuda index. The change was not significant ($p = 0.0952$), which might be attributed to the limited samples size and needs further confirmation in future studies. Additionally, one might assume that the current milk protein hydrolysate containing bioactive peptides may influence whole-body insulin sensitivity secondary to its primary effects on alpha-glucosidase inhibition, which were not obvious after the limited intervention period of six weeks. In addition, the longer intervention period did not strengthen the postprandial effect on glucose response, as iAUC values were comparable to those of single dose intake.

In summary, the study product primarily influenced postprandial glycemia and secondarily influenced insulin sensitivity in the whole body, suggesting rather insulin-independent mechanisms or temporal changes in insulin sensitivity. Moreover, the six week intervention period accentuates the more glycemic and less insulinotropic effect of the current milk protein hydrolysate, as the glycemic marker HbA_{1c} was significantly reduced ($p = 0.0244$). Notably, 52.4% of the subjects completed the study with HbA_{1c} levels $< 5.7\%$ after the six week intervention period, and the significant reduction of HbA_{1c} is worth mentioning in the short time period of six weeks, which has to be confirmed in further studies with longer intervention periods.

Regarding study limitations, the current study was performed in cross-over design to control for inter-individual variability. This variability cannot be estimated from the data, as study products were only provided once to subjects. Nevertheless, data from the open-label single arm phase performed with a daily intake of the low-dosed milk peptide concentration for six weeks suggest minor inter-individual variability and overall confirmed the results of the three-way-cross-over study with single dose intake regimen. Furthermore, one has to take into account that T2DM—and even the prediabetic state—is a heterogeneous disease with multiple pathophysiologies. Both incretins and microbiota in the gastrointestinal tract are known to be affected in prediabetics [47,48], which might have an impact on the postprandial responses. Although these parameters were not assessed in this study, the current results look very promising and should be confirmed in further investigations.

5. Conclusions

The objective of the current study was to assess whether alpha-glucosidase inhibiting bioactive peptides from milk protein hydrolysate might improve postprandial glucose profiles in prediabetic subjects. We demonstrated that low dose milk peptides had a significant impact on postprandial blood glucose profile with more glycemic than insulinotropic properties in prediabetic subjects after a challenge meal high in carbohydrates. This was confirmed after a single dose intake and after a six week intervention period, whereas impacts on postprandial effects were not strengthened by intervention over a longer period. Furthermore, the study product primarily influenced postprandial glycemia and secondarily influenced insulin sensitivity in the whole body, as only a minor increase of the Matsuda index and a slight but significant reduction of HbA_{1C} levels were demonstrated after the six week intervention period.

The investigated hydrolyzed milk-derived bioactive peptides (1.4 g/day) of native whey origin seem to be promising and well-tolerated by prediabetic subjects to control postprandial glucose levels, which should be confirmed in further clinical studies with longer intervention periods to ascertain the benefits for glucose homeostasis.

Supplementary Materials: The following are available online at http://www.mdpi.com/2072-6643/11/7/1700/s1, Table S1: CONSORT 2010 checklist of information to include when reporting a randomized trial.

Author Contributions: Project administration and resources: C.S. and T.D. designed the study; Investigation: the study was undertaken at the study site of BioTeSys GmbH under supervision of A.W., T.D., and C.S. Formal analysis and visualization: C.S. and M.W. planned and performed statistical analysis and created the figures. Writing: T.S. and C.S. drafted the manuscript and made the final approval of the published version. A.B. contributed to discussion and manuscript revision. All authors significantly contributed to results interpretation, critical manuscript revision, and approval of the final manuscript.

Funding: This research was funded by Ingredia S.A., France.

Acknowledgments: The authors would like to thank all subjects who took part in this clinical trial.

Conflicts of Interest: C.S., A.W., T.D., T.S.: employees of contracted research organization. M.W. is an independent statistician supporting with data analysis. The study was financially supported by Ingredia S.A., France. The sponsors contributed to the discussion about the study design and selection of outcome measures prior to the study start. Planning and organization of the study and its realization, data analysis and report generating were independently undertaken solely by BioTeSys GmbH and M.W. The authors from BioTeSys GmbH and M.W. declare no conflict of interest regarding the publication of this paper.

References

1. DeFronzo, R.A.; Bonadonna, R.C.; Ferrannini, E. Pathogenesis of NIDDM: A balanced overview. *Diabetes Care* **1992**, *15*, 318–368. [CrossRef]
2. Fiorentino, T.V.; Marini, M.A.; Andreozzi, F.; Arturi, F.; Succurro, E.; Perticone, M.; Sciacqua, A.; Hribal, M.L.; Perticone, F.; Sesti, G. One-Hour Postload Hyperglycemia Is a Stronger Predictor of Type 2 Diabetes Than Impaired Fasting Glucose. *J. Clin. Endocrinol. Metab.* **2015**, *100*, 3744–3751. [CrossRef]
3. Fiorentino, T.V.; Marini, M.A.; Succurro, E.; Andreozzi, F.; Perticone, M.; Hribal, M.L.; Sciacqua, A.; Perticone, F.; Sesti, G. One-Hour Postload Hyperglycemia: Implications for Prediction and Prevention of Type 2 Diabetes. *J. Clin. Endocrinol. Metab.* **2018**, *103*, 3131–3143. [CrossRef]
4. Kanat, M.; DeFronzo, R.A.; Abdul-Ghani, M.A. Treatment of prediabetes. *World J. Diabetes* **2015**, *6*, 1207–1222. [CrossRef]
5. Jakubowicz, D.; Froy, O. Biochemical and metabolic mechanisms by which dietary whey protein may combat obesity and Type 2 diabetes. *J. Nutr. Biochem.* **2013**, *24*, 1–5. [CrossRef]
6. Li, J.; Janle, E.; Campbell, W.W. Postprandial Glycemic and Insulinemic Responses to Common Breakfast Beverages Consumed with a Standard Meal in Adults Who Are Overweight and Obese. *Nutrients* **2017**, *9*, 32. [CrossRef]
7. Liljeberg, E.H.; Bjorck, I. Milk as a supplement to mixed meals may elevate postprandial insulinaemia. *Eur. J. Clin. Nutr.* **2001**, *55*, 994–999. [CrossRef]
8. Fumeron, F. Produits laitiers et prévention du diabète de type 2. Cholé-Doc 2013. Available online: https://www.cerin.org/fileadmin/user_upload/PDF/Nutrinews-hebdo/NNH_264/264-Les-produits-laitiers-previennent-le-diabete.pdf (accessed on 24 July 2019).
9. Pfeuffer, M.; Schrezenmeir, J. Milk and the metabolic syndrome. *Obes. Rev.* **2007**, *8*, 109–118. [CrossRef]
10. Jauhiainen, T.; Korpela, R. Milk peptides and blood pressure. *J. Nutr.* **2007**, *137*, 825S–829S. [CrossRef]
11. Luhovyy, B.L.; Akhavan, T.; Anderson, G.H. Whey proteins in the regulation of food intake and satiety. *J. Am. Coll. Nutr.* **2007**, *26*, 704S–712S. [CrossRef]
12. Darewicz, M.; Borawska, J.; Pliszka, M. Carp proteins as a source of bioactive peptides-an in silico approach. *Czech J. Food Sci.* **2016**, *34*, 111–117. [CrossRef]
13. Li-Chan, E.C.Y. Bioactive peptides and protein hydrolysates: research trends and challenges for application as nutraceuticals and functional food ingredients. *Curr. Opin. Food Sci.* **2015**, *1*, 28–37. [CrossRef]
14. Bjørnshave, A.; Holst, J.J.; Hermansen, K. Pre-Meal Effect of Whey Proteins on Metabolic Parameters in Subjects with and without Type 2 Diabetes: A Randomized, Crossover Trial. *Nutrients* **2018**, *10*, 122. [CrossRef]
15. Boirie, Y.; Dangin, M.; Gachon, P.; Vasson, M.P.; Maubois, J.L.; Beaufrere, B. Slow and fast dietary proteins differently modulate postprandial protein accretion. *Proc. Natl. Acad. Sci. USA* **1997**, *94*, 14930–14935. [CrossRef]
16. Hamarsland, H.; Laahne, J.A.L.; Paulsen, G.; Cotter, M.; Børsheim, E.; Raastad, T. Native whey induces higher and faster leucinemia than other whey protein supplements and milk: a randomized controlled trial. *BMC Nutr.* **2017**, *3*, 10. [CrossRef]
17. Manninen, A.H. Protein hydrolysates in sports nutrition. *Nutr. Metab. (Lond.)* **2009**, *6*, 38. [CrossRef]
18. Calbet, J.A.L.; Holst, J.J. Gastric emptying, gastric secretion and enterogastrone response after administration of milk proteins or their peptide hydrolysates in humans. *Eur. J. Nutr.* **2004**, *43*, 127–139. [CrossRef]
19. Koopman, R.; Crombach, N.; Gijsen, A.P.; Walrand, S.; Fauquant, J.; Kies, A.K.; Lemosquet, S.; Saris, W.H.M.; Boirie, Y.; van Loon, L.J. Ingestion of a protein hydrolysate is accompanied by an accelerated in vivo digestion and absorption rate when compared with its intact protein. *Am. J. Clin. Nutr.* **2009**, *90*, 106–115. [CrossRef]
20. Grimble, G.K.; Sarda, M.G.; Sessay, H.F.; Marrett, A.L.; Kapadia, S.A.; Bowling, T.E.; Silk, D.B.A. The influence of whey hydrolysate peptide chain length on nitrogen and carbohydrate absorption in the perfused human jejunum. *Clin. Nutr.* **1994**, *13*, 46. [CrossRef]
21. Marcone, S.; Haughton, K.; Simpson, P.J.; Belton, O.; Fitzgerald, D.J. Milk-derived bioactive peptides inhibit human endothelial-monocyte interactions via PPAR-γ dependent regulation of NF-κB. *J. Inflamm. (Lond.)* **2015**, *12*. [CrossRef]
22. Iwaniak, A.; Darewicz, M.; Minkiewicz, P. Peptides Derived from Foods as Supportive Diet Components in the Prevention of Metabolic Syndrome. *Compr. Rev. Food Sci. Food Saf.* **2018**, *17*, 63–81. [CrossRef]

23. Kumar, S.; Narwal, S.; Kumar, V.; Prakash, O. alpha-glucosidase inhibitors from plants: A natural approach to treat diabetes. *Pharmacogn. Rev.* **2011**, *5*, 19–29. [CrossRef]

24. Patil, P.; Mandal, S.; Tomar, S.K.; Anand, S. Food protein-derived bioactive peptides in management of type 2 diabetes. *Eur. J. Nutr.* **2015**, *54*, 863–880. [CrossRef]

25. Hu, R.; Li, Y.; Lv, Q.; Wu, T.; Tong, N. Acarbose Monotherapy and Type 2 Diabetes Prevention in Eastern and Western Prediabetes: An Ethnicity-specific Meta-analysis. *Clin. Ther.* **2015**, *37*, 1798–1812. [CrossRef]

26. Kang, W.; Song, Y.; Gu, X. α-glucosidase inhibitory in vitro and antidiabetic activity in vivo of Osmanthus fragrans. *J. Med. Plants Res.* **2012**, *6*, 2850–2856.

27. Ben Henda, Y.; Labidi, A.; Arnaudin, I.; Bridiau, N.; Delatouche, R.; Maugard, T.; Piot, J.-M.; Sannier, F.; Thiery, V.; Bordenave-Juchereau, S. Measuring angiotensin-I converting enzyme inhibitory activity by micro plate assays: Comp. using marine cryptides and tentative threshold determinations with captopril and losartan. *J. Agric. Food Chem.* **2013**, *61*, 10685–10690. [CrossRef]

28. Boutrou, R.; Gaudichon, C.; Dupont, D.; Jardin, J.; Airinei, G.; Marsset-Baglieri, A.; Benamouzig, R.; Tome, D.; Leonil, J. Sequential release of milk protein-derived bioactive peptides in the jejunum in healthy humans. *Am. J. Clin. Nutr.* **2013**, *97*, 1314–1323. [CrossRef]

29. Höfle, A.S. Effects of Different Milk Proteins on the Metabolic Response in Healthy and Prediabetic Volunteers. Ph.D. Thesis, Technical University of Munich, Munich, Germany, 2015.

30. De Andrade, M.I.S.; Oliveira, J.S.; Leal, V.S.; da Lima, N.M.S.; Costa, E.C.; de Aquino, N.B.; de Lira, P.I.C. Identification of cutoff points for Homeostatic Model Assessment for Insulin Resistance index in adolescents: systematic review. *Revista Paulista Pediatria (Engl. Ed.)* **2016**, *34*, 234–242. [CrossRef]

31. Kita, M.; Obara, K.; Kondo, S.; Umeda, S.; Ano, Y. Effect of Supplementation of a Whey Peptide Rich in Tryptophan-Tyrosine-Related Peptides on Cognitive Performance in Healthy Adults: A Randomized, Double-Blind, Placebo-Controlled Study. *Nutrients* **2018**, *10*, 899. [CrossRef]

32. Babij, K.; Dąbrowska, A.; Szołtysik, M.; Pokora, M.; Zambrowicz, A.; Chrzanowska, J. The Evaluation of Dipeptidyl Peptidase (DPP)-IV, α-Glucosidase and Angiotensin Converting Enzyme (ACE) Inhibitory Activities of Whey Proteins Hydrolyzed with Serine Protease Isolated from Asian Pumpkin (Cucurbita ficifolia). *Int. J. Pept. Res. Ther.* **2014**, *20*, 483–491. [CrossRef]

33. Kung, B.; Anderson, G.H.; Paré, S.; Tucker, A.J.; Vien, S.; Wright, A.J.; Goff, H.D. Effect of milk protein intake and casein-to-whey ratio in breakfast meals on postprandial glucose, satiety ratings, and subsequent meal intake. *J. Dairy Sci.* **2018**, *101*, 8688–8701. [CrossRef]

34. Chakrabarti, S.; Jahandideh, F.; Davidge, S.T.; Wu, J. Milk-Derived Tripeptides IPP (Ile-Pro-Pro) and VPP (Val-Pro-Pro) Enhance Insulin Sensitivity and Prevent Insulin Resistance in 3T3-F442A Preadipocytes. *J. Agric. Food Chem.* **2018**, *66*, 10179–10187. [CrossRef]

35. Sawada, Y.; Sakamoto, Y.; Toh, M.; Ohara, N.; Hatanaka, Y.; Naka, A.; Kishimoto, Y.; Kondo, K.; Iida, K. Milk-derived peptide Val-Pro-Pro (VPP) inhibits obesity-induced adipose inflammation via an angiotensin-converting enzyme (ACE) dependent cascade. *Mol. Nutr. Food Res.* **2015**, *59*, 2502–2510. [CrossRef]

36. Tessari, P.; Kiwanuka, E.; Cristini, M.; Zaramella, M.; Enslen, M.; Zurlo, C.; Garcia-Rodenas, C. Slow versus fast proteins in the stimulation of beta-cell response and the activation of the entero-insular axis in type 2 diabetes. *Diabetes Metab. Res. Rev.* **2007**, *23*, 378–385. [CrossRef]

37. Nilsson, M.; Holst, J.J.; Bjorck, I.M. Metabolic effects of amino acid mixtures and whey protein in healthy subjects: studies using glucose-equivalent drinks. *Am. J. Clin. Nutr.* **2007**, *85*, 996–1004. [CrossRef]

38. Gunnerud, U.J.; Heinzle, C.; Holst, J.J.; Ostman, E.M.; Bjorck, I.M.E. Effects of pre-meal drinks with protein and amino acids on glycemic and metabolic responses at a subsequent composite meal. *PLoS ONE* **2012**, *7*, e44731. [CrossRef]

39. Ostman, E.M.; Liljeberg Elmstahl, H.G.; Bjorck, I.M. Inconsistency between glycemic and insulinemic responses to regular and fermented milk products. *Am. J. Clin. Nutr.* **2001**, *74*, 96–100. [CrossRef]

40. Pasin, G.; Comerford, K.B. Dairy foods and dairy proteins in the management of type 2 diabetes: A systematic review of the clinical evidence. *Adv. Nutr.* **2015**, *6*, 245–259. [CrossRef]

41. Bowen, J.; Noakes, M.; Trenerry, C.; Clifton, P.M. Energy intake, ghrelin, and cholecystokinin after different carbohydrate and protein preloads in overweight men. *J. Clin. Endocrinol. Metab.* **2006**, *91*, 1477–1483. [CrossRef]

42. Hamarsland, H.; Nordengen, A.L.; Nyvik Aas, S.; Holte, K.; Garthe, I.; Paulsen, G.; Cotter, M.; Børsheim, E.; Benestad, H.B.; Raastad, T. Native whey protein with high levels of leucine results in similar post-exercise muscular anabolic responses as regular whey protein: a randomized controlled trial. *J. Int. Soc. Sports Nutr.* **2017**, *14*, 43. [CrossRef]

43. Manders, R.J.F.; Wagenmakers, A.J.M.; Koopman, R.; Zorenc, A.H.G.; Menheere, P.P.C.A.; Schaper, N.C.; Saris, W.H.M.; van Loon, L.J.C. Co-ingestion of a protein hydrolysate and amino acid mixture with carbohydrate improves plasma glucose disposal in patients with type 2 diabetes. *Am. J. Clin. Nutr.* **2005**, *82*, 76–83. [CrossRef] [PubMed]

44. Manders, R.J.; Koopman, R.; Sluijsmans, W.E.; van den Berg, R.; Verbeek, K.; Saris, W.H.; Wagenmakers, A.J.; van Loon, L.J. Co-ingestion of a protein hydrolysate with or without additional leucine effectively reduces postprandial blood glucose excursions in Type 2 diabetic men. *J. Nutr.* **2006**, *136*, 1294–1299. [CrossRef] [PubMed]

45. Manders, R.J.F.; Hansen, D.; Zorenc, A.H.G.; Dendale, P.; Kloek, J.; Saris, W.H.M.; van Loon, L.J. Protein co-ingestion strongly increases postprandial insulin secretion in type 2 diabetes patients. *J. Med. Food* **2014**, *17*, 758–763. [CrossRef] [PubMed]

46. Pal, S.; Ellis, V.; Ho, S. Acute effects of whey protein isolate on cardiovascular risk factors in overweight, post-menopausal women. *Atherosclerosis* **2010**, *212*, 339–344. [CrossRef]

47. Færch, K.; Torekov, S.S.; Vistisen, D.; Johansen, N.B.; Witte, D.R.; Jonsson, A.; Pedersen, O.; Hansen, T.; Lauritzen, T.; Sandbæk, A.; et al. GLP-1 Response to Oral Glucose Is Reduced in Prediabetes, Screen-Detected Type 2 Diabetes, and Obesity and Influenced by Sex: The ADDITION-PRO Study. *Diabetes* **2015**, *64*, 2513–2525. [CrossRef] [PubMed]

48. Allin, K.H.; Tremaroli, V.; Caesar, R.; Jensen, B.A.H.; Damgaard, M.T.F.; Bahl, M.I.; Licht, T.R.; Hansen, T.H.; Nielsen, T.; Dantoft, T.M.; et al. Aberrant intestinal microbiota in individuals with prediabetes. *Diabetologia* **2018**, *61*, 810–820. [CrossRef] [PubMed]

Article

Glucose Response during the Night Is Suppressed by Wheat Albumin in Healthy Participants: A Randomized Controlled Trial

Shinichiro Saito [1,*], Sachiko Oishi [1], Aiko Shudo [2], Yoko Sugiura [2] and Koichi Yasunaga [2]

[1] Biological Research Laboratories, Kao Corporation, 2-1-3 Bunka Sumida-ku, Tokyo 131-8501, Japan; oishi.sachiko@kao.com

[2] Health Care Food Research Laboratories, Kao Corporation, 2-1-3 Bunka Sumida-ku, Tokyo 131-8501, Japan; shudou.aiko@kao.com (A.S.); sugiura.yoko@kao.com (Y.S.); yasunaga.kouichi@kao.com (K.Y.)

* Correspondence: saito.shinichiro@kao.com; Tel.: +81-3-5630-7456

Received: 25 December 2018; Accepted: 15 January 2019; Published: 17 January 2019

Abstract: Postprandial blood glucose excursions are important for achieving optimal glycemic control. In normal-weight individuals, glucose tolerance is diminished in the evening compared to glucose tolerance in the morning. Wheat albumin (WA) has the potential to suppress the postprandial glucose response with a relatively small dose, compared to the dose required when using dietary fiber. In the present study, the effect of WA on glycemic control during the night was investigated after a late evening meal. A randomly assigned crossover trial involving a single oral ingestion in healthy male participants was performed in a double-blind placebo-controlled manner. The participants ingested the placebo (PL) tablets or the WA (1.5 g)-containing tablets 3 min before an evening meal at 22:00 hour, and blood samples were drawn during the night until 07:00 hour using an intravenous cannula. The participants slept from 00:30 hour to 06:30 hour. Glucose response, as a primary outcome during the night, was suppressed significantly by the WA treatment compared to the PL treatment, but the insulin response was not. Plasma glucose-dependent insulinotropic polypeptide concentration during the night was lowered significantly by the WA treatment compared to the PL treatment. In conclusion, WA may be a useful food constituent for glycemic control during the night.

Keywords: glucose; human; night; postprandial; wheat albumin

1. Introduction

An estimated 425 million people worldwide had diabetes in 2017, and this number is projected to reach 700 million by 2045 [1]. Many epidemiologic studies have demonstrated a complex association between glycemia and cardiovascular risk [2,3], with evidence suggesting that an acute increase in glycemia, particularly after a meal, may have direct detrimental effects on the cardiovascular system [4]. Until recently, there has been a strong emphasis on fasting plasma glucose, and the predominant focus of therapy has been on lowering hemoglobin A1c (HbA1c) levels [5]. Although the control of fasting hyperglycemia is necessary, it is usually not sufficient to achieve optimal glycemic control. A growing body of evidence suggests that the reduction of postprandial plasma glucose excursions is as important, or perhaps even more important for achieving HbA1c goals [6,7]. The use of a variety of both non-pharmacologic and pharmacologic therapies is recommended to control postprandial plasma glucose [7]. This is particularly relevant during the night, as glucose tolerance is diminished compared to its level in the morning, even in normal-weight individuals [8–16]. Therefore, even in healthy, non-diabetic people, the use of non-pharmacologic therapies, such as the control of dietary and fitness habits, to protect against impaired glucose tolerance during the night is recommended.

Wheat albumin (WA) has a long history of consumption in humans as a natural food constituent and is a potentially protective agent against postprandial hyperglycemia via its alpha-amylase inhibiting activity with no change in insulin secretion [17], suggesting that WA might improve postprandial insulin sensitivity. Inhibitors of carbohydrate digestion and absorption have been reported to improve blood glucose control with a low risk of hypoglycemia [18], so WA might be a good candidate for night care. Additionally, WA has the potential to suppress the glucose response at a relatively lower dose [17] than other food constituents like wheat or oat fiber (~6 g) [19,20]. Therefore, it could be used in a wide variety of functional food products or incorporated easily into a habitual diet.

Thus, the primary objective of the present study was to investigate the potential effect of WA as a dietary therapy agent to protect against the diminished glucose response during the night in healthy individuals.

2. Materials and Methods

2.1. Ethics Approval and Consent to Participate

This study was performed in accordance with the tenets of the Declaration of Helsinki (2013) and was approved by the Ethical Committee of the Oriental Ueno Kenshin Center (Tokyo, Japan). After receiving a full explanation of the study, all participants provided written informed consent. The study was registered with the University Hospital Medical Information Network (UMIN) clinical registry, prior to the enrollment of the first participant, as UMIN000014533 (registered 15 July 2014 [21].

2.2. Study Design

This was a randomized, double-blind, placebo-controlled, crossover trial with a 1-week washout period, performed under the supervision of a physician in charge. Between the screening and the second visit, the participants were instructed to maintain and record their normal level of physical activity and their normal dietary, alcohol, and smoking habits. As shown in Figure 1, during the study, the participants were free-living, but were prohibited from drinking alcohol the day before the visits to the clinic and from smoking cigarettes for 2 hours before the visits. The participants ate designated meals for dinner in the evening at 21:00 hour one day before the trial, and for breakfast at 08:00 hour and lunch at 12:00 hour on the day of the trial. The participants were not allowed any energy intake after the designated dinner until the trial. The participants visited the clinic at 17:00 hour and were examined by the physician in charge. From 17:00 hour to 22:00 hour, anthropometric parameters were measured, and for the rest of the time participants read books or watched TV for naturalization in a sitting position with no energy consumption. Immediately after obtaining a blood sample, the participants orally ingested a single dose of WA (1.5 g)-containing tablets or placebo (PL) tablets 3 min before ingesting a designated evening meal at 22:00 hour. Blood samples were then drawn every 30 min until 00:00 hour, and then at 01:00 hour, 02:00 hour, and 07:00 hour, using an intravenous cannula. The participants slept from 00:30 hour to 06:30 hour, and their sleep quality was measured using an ActiGraph (ActiGraph, Pensacola, FL, USA). The amount and timing of water consumption during the visits were controlled. The study was conducted at Sumida Hospital, Tokyo, Japan and managed by TES Holdings Co., Ltd. (Tokyo, Japan), a contract research organization (CRO). The CRO managed the random allocation, enrollment, assignment of participants, and blinding of the assignment, and assessed the outcomes under the supervision of the physician in charge. Throughout the study (from screening to finalizing the dataset), the treatment allocation was concealed from the people involved, including the participants, the caregivers, the physicians, the CRO members, the manufacturer of the test tablets, the person in charge of the allocation, and the outcome assessors.

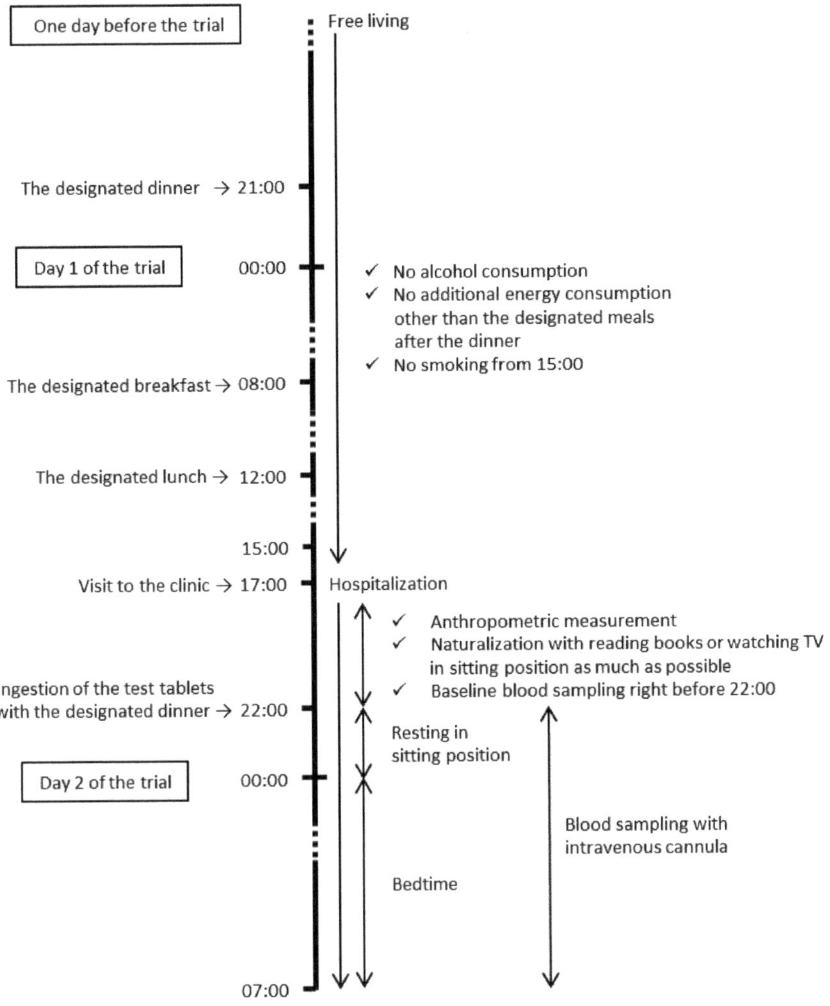

Figure 1. The study protocol.

2.3. Participants

The appropriate sample size for the primary outcome of postprandial blood glucose was estimated to be 20 participants based on the outcomes of our unpublished pilot study (power 0.8 and type I error 0.05). In the present study, potential participants were screened based on the following inclusion criteria: (1) $5.3 \leq$ fasting blood glucose < 7.0 mmol/L, (2) $5.2 \leq$ HbA1c $< 6.5\%$, (3) $23 \leq$ body mass index (BMI) < 30, and (4) $30 \leq$ age < 60 years. Participants were excluded if they met the following exclusion criteria: (1) presence of liver, kidney, or heart disease; respiratory, endocrine, or nervous system disorder; metabolism or consciousness dysfunction; diabetes; or other disease, (2) surgery in the 2 months before the trial, (3) history of gastrectomy or enterectomy, (4) taking medications for hyperglycemia, lipidemia, or hypertension, (5) taking supplements or food for a specific health use authorized by the government, (6) allergies to any constituents in the test meal or tablets, (7) an unpleasant feeling during blood draws, (8) donated 200 mL or more of blood in the month before the trial, (9) habitual breakfast skippers, (10) heavy smokers (>20 cigarettes/day), or (11) shift workers.

The conditions and procedures of the trial were reviewed with all participants before they signed the informed consent form. The participants were randomly assigned to each sequence (ingestion order) with stratified randomization for glucose, hemoglobin A1c, age, and BMI using computer-generated random numbers under blind conditions.

2.4. Test Tablets and Meals

The designated 3 meals (the dinner on the day before the trial, and the breakfast and lunch on the day of the trial) consisted of a Japanese-style diet, such as rice, miso soup, simmered vegetables, grilled meats, and snacks with a total of 9142 kJ (protein = 14%, fat = 22%, and carbohydrate = 64% of the total energy), were provided by the physician in charge before the trial. The test tablets contained 1.5 g WA in 3 tablets for a single oral administration. The PL tablets were prepared using identical ingredients, including flavors and preservatives, but did not contain WA. Each tablet weighed 1.1 g. The energy values were 14.7 kJ per WA tablet and 10.9 kJ per PL tablet. The tablets could not be distinguished by appearance, taste, or odor, and were provided to the CRO after concealment by the manufacturer. The CRO then re-concealed the test tablets and provided them to the participants. The test evening meal on the trial day consisted of curry and rice with a total of 2579 kJ (protein = 7%, fat = 11%, and carbohydrate = 81% of the total energy).

2.5. Laboratory Measurements

The blood samples collected for measuring glucose, insulin, and triglyceride were centrifuged at $1000\times g$ for 15 min at 4 °C to isolate the serum or plasma. The measurements were performed by the Health Sciences Research Institute, Inc. (Yokohama, Kanagawa, Japan). Blood samples for total glucose-dependent insulinotropic polypeptide (GIP) and active glucagon-like peptide-1 (GLP-1) were collected into BD P800 tubes (Becton, Dickinson and Company, Franklin Lakes, NJ, USA) on ice and centrifuged at $1000\times g$ for 15 min at 4 °C to isolate the plasma. Total GIP and active GLP-1 were measured using commercially available enzyme-linked immunosorbent assay kits obtained from Immuno-Biological Laboratories, Co., Ltd. (Fujioka, Gunma, Japan).

2.6. Statistics

The primary outcome of this study was the difference in the area under the curve (AUC) of the blood glucose response during the night (9 hours from 22:00 hour to 07:00 hour) between the treatments assessed with the mixed model, adjusted by sequences and treatments as a fixed effect. As an exploratory assessment, analysis of covariance was also performed on the slope of the changes in the parameters using the linear mixed model, and the p-values for the time and treatment effect and the treatment by time interaction were obtained. In addition, the Bonferroni correction for multiple comparisons was applied to assess statistical difference on each time point between the treatments. A two-sided p-value ≤ 0.05 was considered to indicate statistical significance. All statistical analyses were performed using IBM SPSS Statistics version 19 (IBM Co., Armonk, NY, USA).

3. Results

3.1. Characteristics of the Participants

Eighty-five individuals were screened, and 22 were recruited. Of those recruited, one participant missed the second visit for unknown reasons, and one dropped out due to feeling sick before ingesting the test tablets on the day of the second visit. Based on an examination by the physician in charge, there was no relationship between the reason for dropping out and the test tablets. Thus, 20 participants completed the study and were included in the analyses. The baseline characteristics of these participants are presented in Table 1. From screening until the second visit, no considerable habitual changes were recorded. The percentage of sleep did not differ significantly between the treatments (88.1 $\pm$ 2.3% after PL treatment and 86.3 $\pm$ 3.0% after WA treatment).

Table 1. Baseline characteristics of participants.

Parameter	Value
Number of participants (male/female)	20 (20/0)
Age, years	51 ± 1
Body weight, kg	75.8 ± 1.7
Body mass index, kg/m^2	26.0 ± 0.4
Systolic blood pressure, mmHg	116 ± 2
Diastolic blood pressure, mmHg	76 ± 2
Glucose, mmol/L	5.19 ± 0.05
Insulin, pmol/L	39 ± 4
Triglyceride, mg/dL	1.61 ± 0.16
HbA1c, %	5.6 ± 0.1

Data are mean ± standard error.

3.2. Effects of WA on Blood Glucose, Insulin, and Triglyceride Response during Night

A comparison between the PL and WA treatments revealed that the increase in blood glucose concentration during the night was suppressed significantly by WA, as shown by the AUC. In the exploratory analysis, the treatment by time interaction was also shown to be significant. WA only suppressed the glucose response for 2 hours after the meal, but not after bedtime, suggesting that a hypoglycemic risk by WA treatment was not observed (Figure 2).

Insulin response was not shown to be significantly different in the AUC and interaction assessments, however, a significant treatment effect was observed (Figure 3).

Blood triglyceride was also measured during the night, but no considerable differences between the treatments were observed (data not shown).

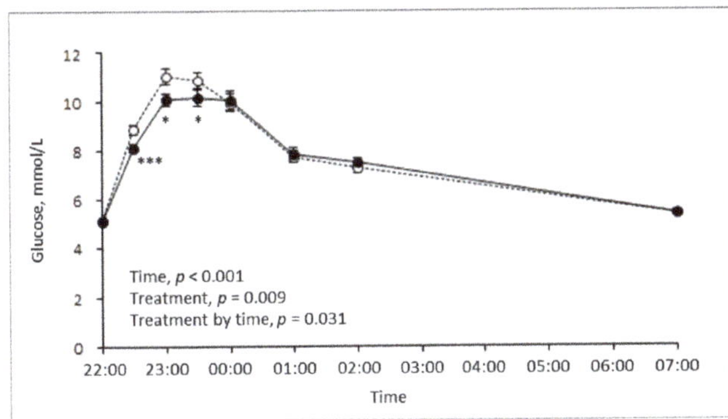
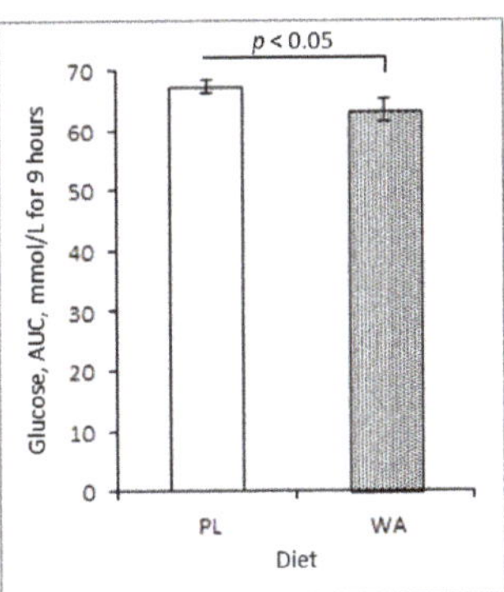

Figure 2. Changes in the blood glucose level during the night following the placebo (PL) (broken line, n = 20) and wheat albumin (WA) treatments (solid line, n = 20). (**Left**) Changes in the blood glucose response. (**Right**) The AUC of the blood glucose level for 9 h from 22:00 hour to 7:00 hour. Data are the mean ± standard error. Significant differences between the treatments: * $p < 0.05$, *** $p < 0.001$.

3.3. Effects of WA on Incretins during the Night

The AUC of the GIP concentration in the blood was significantly lower following the WA treatment than following the PL treatment, but the treatment by time interaction was unchanged. Interestingly, the difference was large at bedtime, but there was no difference in the postprandial state for 2 hours after the meal (Figure 4). There were no significant differences in the blood GLP-1 concentration in any of the assessments (Figure 5).

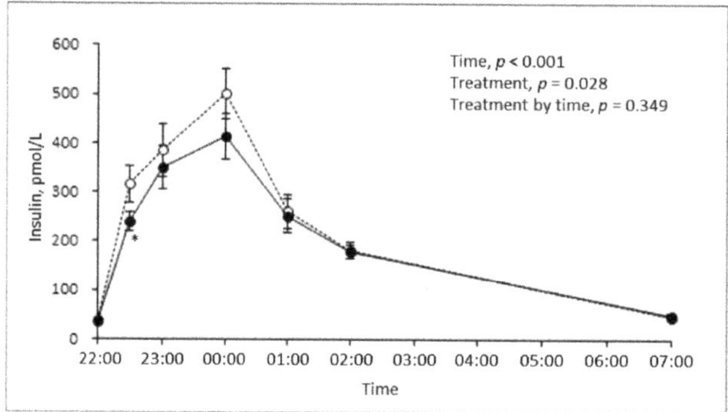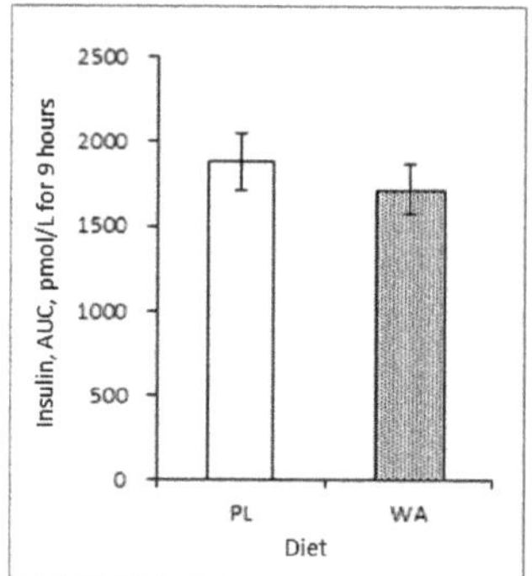

Figure 3. Changes in the blood insulin level during the night between the PL (broken line, *n* = 20) and the WA treatments (solid line, *n* = 20). (**Left**) Changes in the blood insulin response. (**Right**) The AUC of the blood insulin level for 9 h from 22:00 hour to 7:00 hour. Data are the mean ± standard error. Significant differences between the treatments: * $p < 0.05$.

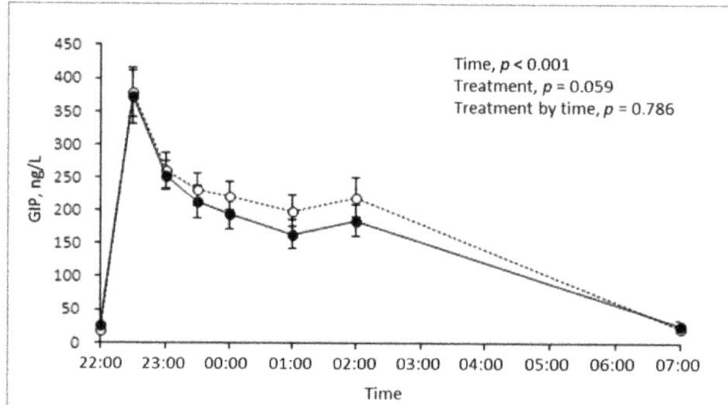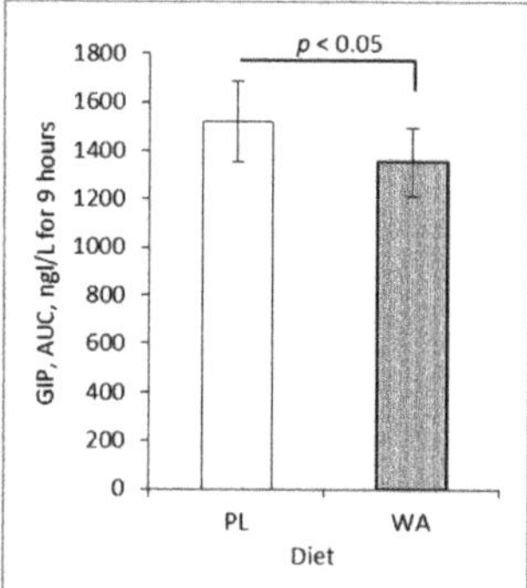

Figure 4. Changes in the blood glucose-dependent insulinotropic polypeptide (GIP) level during the night between the PL (broken line, *n* = 20) and WA treatments (solid line, *n* = 20). (**Left**) Changes in the blood GIP response. (**Right**) The AUC of the blood GIP level for 9 h from 22:00 hour to 7:00 hour. Data are the mean ± standard error.

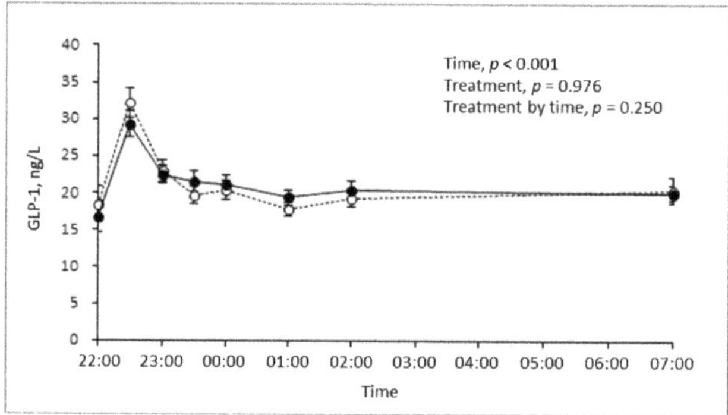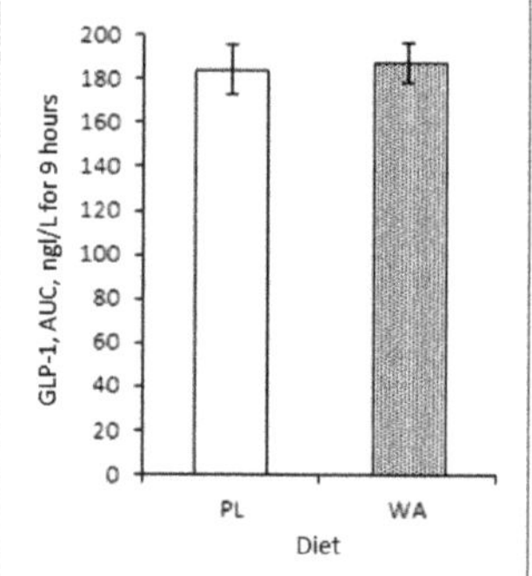

Figure 5. Changes in the blood active glucagon-like peptide-1 (GLP-1) level during the night between the PL (broken line, *n* = 20) and the WA treatments (solid line, *n* = 20). (**Left**) Changes in the blood GLP-1 response. (**Right**) The AUC of the blood GLP-1 level for 9 h from 22:00 hour to 7:00 hour. Data are the mean ± standard error.

4. Discussion

For healthy individuals, the recommended strategies to prevent an elevated glucose level at night are non-pharmaceutical, such as diet and exercise. Dietary therapy is perhaps easier to achieve, and thus may give more continuous efficacy, especially before sleep. Accompanied by the widespread use of a non-invasive glucose monitoring system, nocturnal hypoglycemia and its mortuary risk have been investigated in diabetic patients treated with insulin therapy [22]. However, in healthy individuals and individuals with borderline high values, the risk of a nocturnal hypoglycemic state is also best avoided. Therefore, the use of a food constituent with a moderate efficacy may be an alternative strategy for night care. As previously reported, WA has a suppressive effect on the postprandial glucose level via its inhibitory action on alpha-amylase activity [17], suggesting its potential as a low risk strategy for glycemic control during the night. Therefore, this study investigated the effect of WA on glucose response during the night in healthy individuals. As expected, WA contributed to a lower glucose response during the night but showed no hypoglycemic effect when compared to the PL (Figure 2). Thus, WA is a good candidate as a strategy for bedtime glycemic control in healthy individuals, and perhaps in diabetic patients alongside insulin therapy; however, additional studies are required for these patients to assess safety concerns.

An epidemiology study showed that the habitual intake of late-evening meals was associated with a higher BMI and metabolic risk factors such as high triglycerides and lower high-density lipoprotein cholesterol [23]. A potential explanation for this is the reduced metabolic rate and fuel utilization that occur during sleep [24]. GIP, an incretin that stimulates insulin secretion from pancreatic beta-cells, was reported to have a higher concentration postprandially that was directly associated with a lower metabolic rate [25] and the stimulation of fat accumulation in adipose tissue as an exopancreatic function. In contrast, GLP-1 did not show associations with these factors [26,27]. Moreover, higher blood GIP levels induced by chronic GIP treatment reduced fat utilization in high-fat diet-fed mice [28]. Interestingly, a bigger difference in the total GIP response between the PL and the WA treatments occurred after sleep than before sleep (Figure 4), with a moderate effect on insulin (Figure 3). These findings indicate that WA affects the later phase of the blood GIP level or its secretion, rather than the earlier phase. This might induce metabolic differences in the extrapancreatic actions of GIP, as mentioned above. Therefore, WA might improve not only glucose metabolism, but also fat utilization as energy by lowering the extrapancreatic actions of GIP during sleep. However, further investigation of the effects of WA on metabolic rate or fuel utilization during the night is required.

Dietary intake in the late evening has been suggested to alter clock genes such as *Bmal1* and *Clock* [29]. *Bmal1* knockout mice showed a higher blood glucose concentration than a wild type [30,31]. Circadian mutant mice, and both *Clock* [32,33] and *Bmal1* [34] mutants, showed impaired glucose tolerance, reduced insulin secretion, and defects in the size and proliferation of pancreatic islets that worsened with age. Thus, habitual late-evening meals may induce impaired glucose tolerance and the development of diabetes. As shown in our study, WA has the potential to reduce factors relating to the development of metabolic disorders, such as higher glucose and GIP responses during the night, through consumption in an evening meal, so WA can be expected to rearrange disordered circadian rhythms. Thus, a further investigation into the effect of WA on glucose tolerance with a focus on circadian rhythms or clock gene expression would be of great interest.

Overall, further studies to investigate the effects of WA on glycemic control during the night in diabetic patients, on energy metabolism, and on circadian clock genes would be of considerable interest.

The limitations and potential biases in this study were the imbalanced gender of the participants (only men), the use of a single race (Japanese), and the inadequate sample size for stratified analyses based on the participants' characteristics, such as BMI and fasting glucose level. In addition, all authors in this study are employees of the manufacturer of the studied ingredient.

5. Conclusions

WA might be a useful food constituent for glycemic control during the night.

Author Contributions: S.S. was responsible for the conception and design of the study, and the drafting of the manuscript. S.O. made substantial contributions to the conception and the design. A.S. and Y.S. made substantial contributions to the data collection. K.Y. approved the version to be published and was a general supervisor of the study treatment.

Funding: This study was conducted by TES Holdings Co., Ltd., a CRO company under financial support from the Kao Corporation.

Acknowledgments: We thank Nozomu Higo from Sumida Hospital who was the physician in charge.

Conflicts of Interest: Authors are employees of the Kao Corporation. The test tablets were prepared by the Kao Corporation.

References

1. International Diabetes Federation. *IDF Diabetes Atlas*; International Diabetes Federation: Brussels, Belgium, 2017.

2. Gerstein, H.C. Dysglycemia, not just diabetes, is a continuous risk factor for cardiovascular disease. *Evid.-Based Cardiovasc. Med.* **1997**, *1*, 87–88. [CrossRef]

3. Punthakee, Z.; Werstuck, G.H.; Gerstein, H.C. Diabetes and cardiovascular disease: Explaining the relationship. *Rev. Cardiovasc. Med.* **2007**, *8*, 145–153.

4. Ceriello, A.; Hanefeld, M.; Leiter, L.; Monnier, L.; Moses, A.; Owens, D.; Tajima, N.; Tuomilehto, J. The International Prandial Glucose Regulation (PGR) Study Group. Postprandial Glucose Regulation and Diabetic Complications. *Arch. Intern. Med.* **2004**, *164*, 2090–2095. [CrossRef] [PubMed]

5. Nathan, D.M.; Buse, J.B.; Davidson, M.B.; Hein, R.J.; Holman, R.R.; Sherwin, R.; Zinman, B. Management of hyperglycemia in type 2 diabetes: A consensus algorithm for the initiation and adjustment of therapy: A consensus statement from the American Diabetes Association and the European Association for the Study of Diabetes. *Diabetes Care* **2006**, *29*, 1963–1972. [CrossRef] [PubMed]

6. Ceriello, A. The glucose triad and its role in comprehensive glycaemic control: Current status, future management. *Int. J. Clin. Pract.* **2010**, *64*, 1705–1711. [CrossRef] [PubMed]

7. International Diabetes Federation. *Guideline for Management of PostMeal Glucose in Diabetes*; International Diabetes Federation: Brussels, Belgium, 2011.

8. Jarrett, R.J.; Baker, I.A.; Keen, H.; Oakly, N.W. Diurnal variation in oral glucose tolerance, blood sugar, and plasma insulin levels in the morning, afternoon, and evening. *Br. Med. J.* **1972**, *1*, 199–201. [CrossRef] [PubMed]

9. Carroll, K.F.; Nestel, P.L. Diurnal variation in glucose tolerance and in insulin secretion in man. *Diabetes* **1973**, *22*, 333–348. [CrossRef] [PubMed]

10. Zimmet, P.Z.; Well, J.R.; Rome, R.; Stimmler, L.; Jarrett, J.R. Diurnal variation in glucose tolerance and associated changes in plasma insulin, growth hormone, and esterified fatty acids. *Br. Med. J.* **1974**, *1*, 485–488. [CrossRef] [PubMed]

11. Aparicio, N.J.; Puchulu, F.E.; Gagliardino, J.J.; Ruiz, M.; Llorens, J.M.; Ruiz, J.; Lamas, A.; De Miguel, R. Circadian variation of the blood glucose, plasma insulin, and human growth hormone levels in response to an oral glucose load in normal subjects. *Diabetes* **1974**, *23*, 132–137. [CrossRef]

12. Whichelow, M.J.; Stuge, R.A.; Keen, H.; Jarrett, R.T.; Stimmler, L.; Grainger, S. Diurnal variation in response to intravenous glucose. *Br. Med. J.* **1974**, *1*, 488–491. [CrossRef]

13. Sensi, S. Some aspects of circadian variation of carbohydrate metabolism and related hormones in man. *Chronobiologia* **1974**, *1*, 396–399. [PubMed]

14. Jarrett, R.J.; Keen, H. Further observations on the diurnal variation in oral glucose tolerance. *Br. Med. J.* **1970**, *4*, 334–337. [CrossRef] [PubMed]

15. Baker, I.A.; Jarrett, R.J. Diurnal variation in the blood glucose and plasma insulin response to tolbutamide. *Lancet* **1972**, *1*, 945–947. [CrossRef]

16. Gibson, T.; Jarrett, R.J. Diurnal variation in insulin sensitivity. *Lancet* **1972**, *1*, 947–948. [CrossRef]

17. Kodama, T.; Miyazaki, T.; Kitamura, I.; Suzuki, Y.; Namba, Y.; Sakurai, J.; Torikai, Y.; Inoue, S. Effects of single and long-term administration of wheat albumin on blood glucose control: Randomized controlled clinical trials. *Eur. J. Clin. Nutr.* **2005**, *59*, 384–392. [CrossRef]

18. Josse, R.G.; Chiasson, J.L.; Ryan, E.A.; Lau, D.C.W.; Ross, S.A.; Yale, J.F.; Leiter, L.A.; Maheux, P.; Tessier, D.; Wolever, T.M.S.; et al. Acarbose in the treatment of elderly patients with type 2 diabetes. *Diabetes Res. Clin. Pract.* **2003**, *59*, 37–42. [CrossRef]

19. Lu, Z.X.; Walker, K.Z.; Muir, J.G.; Mascara, T.; O'Dea, K. Arabinoxylan fiber, a byproduct of wheat flour processing, reduces the postprandial glucose response in normoglycemic subjects. *Am. J. Clin. Nutr.* **2000**, *71*, 1123–1128. [CrossRef]

20. Maki, K.C.; Davidson, M.H.; Witchger, M.S.; Dicklin, M.R.; Subbaiah, P.V. Effects of high-fiber oat and wheat cereals on postprandial glucose and lipid responses in healthy men. *Int. J. Vitam. Nutr. Res.* **2007**, *77*, 347–356. [CrossRef]

21. The University Hospital Medical Information Network (UMIN) Clinical Registry. Available online: http://www.umin.ac.jp/ctr/index.htm (accessed on 25 December 2018).

22. Allen, K.V.; Frier, B.M. Nocturnal hypoglycemia: Clinical manifestations and therapeutic strategies toward prevention. *Endocr. Pract.* **2003**, *9*, 530–543. [CrossRef] [PubMed]

23. Oshida, H.; Kutsuma, A.; Nakajima, K. Associations of eating a late-evening meal before bedtime with low serum amylase and unhealthy conditions. *J. Diabetes Metab. Disord.* **2013**, *12*, 53–56. [CrossRef] [PubMed]

24. Katayose, Y.; Tasaki, M.; Ogata, H.; Nakata, Y.; Tokuyama, K.; Satoh, M. Metabolic rate and fuel utilization during sleep assessed by whole-body indirect calorimetry. *Metabolism* **2009**, *58*, 920–926. [CrossRef]

25. Zhou, H.; Yamada, Y.; Tsukiyama, K.; Miyawaki, K.; Hosokawa, M.; Nagashima, K.; Toyoda, K.; Naitoh, R.; Mizunoya, W.; Fushiki, T.; et al. Gastric inhibitory polypeptide modulates adiposity and fat oxidation under diminished insulin action. *Biochem. Biophys. Res. Commun.* **2005**, *335*, 937–942. [CrossRef] [PubMed]

26. Seino, Y.; Fukushima, M.; Yabe, D. GIP and GLP-1, the two incretin hormones: Similarities and differences. *J. Diabetes Investig.* **2010**, *1*, 8–23. [CrossRef] [PubMed]

27. Seino, Y.; Yabe, D. Glucose-dependent insulinotropic polypeptide and glucagon-like peptide-1: Incretin actions beyond the pancreas. *J. Diabetes Investig.* **2013**, *4*, 108–130. [CrossRef] [PubMed]

28. Shimotoyodome, A.; Suzuki, J.; Fukuoka, D.; Tokimitsu, I.; Hase, T. RS4-type resistant starch prevents high-fat diet-induced obesity via increased hepatic fatty acid oxidation and decreased postprandial GIP in C57BL/6J mice. *Am. J. Physiol. Endocrinol. Metab.* **2010**, *298*, E652–E662. [CrossRef] [PubMed]

29. Wang, X.; Xue, J.; Yang, J.; Xie, M. Timed high-fat diet in the evening affects the hepatic circadian clock and PPARα-mediated lipogenic gene expressions in mice. *Genes Nutr.* **2013**, *8*, 457–463. [CrossRef] [PubMed]

30. Shimba, S.; Ogawa, T.; Hitosugi, S.; Ichihashi, Y.; Nakadaira, Y.; Kobayashi, M.; Tezuka, M.; Kosuge, Y.; Ishige, K.; Ito, Y.; et al. Deficient of a clock gene, brain and muscle Arnt-like protein-1 (BMAL1), induces dyslipidemia and ectopic fat formation. *PLoS ONE* **2011**, *6*, e25231. [CrossRef]

31. Marcheva, B.; Ramsey, K.M.; Buhr, E.D.; Kobayashi, Y.; Su, H.; Ko, C.H.; Ivanova, G.; Omura, C.; Mo, S.; Vitaterna, M.H.; et al. Disruption of the clock components CLOCK and BMAL1 leads to hypoinsulinaemia and diabetes. *Nature* **2010**, *466*, 627–631. [CrossRef]

32. King, D.P.; Zhao, Y.; Sangoram, A.M.; Wilsbacher, L.D.; Tanaka, M.; Antoch, M.P.; Steeves, T.D.; Vitaterna, M.H.; Kornhauser, J.M.; Lowrey, P.L.; et al. Positional cloning of the mouse circadian clock gene. *Cell* **1997**, *89*, 641–653. [CrossRef]

33. Turek, F.W.; Joshu, C.; Kohsaka, A.; Lin, E.; Ivanova, G.; McDearmon, E.; Laposky, A.; Losee-Olson, S.; Easton, A.; Jensen, D.R.; et al. Obesity and metabolic syndrome in circadian Clock mutant mice. *Science* **2005**, *308*, 1043–1045. [CrossRef]

34. Bunger, M.K.; Wilsbacher, L.D.; Moran, S.M.; Clendenin, C.; Radcliffe, L.A.; Hogenesch, J.B.; Simon, M.C.; Takahashi, J.S.; Bradfield, C.A. Mop3 is an essential component of the master circadian pacemaker in mammals. *Cell* **2000**, *103*, 1009–1017. [CrossRef]

Article

Effects of Dietary Protein and Fat Content on Intrahepatocellular and Intramyocellular Lipids during a 6-Day Hypercaloric, High Sucrose Diet: A Randomized Controlled Trial in Normal Weight Healthy Subjects

Anna Surowska [1,†], Prasanthi Jegatheesan [1,†], Vanessa Campos [1], Anne-Sophie Marques [1], Léonie Egli [1], Jérémy Cros [1], Robin Rosset [1], Virgile Lecoultre [1], Roland Kreis [2], Chris Boesch [2], Bertrand Pouymayou [2], Philippe Schneiter [1] and Luc Tappy [1,*]

[1] Department of Physiology, University of Lausanne, 1005 Lausanne, Switzerland; anna.surowska@unil.ch (A.S.); pira_jegatheesan@hotmail.com (P.J.); vanessacaroline.campos@rdls.nestle.com (V.C.); asophie.marques@gmail.com (A.-S.M.); Leonie.Egli@rdls.nestle.com (L.E.); Jeremy.cros@unil.ch (J.C.); rosset.robin@gmail.com (R.R.); virgile.lecoultre@hibroye.ch (V.L.); philippe.schneiter@unil.ch (P.S.)

[2] Department for Biomedical Research, University of Bern and Institute of Diagnostic Interventional and Pediatric Radiology, University Hospital, 3012 Bern, Switzerland; roland.kreis@insel.ch (R.K.); Chris.Boesch@insel.ch (C.B.); bertrand.pouymayou@insel.ch (B.P.)

* Correspondence: luc.tappy@unil.ch; Tel.: +41-21-692-55-41

† These authors contributed equally to this work.

Received: 14 November 2018; Accepted: 15 January 2019; Published: 21 January 2019

Abstract: Sucrose overfeeding increases intrahepatocellular (IHCL) and intramyocellular (IMCL) lipid concentrations in healthy subjects. We hypothesized that these effects would be modulated by diet protein/fat content. Twelve healthy men and women were studied on two occasions in a randomized, cross-over trial. On each occasion, they received a 3-day 12% protein weight maintenance diet (WM) followed by a 6-day hypercaloric high sucrose diet (150% energy requirements). On one occasion the hypercaloric diet contained 5% protein and 25% fat (low protein-high fat, LP-HF), on the other occasion it contained 20% protein and 10% fat (high protein-low fat, HP-LF). IHCL and IMCL concentrations (magnetic resonance spectroscopy) and energy expenditure (indirect calorimetry) were measured after WM, and again after HP-LF/LP-HF. IHCL increased from 25.0 ± 3.6 after WM to 147.1 ± 26.9 mmol/kg wet weight (ww) after LP-HF and from 30.3 ± 7.7 to 57.8 ± 14.8 after HP-LF (two-way ANOVA with interaction: $p < 0.001$ overfeeding x protein/fat content). IMCL increased from 7.1 ± 0.6 to 8.8 ± 0.7 mmol/kg ww after LP-HF and from 6.2 ± 0.6 to 6.9 ± 0.6 after HP-LF, ($p < 0.002$). These results indicate that liver and muscle fat deposition is enhanced when sucrose overfeeding is associated with a low protein, high fat diet compared to a high protein, low fat diet.

Keywords: sucrose overfeeding; hepatic steatosis; intramyocellular lipids; intrahepatocellular lipids; dietary protein content; dietary fat content; energy expenditure; plasma triglyceride

1. Introduction

Consumption of hypercaloric high-fructose or high-sucrose diets can lead to the deposition of fat in ectopic sites such as visceral adipose tissue, the liver (intrahepatocellular lipids, IHCL), skeletal muscle (intramyocellular lipids, IMCL), the heart, and the pancreas [1]. Such ectopic fat deposition has been associated with insulin resistance and increased risk of cardiovascular and hepatic disorders [2,3]. In addition, hypercaloric high-fructose diets have been shown to impair hepatic insulin sensitivity [4,5], to increase fasting and postprandial blood triglycerides [6,7] and uric acid [8] concentrations, and may therefore be associated with a particularly ominous constellation of cardiometabolic risk factors.

Most studies that have documented metabolic effects of fructose or sucrose overfeeding have involved either the addition of fructose or sucrose to a weight maintenance diet, or the substitution of fructose or sucrose for dietary starch. In real life conditions, however, the addition of sucrose to an *ad libitum* diet is expected to impact habitual food consumption and hence to alter both total energy intake and the dietary macronutrient composition. It has indeed been reported that the addition of fructose-sweetened beverages to the spontaneous diet of overweight subjects was associated with a partial suppression of dietary fat and protein intake from solid foods [9]. One may therefore hypothesize that the metabolic effects of overfeeding depend not only on the amount of excess sucrose, but also on how it impacts other dietary macronutrient intake. Dietary sucrose and fat content may have additive effects on IHCL [10]. Interactions between dietary sucrose and protein are also relevant, since dietary protein intake has been shown to modulate overfeeding-induced ectopic lipid storage: in rodents fed a high fructose diet, the increase in IHCL was lower when excess dietary fructose was associated with a high, compared to a low, protein intake [11,12]. Similar observations were reported for humans overfed with lipids and protein compared to lipids alone [13–15], and with fructose and essential amino-acids compared with fructose alone [16]. In addition, a high protein intake is associated with an increase in energy expenditure, and may thus reduce energy storage [17]. We therefore hypothesized that, in normal weight human subjects, a short-term sucrose overfeeding associated with a high-protein, low-fat intake would blunt intrahepatocellular and intramyocellular lipid storage compared to the same sucrose overfeeding associated with a low-protein, high-fat diet. To assess this hypothesis, we carried out a randomized, cross-over controlled trial in 12 healthy male and female subjects. We monitored IHCL and IMCL, postprandial energy expenditure (EE), and blood metabolite concentrations at baseline, i.e. after 3 days on a 10% sucrose weight maintenance diet (WM), and after 6-days overfeeding with 50% extra-energy added as 40% sucrose and 10% lactose with either a high protein-low fat (HP-LF) or a low protein-high fat (LP-HF) content.

2. Materials and Methods

2.1. Subjects

Twelve healthy and non-obese volunteers (6 males, mean age 21 ± 1 years, weight 71.6 ± 2.3 kg, BMI 22.5 ± 0.8 kg/m^2; 6 females mean age 23 ± 1 years, weight 57.3 ± 0.8 kg, BMI 21.2 ± 0.7 kg/m^2) were included in this study. Volunteers were recruited through advertisements posted at the University of Lausanne and the Lausanne University Hospital. All volunteers were sedentary (less than 2 h of strenuous physical activity per week), were nonsmokers, had no lactose intolerance as documented by a lactose hydrogen breath test [18], and did not take any medication, (except for contraceptive agents which were used by all female participants). They all provided informed written consent.

2.2. Experimental Protocol

The experimental protocol was approved by the ethical committee (Commission d'éthique pour la recherche humaine de l'Etat de Vaud, Switzerland), and was registered at clinicaltrials.gov (NCT02168218). All procedures were performed in accordance with the 1983 revision of the Declaration of Helsinki. The primary outcome of the study was whole body protein turnover using labelled leucine, and will be reported separately. IHCL, IMCL and EE, which are the main focus of this paper, were all secondary outcomes. The experimental protocol is presented in Figure 1.

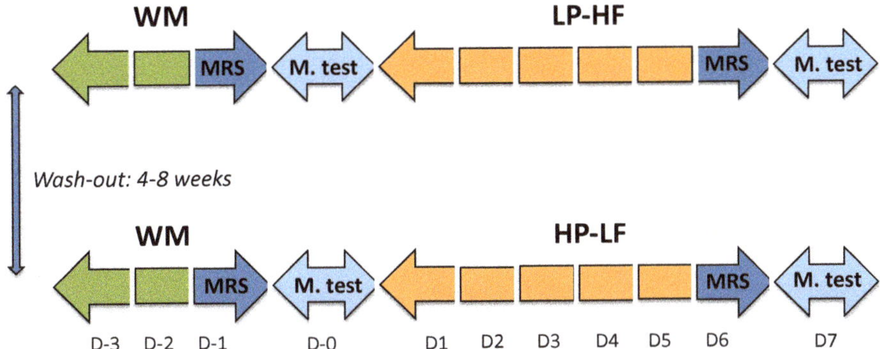

Figure 1. Experimental protocol. Each participant took part in two overfeeding periods according to a randomized, cross-over design. WM: weight maintenance diet, LP-HF: hypercaloric (150% energy requirement high-sucrose, low protein-high fat); HP-LF: hypercaloric (150% energy requirement high-sucrose, high protein-low fat); MRS: magnetic resonance spectroscopy for measurement of IHCL and IMCL; M. test: metabolic test, consisting of measurements of energy expenditure, plasma hormones, and substrate concentrations after ingestion of WM meal providing 40% of total energy requirements (D0), or LP-HF/HP-LF meals providing 60% of total energy requirements.

2.3. Dietary Interventions

All participants were studied on two occasions, each one consisting of a 3-day (D-3–D-1) weight-maintenance (WM), low sucrose diet followed by 6-day of sucrose + lactose overfeeding (D1–D6). On one occasion this overfeeding consisted of a 5% dietary protein and 25% fat content; on the other occasion, it was comprised of 20% dietary protein and 10% fat content. The dietary conditions were applied according to a randomized, cross-over design (Figure 1). Randomization was performed according to a pre-defined sequence, which was generated using R, version 3.0.1. (R Foundation for Statistical Computing, Vienna, Austria). The intervention was not blinded due to the nature of the drinks consumed. The two interventions were separated by a washout period of four to eight weeks.

WM diets were prepared from market foods and provided 100% of energy requirements (estimated from basal energy expenditure, calculated with the Harris-Benedict equation, times a physical activity level of 1.5). Food intake was partitioned into 3 meals/day and 2 snacks/day. It contained 45% of total energy as starch, 10% as sucrose, 33% as lipid, and 12% as protein and 22.6 ± 0.9 g dietary fiber/day; beverages were provided *ad libitum* as water. Overfeeding was attained by adding an extra 50% energy to the weight-maintenance energy requirements, in the form of six drinks per day. Drinks were prepared with skimmed milk and sucrose for the HP-LF condition or with water, lactose, and sucrose for the LP-HF condition, and had a volume of 218 ± 52 ml each. Solid diets were adjusted to obtain the same total energy (150% energy requirement): starch (29%), sucrose (34%) and lactose (7%) in both diets, with 20% protein (2.7 g/kg/day) and 10% fat in HP-LF or 5% protein (0.8 g/kg/day) and 25% fat in LP-HF. The addition of fat in LP-HF was mainly achieved by the addition of olive oil, butter, sauces, and cereals bars. Water consumption was left *ad libitum*. The detailed compositions of all three diets are shown in Table 1.

During each intervention, participants came to the metabolic unit of the Physiology Department of the University of Lausanne to consume their breakfasts, lunches, dinners, and three supplemental drinks under supervision. Every day, they also received two packages of snacks, together with three supplemental drinks during the overfeeding periods to consume between main meals, and were instructed not to consume any other food or drinks except plain water.

Table 1. Energy content and macronutrient composition of WM, LP-HF and HP-LF.

Diet Composition	WM		LP-HF			HP-LF		
	Solid Diet kcal/day (%)	Beverages kcal/day (%)	Solid Diet kcal/day (%)	Beverages kcal/day (%)	Total LP-HF kcal/day (%)	Solid Diet kcal/day (%)	Beverages kcal/day (%)	Total HP-LF kcal/day (%)
Starch	1061 (45)	-	1054	-	1054 (29)	1043	-	1043 (29)
Sucrose	249 (10)	-	241	965	1206 (34)	246	964	1210 (34)
Lactose	-	-	-	245	245 (7)	-	246	246 (7)
Protein	274 (12)	-	194	-	194 (5)	514	178	692 (20)
Fat	781 (33)	-	886	-	886 (25)	357	12	369 (10)
SFA	263 (34)	-	313	-	313 (35)	184	-	184 (52)
MUFA	280 (36)	-	389	-	389 (44)	102	-	102 (29)
PUFA	202 (26)	-	168	-	168 (19)	55	-	55 (15)
Total kcal	2365	-	2375	1210	3585	2160	1400	3560

WM: weight maintenance diet; LP-HF: high-sucrose, low-protein; HP-LF: high-sucrose, high-protein. Data are expressed as kcal/day; values into bracket represent % of total energy intake. For SFA, MUFA and PUFA, values in () are given as % total fat intake.

2.4. Measurements of IHCL and IMCL

For each intervention, IHCL and IMCL were measured at 4:00 pm on the 3rd day (D-1) on the WM diet (WM_{LP-HF} and WM_{HP-LF}) and on the 6th day (D6) on the hypercaloric diets (HP-LF and LP-HF). IHCL and IMCL content were determined by ^{1}H-MRS using a clinical 3T MR system (Verio, Siemens Medical, Germany) using methods similar to those described previously for IMCL [19,20] and for IHCL [21]. For the latter, quantification was based on the unsuppressed water signal corrected for transverse relaxation (characterized by the T_2 value) as determined in each subject individually. Since T_2 values were found to be significantly different before (WM_{LP-HF}, WM_{HP-LF}) versus after the diets (LP-HF, HP-LF), but did not differ between diets (LP-HF vs. HP-LF), individually averaged T_2 values for pre- and post-diet sessions were used for IHCL quantification. Results were expressed as mmol/kg ww.

2.5. Metabolic Tests

On days following IHCL and IMCL measurements (D0 and D7), participants were asked to arrive in the fasting state at the Metabolism, Nutrition and Physical Activity Research Center of the Department of Physiology of the University of Lausanne at 7:00 am for a metabolic test (schema shown in Figure 2). They had performed a 24-h urine collection the day before.

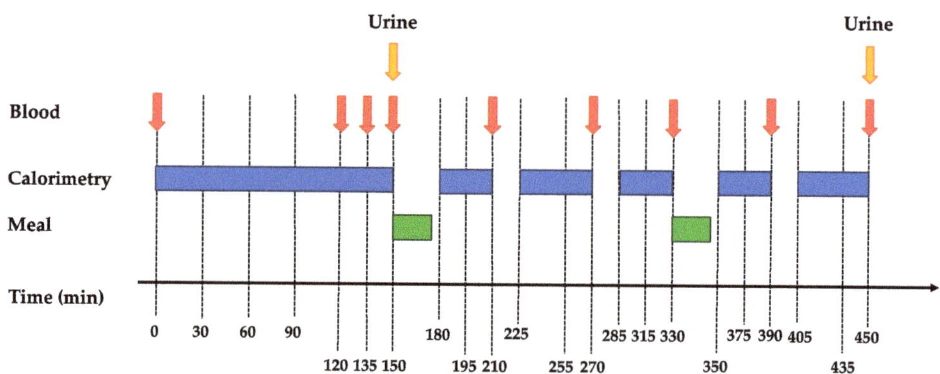

Figure 2. Schema of metabolic tests at D0 and D7.

This metabolic test aimed at comparing their fasting and postprandial energy expenditure, plasma hormones, and substrate profiles during periods of weight maintenance and overfeeding. At their arrival, participants were asked to void and discard their urine. They were then weighed and transferred to a bed where they remained in a semi-recumbent position for the next 7.5 h. A catheter was inserted into an antecubital vein for blood collection. Subjects remained fasted for the initial 2.5 h.

Four fasting blood samples and a urine collection were obtained during this period. Thereafter, they received two meals, one at 150 min and the second one at 330 min. Meal composition corresponded to the current intervention (i.e., WM on D0 and either HP-LF or LP-HF on D7). The sum of these two meals contained 40% (30% in first and 10% in the second meal) of total daily energy intake, which corresponded to 40% of energy requirements with WM, and to 60% of daily energy requirements during overfeeding periods (HP-LF and LP-HF). Postprandial blood samples were collected at the times 210 min, 270 min, 330 min, 390 min, and 450 min. Respiratory gas exchanges were monitored throughout the experiment by open-circuit indirect calorimetry (Quark RMR, version 9.1b, Cosmed, Rome, Italy), except for brief interruptions during meals. A second urine collection was obtained at the end of the test (time 450 min). Energy expenditure (EE) was calculated using the equations of Livesey and Elia [22].

2.6. Analytical Procedures

Plasma glucose, triglycerides (TG), lactate, and urine urea were measured by enzymatic methods (Randox Laboratories, Crumlin, County Antrim, UK). Plasma fructose concentrations were measured by GC–MS apparatus (Agilent Technologies, Santa Clara, CA, USA) [23]. Insulin and glucagon were assessed by radioimmunoassays (Millipore, Billerica, MA, USA). Plasma lipoprotein subfractions were separated by ultracentrifugation [24].

2.7. Statistical Analysis

All results are expressed as means $\pm$ SEMs. Postprandial results for all parameters (except for IGF1 and glucagon, which were determined in fasting conditions at only 2-time points postprandial) were expressed as the incremental area under the curve (iAUC $_{(0\text{-}300 \text{ min})}$), which was obtained using the trapezoidal method by subtracting the fasting value. As a preliminary analysis, the normality of data was checked with Shapiro-Wilk tests for all parameters analyzed. Non-normally distributed data were log-transformed (IHCL, fasting insulin, glucagon, TG, and postprandial glucagon). Two-way ANOVA assessed the effects of overfeeding, protein/fat content (HP-LF vs. LP-HF), and interaction between overfeeding x protein/fat content with repeated measures. Tukey post hoc tests were performed to compare individuals when needed. All statistical analyses were performed using Prism 7 (GraphPad Software, Inc., La Jolla, USA). The number of subjects included in the study was based on a power analysis related to whole body protein turnover (not reported here).

3. Results

The recruitment and follow up of subjects took place between June 2013 and April 2016. All volunteers completed the investigation and reported that they did not take any additional caloric drinks and food during the study. One volunteer was not included in the calculation of postprandial fructose due to missing plasma samples. Two volunteers were excluded from 24 h urinary concentration, excretion, and clearance calculation due to missing urine collections. All other calculations were performed with all 12 volunteers.

3.1. Fasting Condition

Fasting parameters are shown in Table 2. All fasting parameters were not significantly different after WM$_{\text{LP-HF}}$ and WM$_{\text{HP-LF}}$. Body weight increased by 0.7 $\pm$ 0.1 kg (males 0.9 $\pm$ 0.2 kg, females 0.6 $\pm$ 0.1 kg) between D0 and D7 after LP-HF and by 1.4 $\pm$ 0.2 kg after HP-LF (males 1.8 $\pm$ 0.2 kg, females 0.9 $\pm$ 0.1 kg) (for the whole group: $p < 0.001$ for overfeeding, $p > 0.999$ for protein/fat content, $p = 0.009$ for overfeeding $\times$ protein/fat content). Fasting EE increased from 1.11 $\pm$ 0.06 kcal/min (WM$_{\text{LP-HF}}$) to 1.12 $\pm$ 0.05 kcal/min (LP-HF), and from 1.10 $\pm$ 0.05 kcal/min (WM$_{\text{HP-LF}}$) to 1.18 $\pm$ 0.05 kcal/min (HP-LF), ($p = 0.018$ for overfeeding, $p = 0.126$ for protein/fat content, $p = 0.024$ overfeeding $\times$ protein/fat content). Fasting plasma glucose, fructose, lactate, TG, and insulin all increased to the same extent with HP-LF and LP-HF (Table 2). Fasting plasma NEFA decreased to the same extent with

HP-LF and LP-HF. In contrast, fasting glucagon concentration and IGF-1 concentrations increased with HP-LF, but remained stable (glucagon) or slightly decreased (IGF-1) with LP-HF.

Table 2. Fasting plasma metabolites and hormones concentrations.

Fasting	WM (LP-HF)	LP-HF	WM (HP-LF)	HP-LF	*p* Value		
					Overfeeding	Protein/Fat Content	OxP
Glucose (mmol/L)	4.56 ± 0.07	4.78 ± 0.07	4.46 ± 0.11	4.76 ± 0.09	<0.001	0.444	0.383
Fructose (µmol/L)	25.95 ± 1.41	27.15 ± 1.49	26.35 ± 1.37	28.0 ± 1.32	0.022	0.611	0.750
Lactate (mmol/L)	0.70 ± 0.06	1.22 ± 0.07	0.64 ± 0.04	1.16 ± 0.09	<0.001	0.107	0.935
Uric acid (mmol/L)	0.38 ± 0.02	0.38 ± 0.03	0.39 ± 0.02	0.30 ± 0.02	<0.001	<0.001	<0.001
TG (mmol/L)	0.68 ± 0.07	1.54 ± 0.22	0.66 ± 0.08	1.68 ± 0.19	<0.001	0.429	0.119
NEFA (mmol/L)	0.72 ± 0.05	0.44 ± 0.09	0.77 ± 0.04	0.37 ± 0.07	<0.001	0.779	0.097
Insulin (µU/mL)	8.42 ± 0.83	10.95 ± 1.02	7.82 ± 0.79	11.73 ± 1.54	<0.001	0.744	0.295
Glucagon (pg/mL)	72.42 ± 4.83	72.49 ± 4.94	68.67 ± 4.03	79.27 ± 5.35	0.059	0.276	0.036
IGF-1 (ng/mL)	212 ± 13	176 ± 12	174 ± 18	208 ± 13	0.901	0.712	<0.001

WM: weight maintenance diet; LP-HF: high-sucrose, low-protein; HP-LF: high-sucrose, high-protein. All values are mean ± SEM, *n* = 12. A significant difference in each condition, $p < 0.05$ (2-way ANOVA with repeated measures). OxP: Overfeeding x protein/fat content.

3.2. IHCL and IMCL Concentrations

IHCL and IMCL concentrations after WM and after LP-HF and HP-LF are shown in Figure 3. No statistically significant difference was observed between WM_{LP-HF} and WM_{HP-LF}. Compared to WM conditions, IHCL and IMCL concentrations increased significantly with both LP-HF and HP-LF overfeeding. However, IHCL increased more importantly with LP-HF than with HP-LF ($p < 0.001$ for effect of overfeeding, $p < 0.001$ for effect of dietary protein/fat content, and $p < 0.001$ for interaction overfeeding x protein/fat content). IMCL also increased more with LP-HF than with HP-LF ($p < 0.001$ for overfeeding, $p = 0.025$ for protein/fat content, and $p = 0.002$ for overfeeding x protein/fat content).

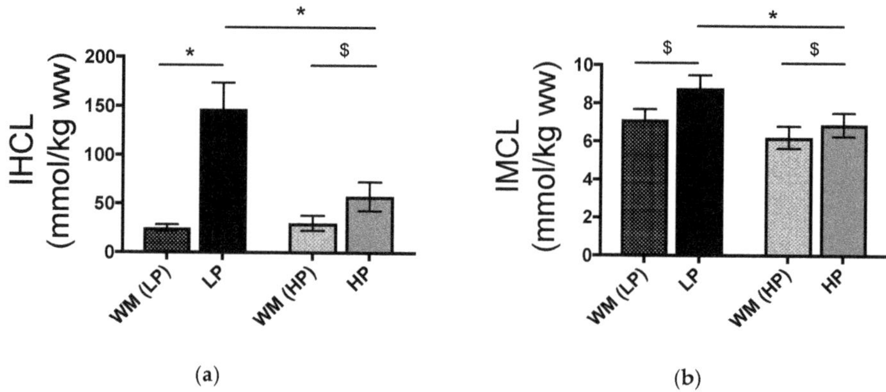

(a) (b)

Figure 3. Intrahepatocellular (IHCL) (**a**) and intramyocellular (IMCL) lipids (**b**) in response to weight maintaining diet (WM_{LP-HF} and WM_{HP-LF}) and overfeeding with LP-HF and HP-LF. *n* = 12; significant responses from WM_{LP-HF} and WM_{HP-LF} were measured by 2-way ANOVA for repeated measures with interaction. *: $p < 0.001$, interaction overfeeding × protein/fat content. $: $p < 0.005$, Tukey post hoc tests.

3.3. Postprandial Parameters

Postprandial metabolic parameters were not significantly different after WM diets. Postprandial EE and diet-induced thermogenesis were both significantly higher with LP-HF and HP-LF than under their respective WM conditions. Furthermore, EE increased more after HP-LF (from 1.23 ± 0.05

to 1.55 ± 0.06 kcal/min) than after LP-HF (from 1.24 ± 0.05 to 1.41 ± 0.06 kcal/min) ($p < 0.001$ for overfeeding, $p = 0.013$ for protein/fat content, and $p < 0.001$ for overfeeding x protein/fat content).

The postprandial iAUCs for blood metabolites and hormones are shown in Table 3. Postprandial blood glucose did not significantly change with HP-LF and LP-HF compared to their respectively WM conditions. Postprandial fructose, lactate, TG, and insulin iAUC were significantly higher in HP-LF and LP-HF than in the respective WM conditions.

Table 3. Metabolites and hormones at postprandial states.

Postprandial	WM (LP-HF)	LP-HF	WM (HP-LF)	HP-LF	*p* Value		
					Overfeeding	Protein/Fat Content	OxP
iAUC Glucose (mmol/L*300min)	504.0 ± 40.5	495.3 ± 69.2	560.3 ± 43.7	471.4 ± 57.2	0.242	0.616	0.189
iAUC Fructose (mmol/L*300min)	4.2 ± 0.3	30.3 ± 2.9	4.8 ± 0.5	23.4 ± 2.2	<0.001	0.005	0.003
iAUC Lactate (mmol/L*300min)	78.8 ± 12.8	239.8 ± 24.5	92.2 ± 15.1	139.3 ± 15.7	<0.001	0.001	0.001
iAUC TG (mmol/L*300min)	29.3 ± 6.8	121.3 ± 15.3	24.2 ± 8.0	126.7 ± 16.9	<0.001	0.986	0.471
iAUC NEFA (mmol/L*300min)	−162.6 ± 12.5	−86.7 ± 23.2	−173.1 ± 11.0	−76.6 ± 17.2	<0.001	0.984	0.051
iAUC Insulin (μU/ml*300min)	11378 ± 1232	19228 ± 1708	11138 ± 1488	24123 ± 2790	<0.001	0.061	0.028

WM: weight maintenance diet; LP-HF: high-sucrose, low-protein; HP-LF: high-sucrose, high-protein. All values are mean ± SEM, $n = 12$. A significant difference in each condition, $p < 0.05$ (2-way ANOVA, with repeated measures). OxP: Overfeeding x protein/fat content. In the calculation of iAUC fructose ($n = 11$) one volunteer was excluded for reason of missing plasma data.

HP-LF and LP-HF nonetheless differentially altered postprandial insulin, fructose, and lactate concentrations: HP-LF increased postprandial insulin concentrations more than LP-HF, but decreased postprandial fructose and lactate (see Table 3 for detailed statistics). Postprandial plasma uric acid concentration, measured at time 450 min, decreased from 0.38 ± 0.02 (WM) to 0.30 ± 0.02 mmol/L with HP-LF, but increased from 0.38 ± 0.03 (WM) to 0.42 ± 0.04 mmol/L with LP-HF ($p = 0.283$ for overfeeding, $p = 0.001$ for diet, $p < 0.001$ for overfeeding × protein/fat content). Plasma glucagon, measured at time 450 min, increased from 62.6 ± 3.3 to 87.9 ± 8.4 pg/mLwith HP-LF, but did not change with LP-HF: 65.1 ± 4.4 vs. LP-HF: 70.6 ± 4.8 pg/mL, ($p < 0.001$ for overfeeding, $p = 0.026$ for protein/fat content, and $p = 0.001$ for overfeeding x protein/fat content).

24-h urinary excretion and clearance of creatinine and uric acid are shown in Table 4. LP-HF and HP-LF did not significantly change 24-h urinary excretion and clearance of creatinine. HP-LF increased urinary excretion of uric acid and uric acid clearance while LP-HF decreased it. Compared to LP-HF, HP-LF significantly increased urinary creatinine and uric acid clearance; it also increased total 24-h uric acid excretion.

Table 4. 24-h urinary creatinine and uric acid excretion and clearance.

	WM (LP-HF)	LP-HF	WM (HP-LF)	HP-LF	*p* Value		
					Overfeeding	Protein/Fat Content	OxP
24-h urinary excretion							
Creatinine (mmol/24h)	13.6 ± 1.8	12.6 ± 1.2	13.0 ± 0.8	12.7 ± 1.1	0.264	0.450	0.638
Uric acid (mmol/24h)	3.5 ± 0.2	3.3 ± 0.2	3.3 ± 0.2	4.1 ± 0.4	0.049	0.238	0.022
Urinary clearance rate							
Creatinine (ml/min)	129.8 ± 9.4	133.7 ± 9.7	135.1 ± 10.8	153.0 ± 13.0	0.279	0.309	0.282
Uric acid (ml/min)	6.9 ± 0.6	6.5 ± 0.6	6.1 ± 0.4	10.0 ± 1.4	0.005	0.015	0.004

WM: weight maintenance diet; LP-HF: high-sucrose, low-protein; HP-LF: high-sucrose, high-protein. All values are mean ± SEM, $n = 10$ as two volunteers were excluded because of missing samples. A significant difference in each condition, $p < 0.05$ (2-way ANOVA, with repeated measures). OxP: Overfeeding x protein/fat content.

4. Discussion

This study was designed to assess whether the consequences of sucrose overfeeding differ according to concomitant changes in daily protein and fat intake. Our main findings were that both HP-LF and LP-HF increased IHCL, IMCL, and blood triglycerides concentrations, but increments were reduced on average by 78% for IHCL and by 59% for IMCL with HP-LF compared to LP-HF.

In addition, fasting and postprandial EE were significantly higher with HP-LF than LP-HF. However, blood triglyceride concentrations were not significantly different with HP-LF and LP-HF. Finally, blood uric acid concentrations were increased with LP-HF, but decreased with HP-LF.

Our experimental design compared the effects of two hypercaloric high sucrose diets, one with a high protein-low fat content and the other with a low protein-high fat content, to that of a weight maintenance control diet. All three diets contained an amount of starch equivalent to approximately 45% total energy requirements, and the two hypercaloric diets contained 150% of daily energy requirements, with about 50% of energy requirements as sucrose, and 7% of energy requirements as lactose. Lactose intake was higher in HP-LF than in WM because of a high milk protein intake and was balanced by lactose addition in LP-HF in order to have equal carbohydrate amounts and composition in both diets. Dietary saturated-monounsaturated and polyunsaturated fatty acid proportions were also different in each diet.

The dietary composition had a profound effect on the amount of ectopic lipids being deposited during overfeeding. HP-LF and LP-HF both increased lipid storage in the liver and muscle, two sites in which ectopic lipid deposition is known to be associated with adverse long-term effects [1]. Several short-term studies had previously documented that excess energy intake from fructose or glucose increased IHCL [10,25,26] and IMCL [26–28]. In our study, this effect was most notable in the liver, where IHCL increased by 542 ± 105% after LP-HF. It was milder in skeletal muscle, where we nonetheless observed a significant increase of +24 ± 3% after LP-HF. In both sites, the increases induced by HP-LF were significantly lower than those induced by LP-HF. Excess energy intake from sugars is thought to increase IHCL by enhancing hepatic *de novo* lipogenesis and inhibiting intrahepatic lipid oxidation [29]. Several hypotheses can be proposed to account for the differential effects of HP-LF and LP-HF. First, LP-HF contained more lipids than HP-LF. Previous experiments have shown that fat overfeeding increases IHCL synthesis from intestinally derived TG-rich lipoprotein particles and/or circulating NEFA [13,30,31]. It has also been shown that fructose and fat have additive effects on IHCL during combined fructose-fat overfeeding [10]. It is therefore likely that, with LP-HF, the high dietary sugar and fat intake had additive effects on IHCL. Second, dietary protein may decrease IHCL independently of dietary fat or energy intake. In support of this hypothesis, a former study reported that IHCL were increased in healthy subjects fed a hypercaloric, high fat diet containing 130% energy requirements. However, the addition of protein to this high fat diet resulted in a similar daily fat and carbohydrate intake, but also in a higher total energy and protein intake with significantly reduced IHCL [13]. The mechanisms by which an increased protein intake may reduce IHCL remain unknown. Inhibition of *de novo* lipogenesis has been postulated [13], but fractional hepatic *de novo* lipogenesis was stimulated to the same extent in healthy subjects overfed with fructose alone or with fructose and proteins [16]. A stimulation of hepatic VLDL-TG secretion and extrahepatic VLDL-TG clearance [16], or a protein-induced increase in plasma bile acid concentrations [13] have also been proposed to play a role. In contrast, no effect of dietary protein intake on IMCL has been reported to our knowledge. Finally, changes in dietary fatty acids composition may modulate diet-induced hepatic fat deposition (reviewed in reference [32]). Hepatic steatosis in animal models is readily produced by consumption of a high saturated fat diet with low PUFA content. In contrast, there is evidence that PUFA or oleic acid supplementation may actually blunt diet-induced hepatic steatosis [32]. In the present study, dietary protein intake in HP-LF was increased through the consumption of skimmed dairy products to avoid an increase in SFA, and dietary fat intake in LP-HF was increased by consumption of vegetable oils (mainly olive oil). As a result, total daily SFA intake was only slightly higher in LP-HF than in HP-LF (34.7 ± 1.5 vs. 20.4 ± 0.9 g/day) while MUFA+PUFA intake was markedly increased. It is therefore unlikely that the higher IHCL observed with LP-HF can be explained by the differences in dietary fat composition.

The postprandial increases in plasma TG concentrations were 5-fold higher with HP-LF and 4-fold higher with LP-HF than with WM. Several studies have reported that fructose and sucrose overfeeding increases fasting and postprandial blood triglyceride by increasing hepatic *de novo* lipogenesis and

VLDL-TG secretion and by decreasing the postprandial clearance of triglyceride-rich lipoprotein particles [27,33,34]. It is therefore likely that an upregulation of lipogenic enzymes with sucrose overfeeding contributed to this hypertriglyceridemia. However, the meals administered during the metabolic tests contained 50% more total energy in overfeeding than in weight-maintenance control conditions, and, therefore, contained also more sucrose and fat, which makes it difficult to sort out the relative role of sucrose and other macronutrients. Globally, the increase in postprandial TG concentrations was not significantly different in HP-LF and LP-HF.

The effect of overfeeding on energy expenditure was also markedly dependent on dietary composition. Postprandial EE increased significantly with both HP-LF and LP-HF, mainly due to the fact that the test meals ingested in both conditions had a caloric content 50% higher than in the control weight-maintenance condition. Postprandial EE increased more with HP-LF than LP-HF. This is most likely explained by the high energy cost of amino-acid metabolism [35].

We also assessed whether dietary composition had significant effects on postprandial blood metabolic markers during overfeeding. The total carbohydrate and sucrose content of meals ingested during the metabolic tests were higher in overfeeding than in the WM control condition, and postprandial increments in blood fructose, lactate, and insulin were accordingly enhanced. Similarly, postprandial NEFA was decreased to lower levels in overfeeding than in WM conditions. However, postprandial blood glucose responses were not significantly altered. Most postprandial parameters were not significantly different in HP-LF and LP-HF overfeeding. However, postprandial glucagon increased more with HP-LF than with LP-HF, as expected due to the well-known stimulation of glucagon secretion by circulating amino-acids after protein ingestion [36]. Surprisingly, blood fructose and lactate concentration increased less with HP-LF than LP-HF. It is possible that the lower lactate concentration was secondary to glucagon stimulating hepatic lactate uptake [37]. The lower fructose response was unexpected, however, and may suggest that hepatic fructose extraction was enhanced when consumed with proteins. Nutrient- or glucagon-mediated changes in portal blood flow may also be implicated [38]. Alternatively, it is possible that gastric emptying was delayed with HP-LF meals, thus accounting for a slower fructose absorption [39]. Finally, compared to WM, postprandial increases in uric acid were higher with LP-HF, but lower with HP-LF, while urinary uric acid excretion and uric acid clearance were significantly increased with HP-LF. This suggests that both HP-LF and LP-HF increased uric acid production, possibly due to the fructose component of sucrose [40], and that an increase in glomerular filtration rate, possibly mediated by glucagon [41], increased uric acid excretion, thus preventing an increase in blood uric acid. Elevated lactate concentrations are also known to impair renal uric acid clearance [42], and it is, therefore, possible that lower lactate concentrations during HP-LF than LP-HF overfeeding also played a role. Our data, however, do not allow accurate comparisons of uric acid production and excretion between HP-LF and LP-HF.

The present study limitations need to be acknowledged. First, we did not include isotopic measurements of *de novo* lipogenesis and VLDL-TG kinetics, and therefore cannot identify the mechanisms by which HP-LF decreased IHCL and IMCL compared to LP-HF. Second, not only total dietary fat intake, but also the proportions of SFA-MUFA-PUFA were different between diets, and we cannot exclude the possibility that this may have impacted IHCL or IMCL storage. Third, in HP-LF condition, dietary protein content was increased by addition of dairy products; whether the observed effects are generic to dietary proteins or specific to dairy products remains to be evaluated. Finally, our study was of short duration and was limited to a small group of healthy male and female subjects, and results may not apply to other subgroups of the population (e.g., overweight subjects or subjects with the metabolic syndrome).

5. Conclusions

In summary, our data indicate that overfeeding with a high sucrose, high protein/low-fat diet markedly reduces ectopic fat accumulation in the liver and muscle, and increases energy expenditure, compared to an isocaloric overfeeding with high sucrose, low protein/high-fat diet. This may be due to an additive effect of sucrose and dietary fat and/or a protective effect of dietary protein on ectopic fat accumulation.

Author Contributions: Conceptualization, L.E., C.B., R.K., and L.T.; methodology, P.S., C.B., R.K, V.C.; validation, V.C., A.S., P.J.; formal analysis, A.S.; investigation, A.S., P.J., V.C., L.E., A.-S.M., R.K., V.L., R.R., B.P, J.C.; writing—original draft preparation, A.S.; writing—review and editing, all.; visualization, A.S., P.J.; project administration, A.S.; funding acquisition, L.T., C.B., R.K.

Funding: This research was funded by grant from the Swiss National Foundation for science 32003B_156167, and by a grant from the Institute Benjamin Delessert Foundation to P.J.

Acknowledgments: We thank the staff of the Department of Physiology of Lausanne for their great assistance, Shawna McCallin for language editing, and all the volunteers for their participation and commitment.

Conflicts of Interest: L.T. has received research support from Soremartec Italia srl for projects unrelated to this report, and speakers' fees from Soremartec Italia srl, Nestlé AG, Switzerland, and the Gatorade Sport Science Institute, USA. L.E. and V.C. are presently employed by Nestec SA, Switzerland. Other authors declare no conflict of interest.

References

1. Szendroedi, J.; Roden, M. Ectopic lipids and organ function. *Curr. Opin. Lipidol.* **2009**, *20*, 50–56. [CrossRef]
2. Morelli, M.; Gaggini, M.; Daniele, G.; Marraccini, P.; Sicari, R.; Gastaldelli, A. Ectopic fat: The true culprit linking obesity and cardiovascular disease? *Thromb. Haemost.* **2013**, *110*, 651–660. [CrossRef] [PubMed]
3. Britton, K.A.; Fox, C.S. Ectopic fat depots and cardiovascular disease. *Circulation* **2011**, *124*, e837–e841. [CrossRef] [PubMed]
4. Stanhope, K.L.; Havel, P.J. Fructose consumption: Considerations for future research on its effects on adipose distribution, lipid metabolism, and insulin sensitivity in humans. *J. Nutr.* **2009**, *139*, 1236S–1241S. [CrossRef] [PubMed]
5. Aeberli, I.; Hochuli, M.; Gerber, P.A.; Sze, L.; Murer, S.B.; Tappy, L.; Spinas, G.A.; Berneis, K. Moderate amounts of fructose consumption impair insulin sensitivity in healthy young men: A randomized controlled trial. *Diabetes Care* **2013**, *36*, 150–156. [CrossRef] [PubMed]
6. Stanhope, K.L.; Havel, P.J. Fructose consumption: Potential mechanisms for its effects to increase visceral adiposity and induce dyslipidemia and insulin resistance. *Curr. Opin. Lipidol.* **2008**, *19*, 16–24. [CrossRef]
7. Stanhope, K.L.; Schwarz, J.M.; Keim, N.L.; Griffen, S.C.; Bremer, A.A.; Graham, J.L.; Hatcher, B.; Cox, C.L.; Dyachenko, A.; Zhang, W.; et al. Consuming fructose-sweetened, not glucose-sweetened, beverages increases visceral adiposity and lipids and decreases insulin sensitivity in overweight/obese humans. *J. Clin. Investig.* **2009**, *119*, 1322–1334. [CrossRef]
8. Wang, D.D.; Sievenpiper, J.L.; de Souza, R.J.; Chiavaroli, L.; Ha, V.; Cozma, A.I.; Mirrahimi, A.; Yu, M.E.; Carleton, A.J.; Di Buono, M.; et al. The effects of fructose intake on serum uric acid vary among controlled dietary trials. *J. Nutr.* **2012**, *142*, 916–923. [CrossRef]
9. Taskinen, M.R.; Soderlund, S.; Bogl, L.H.; Hakkarainen, A.; Matikainen, N.; Pietilainen, K.H.; Rasanen, S.; Lundbom, N.; Bjornson, E.; Eliasson, B.; et al. Adverse effects of fructose on cardiometabolic risk factors and hepatic lipid metabolism in subjects with abdominal obesity. *J. Intern. Med.* **2017**, *282*, 187–201. [CrossRef]
10. Sobrecases, H.; Le, K.A.; Bortolotti, M.; Schneiter, P.; Ith, M.; Kreis, R.; Boesch, C.; Tappy, L. Effects of short-term overfeeding with fructose, fat and fructose plus fat on plasma and hepatic lipids in healthy men. *Diabetes Metab.* **2010**, *36*, 244–246. [CrossRef]
11. Hamad, E.M.; Taha, S.H.; Abou Dawood, A.G.; Sitohy, M.Z.; Abdel-Hamid, M. Protective effect of whey proteins against nonalcoholic fatty liver in rats. *Lipids Health Dis.* **2011**, *10*, 57. [CrossRef] [PubMed]
12. Chaumontet, C.; Even, P.C.; Schwarz, J.; Simonin-Foucault, A.; Piedcoq, J.; Fromentin, G.; Azzout-Marniche, D.; Tome, D. High dietary protein decreases fat deposition induced by high-fat and high-sucrose diet in rats. *Br. J. Nutr.* **2015**, *114*, 1132–1142. [CrossRef] [PubMed]
13. Bortolotti, M.; Kreis, R.; Debard, C.; Cariou, B.; Faeh, D.; Chetiveaux, M.; Ith, M.; Vermathen, P.; Stefanoni, N.; Le, K.A.; et al. High protein intake reduces intrahepatocellular lipid deposition in humans. *Am. J. Clin. Nutr.* **2009**, *90*, 1002–1010. [CrossRef] [PubMed]

14. Martens, E.A.; Gatta-Cherifi, B.; Gonnissen, H.K.; Westerterp-Plantenga, M.S. The potential of a high protein-low carbohydrate diet to preserve intrahepatic triglyceride content in healthy humans. *PLoS ONE* **2014**, *9*, e109617. [CrossRef] [PubMed]

15. Rietman, A.; Schwarz, J.; Blokker, B.A.; Siebelink, E.; Kok, F.J.; Afman, L.A.; Tome, D.; Mensink, M. Increasing protein intake modulates lipid metabolism in healthy young men and women consuming a high-fat hypercaloric diet. *J. Nutr.* **2014**, *144*, 1174–1180. [CrossRef] [PubMed]

16. Theytaz, F.; Noguchi, Y.; Egli, L.; Campos, V.; Buehler, T.; Hodson, L.; Patterson, B.W.; Nishikata, N.; Kreis, R.; Mittendorfer, B.; et al. Effects of supplementation with essential amino acids on intrahepatic lipid concentrations during fructose overfeeding in humans. *Am. J. Clin. Nutr.* **2012**, *96*, 1008–1016. [CrossRef] [PubMed]

17. Bray, G.A.; Smith, S.R.; de Jonge, L.; Xie, H.; Rood, J.; Martin, C.K.; Most, M.; Brock, C.; Mancuso, S.; Redman, L.M. Effect of dietary protein content on weight gain, energy expenditure, and body composition during overeating: A randomized controlled trial. *JAMA* **2012**, *307*, 47–55. [CrossRef] [PubMed]

18. Eisenmann, A.; Amann, A.; Said, M.; Datta, B.; Ledochowski, M. Implementation and interpretation of hydrogen breath tests. *J. Breath Res.* **2008**, *2*, 046002. [CrossRef]

19. Le, K.A.; Faeh, D.; Stettler, R.; Ith, M.; Kreis, R.; Vermathen, P.; Boesch, C.; Ravussin, E.; Tappy, L. A 4-wk high-fructose diet alters lipid metabolism without affecting insulin sensitivity or ectopic lipids in healthy humans. *Am. J. Clin. Nutr.* **2006**, *84*, 1374–1379. [CrossRef]

20. Boesch, C.; Kreis, R. Observation of intramyocellular lipids by 1h-magnetic resonance spectroscopy. *Ann. N. Y. Acad. Sci.* **2000**, *904*, 25–31. [CrossRef]

21. Cros, J.; Pianezzi, E.; Rosset, R.; Egli, L.; Schneiter, P.; Cornette, F.; Pouymayou, B.; Heinzer, R.; Tappy, L.; Kreis, R.; et al. Impact of sleep restriction on metabolic outcomes induced by overfeeding: A randomized controlled trial in healthy individuals. *Am. J. Clin. Nutr.* **2019**. [CrossRef] [PubMed]

22. Elia, M.; Livesey, G. Energy expenditure and fuel selection in biological systems: The theory and practice of calculations based on indirect calorimetry and tracer methods. *World Rev. Nutr. Diet.* **1992**, *70*, 68–131. [PubMed]

23. Tran, C.; Jacot-Descombes, D.; Lecoultre, V.; Fielding, B.A.; Carrel, G.; Le, K.A.; Schneiter, P.; Bortolotti, M.; Frayn, K.N.; Tappy, L. Sex differences in lipid and glucose kinetics after ingestion of an acute oral fructose load. *Br. J. Nutr.* **2010**, *104*, 1139–1147. [CrossRef] [PubMed]

24. Karpe, F.; Steiner, G.; Olivecrona, T.; Carlson, L.A.; Hamsten, A. Metabolism of triglyceride-rich lipoproteins during alimentary lipemia. *J. Clin. Investig.* **1993**, *91*, 748–758. [CrossRef] [PubMed]

25. Lecoultre, V.; Egli, L.; Carrel, G.; Theytaz, F.; Kreis, R.; Schneiter, P.; Boss, A.; Zwygart, K.; Le, K.A.; Bortolotti, M.; et al. Effects of fructose and glucose overfeeding on hepatic insulin sensitivity and intrahepatic lipids in healthy humans. *Obesity* **2013**, *21*, 782–785. [CrossRef] [PubMed]

26. Johnston, R.D.; Stephenson, M.C.; Crossland, H.; Cordon, S.M.; Palcidi, E.; Cox, E.F.; Taylor, M.A.; Aithal, G.P.; Macdonald, I.A. No difference between high-fructose and high-glucose diets on liver triacylglycerol or biochemistry in healthy overweight men. *Gastroenterology* **2013**, *145*, 1016–1025.e2. [CrossRef]

27. Le, K.A.; Ith, M.; Kreis, R.; Faeh, D.; Bortolotti, M.; Tran, C.; Boesch, C.; Tappy, L. Fructose overconsumption causes dyslipidemia and ectopic lipid deposition in healthy subjects with and without a family history of type 2 diabetes. *Am. J. Clin. Nutr.* **2009**, *89*, 1760–1765. [CrossRef]

28. Ngo Sock, E.T.; Le, K.A.; Ith, M.; Kreis, R.; Boesch, C.; Tappy, L. Effects of a short-term overfeeding with fructose or glucose in healthy young males. *Br. J. Nutr.* **2010**, *103*, 939–943. [CrossRef]

29. Tappy, L.; Le, K.A. Does fructose consumption contribute to non-alcoholic fatty liver disease? *Clin. Res. Hepatol. Gastroenterol.* **2012**, *36*, 554–560. [CrossRef]

30. Donnelly, K.L.; Smith, C.I.; Schwarzenberg, S.J.; Jessurun, J.; Boldt, M.D.; Parks, E.J. Sources of fatty acids stored in liver and secreted via lipoproteins in patients with nonalcoholic fatty liver disease. *J. Clin. Investig.* **2005**, *115*, 1343–1351. [CrossRef]

31. Kotronen, A.; Yki-Jarvinen, H. Fatty liver: A novel component of the metabolic syndrome. *Arterioscler. Thromb. Vasc. Biol.* **2008**, *28*, 27–38. [CrossRef] [PubMed]

32. Ferramosca, A.; Zara, V. Modulation of hepatic steatosis by dietary fatty acids. *World J. Gastroenterol.* **2014**, *20*, 1746–1755. [CrossRef]

33. Teff, K.L.; Elliott, S.S.; Tschop, M.; Kieffer, T.J.; Rader, D.; Heiman, M.; Townsend, R.R.; Keim, N.L.; D'Alessio, D.; Havel, P.J. Dietary fructose reduces circulating insulin and leptin, attenuates postprandial

suppression of ghrelin, and increases triglycerides in women. *J. Clin. Endocrinol. Metab.* **2004**, *89*, 2963–2972. [CrossRef] [PubMed]

34. Stanhope, K.L.; Bremer, A.A.; Medici, V.; Nakajima, K.; Ito, Y.; Nakano, T.; Chen, G.; Fong, T.H.; Lee, V.; Menorca, R.I.; et al. Consumption of fructose and high fructose corn syrup increase postprandial triglycerides, ldl-cholesterol, and apolipoprotein-b in young men and women. *J. Clin. Endocrinol. Metab.* **2011**, *96*, E1596–E1605. [CrossRef] [PubMed]

35. Tappy, L. Thermic effect of food and sympathetic nervous system activity in humans. *Reprod. Nutr. Dev.* **1996**, *36*, 391–397. [CrossRef] [PubMed]

36. Calbet, J.A.; MacLean, D.A. Plasma glucagon and insulin responses depend on the rate of appearance of amino acids after ingestion of different protein solutions in humans. *J. Nutr.* **2002**, *132*, 2174–2182. [CrossRef] [PubMed]

37. Ramnanan, C.J.; Edgerton, D.S.; Kraft, G.; Cherrington, A.D. Physiologic action of glucagon on liver glucose metabolism. *Diabetes Obes. Metab.* **2011**, *13* (Suppl. 1), 118–125. [CrossRef]

38. Granger, D.N.; Richardson, P.D.; Kvietys, P.R.; Mortillaro, N.A. Intestinal blood flow. *Gastroenterology* **1980**, *78*, 837–863. [PubMed]

39. Ma, J.; Stevens, J.E.; Cukier, K.; Maddox, A.F.; Wishart, J.M.; Jones, K.L.; Clifton, P.M.; Horowitz, M.; Rayner, C.K. Effects of a protein preload on gastric emptying, glycemia, and gut hormones after a carbohydrate meal in diet-controlled type 2 diabetes. *Diabetes Care* **2009**, *32*, 1600–1602. [CrossRef] [PubMed]

40. Le, M.T.; Frye, R.F.; Rivard, C.J.; Cheng, J.; McFann, K.K.; Segal, M.S.; Johnson, R.J.; Johnson, J.A. Effects of high-fructose corn syrup and sucrose on the pharmacokinetics of fructose and acute metabolic and hemodynamic responses in healthy subjects. *Metabolism* **2012**, *61*, 641–651. [CrossRef] [PubMed]

41. Ahloulay, M.; Dechaux, M.; Laborde, K.; Bankir, L. Influence of glucagon on gfr and on urea and electrolyte excretion: Direct and indirect effects. *Am. J. Physiol.* **1995**, *269*, F225–F235. [CrossRef] [PubMed]

42. Yu, T.F.; Sirota, J.H.; Berger, L.; Halpern, M.; Gutman, A.B. Effect of sodium lactate infusion on urate clearance in man. *Proc. Soc. Exp. Biol. Med.* **1957**, *96*, 809–813. [CrossRef] [PubMed]

MDPI
St. Alban-Anlage 66
4052 Basel
Switzerland
Tel. +41 61 683 77 34
Fax +41 61 302 89 18
www.mdpi.com

Nutrients Editorial Office
E-mail: nutrients@mdpi.com
www.mdpi.com/journal/nutrients